# BIOLOGY AND CULTURE OF PORTUNID CRABS OF WORLD SEAS

*Biology and Ecology of Marine Life*

# BIOLOGY AND CULTURE OF PORTUNID CRABS OF WORLD SEAS

Ramasamy Santhanam, PhD

Apple Academic Press Inc.
3333 Mistwell Crescent
Oakville, ON L6L 0A2 Canada

Apple Academic Press Inc.
9 Spinnaker Way
Waretown, NJ 08758 USA

© 2018 by Apple Academic Press, Inc.

First issued in paperback 2021

*Exclusive worldwide distribution by CRC Press, a member of Taylor & Francis Group*
No claim to original U.S. Government works

ISBN 13: 978-1-77-463646-6 (pbk)
ISBN 13: 978-1-77-188590-4 (hbk)

All rights reserved. No part of this work may be reprinted or reproduced or utilized in any form or by any electric, mechanical or other means, now known or hereafter invented, including photocopying and recording, or in any information storage or retrieval system, without permission in writing from the publisher or its distributor, except in the case of brief excerpts or quotations for use in reviews or critical articles.

This book contains information obtained from authentic and highly regarded sources. Reprinted material is quoted with permission and sources are indicated. Copyright for individual articles remains with the authors as indicated. A wide variety of references are listed. Reasonable efforts have been made to publish reliable data and information, but the authors, editors, and the publisher cannot assume responsibility for the validity of all materials or the consequences of their use. The authors, editors, and the publisher have attempted to trace the copyright holders of all material reproduced in this publication and apologize to copyright holders if permission to publish in this form has not been obtained. If any copyright material has not been acknowledged, please write and let us know so we may rectify in any future reprint.

**Trademark Notice:** Registered trademark of products or corporate names are used only for explanation and identification without intent to infringe.

---

### Library and Archives Canada Cataloguing in Publication

Santhanam, Ramasamy, 1946-, author
Biology and culture of portunid crabs of world seas / Ramasamy Santhanam, PhD.
(Biology and ecology of marine life book series)
Includes bibliographical references and index.
Issued in print and electronic formats.
ISBN 978-1-77188-590-4 (hardcover).--ISBN 978-1-315-20581-6 (PDF)
1. Swimming crabs. I. Title. II. Series: Santhanam, Ramasamy, 1946- Biology and ecology of marine life book series
QL444.M33S26 2017          595.3'84   2017-905051-6     C2017-905052-4

---

### Library of Congress Cataloging-in-Publication Data

Names: Santhanam, Ramasamy, 1946- author.
Title: Biology and culture of portunid crabs of world seas / Ramasamy Santhanam, PhD.
Description: Waretown, NJ : Apple Academic Press, 2017. | Series: Biology and   ecology of marine life | Includes bibliographical references and index.
Identifiers: LCCN 2017034068 (print) | LCCN 2017034334 (ebook) | ISBN   9781315205816 (ebook) | ISBN 9781771885904 (hardcover : alk. paper)
Subjects: LCSH: Swimming crabs.
Classification: LCC QL444.M33 (ebook) | LCC QL444.M33 S253 2017 (print) | DDC 595.3/86--dc23
LC record available at https://lccn.loc.gov/2017034068

---

Apple Academic Press also publishes its books in a variety of electronic formats. Some content that appears in print may not be available in electronic format. For information about Apple Academic Press products, visit our website at **www.appleacademicpress.com** and the CRC Press website at **www.crcpress.com**

# ABOUT THE AUTHOR

**Ramasamy Santhanam, PhD**
*Former Dean, Fisheries College and Research Institute, Tamilnadu Veterinary and Animal Sciences University, Thoothukudi, India*

Dr. Ramasamy Santhanam is the former Dean of the Fisheries College and Research Institute at the Tamilnadu Veterinary and Animal Sciences University in Thoothukudi, India. His fields of specialization are marine biology and fisheries environment. Presently he is serving as a resource person for various universities in India. He has also served as an expert for the Environment Management Capacity Building, a World Bank-aided project of the Department of Ocean Development, Govt. of India. He was a member of the American Fisheries Society, the World Aquaculture Society, the Global Fisheries Ecosystem Management Network (GFEMN), and the International Union for Conservation of Nature's (IUCN) Commission on Ecosystem Management, Switzerland. To his credit, Dr. Santhanam has 18 books on fisheries science/marine biology as well as 70 research papers.

# BIOLOGY AND ECOLOGY OF MARINE LIFE BOOK SERIES

**Series Author:**

Ramasamy Santhanam, PhD
Former Dean, Fisheries College and Research Institute,
Tamil Nadu Veterinary and Animal Sciences University,
Thoothukkudi-628 008, India
Email: rsanthaanamin@yahoo.co.in

**Books in the Series:**

- Biology and Culture of Portunid Crabs of the World Seas
- Biology and Ecology of Edible Marine Bivalve Molluscs
- Biology and Ecology of Edible Marine Gastropod Molluscs
- Biology and Ecology of Venomous Marine Snails
- Biology and Ecology of Venomous Stingrays
- Biology and Ecology of Toxic Pufferfish

# CONTENTS

*List of Abbreviations* ............................................................. *ix*

*Preface* ............................................................................. *xi*

1. **Introduction** .............................................................. 1
2. **Biology and Ecology of Portunid Crabs** ......................... 9
3. **Profile of Portunid Crabs (Family: Portunidae)** .............. 23
4. **Aquaculture of Portunid Crabs** ................................... 273
5. **Nutritional Values of Portunid Crabs** ........................... 307
6. **Biomedical, Food, and Industrial Applications of Portunid Crab Wastes** .......................................... 329
7. **Diseases and Parasites of Portunid Crabs** ...................... 349

*References* ..................................................................... *375*

*Index* ............................................................................ *389*

# LIST OF ABBREVIATIONS

| | |
|---|---|
| BCD | bitter crab disease |
| BCS | bitter crab syndrome |
| BW | bodyweight |
| CB | carapace is broad |
| CL | carapace length |
| COD | chemical oxygen demand |
| CPUE | catch-per-unit-effort |
| CW | carapace width |
| GFEMN | Global Fisheries Ecosystem Management Network |
| HA | haemagglutinin |
| HDL | high density lipoprotein |
| HLV | herpeslike virus |
| ICW | internal carapace width |
| IUCN | International Union for Conservation of Nature |
| MCRV | mud crab reovirus |
| NOCC | $N,O$-carboxymethylchitosan |
| PCB | polychlorinated biphenyls |
| PCD | pink crab disease |
| PL | postlarvae |
| PRT | plasma recalcification time |
| SDS | shell disease syndrome |
| SGR | specific growth rate |
| TBA | thiobarbituric acid |
| TWG | total weight gained |
| USEPA | United States Environmental Protection Agency |
| USFDA | United States Food and Drug Administration |
| WHBCT | whole blood clotting time |
| WOF | warmed-over favor |
| WSSV | white spot syndrome virus |

# PREFACE

Marine crabs play a vital role directly or indirectly in the livelihood of millions of people around the world. They have been reported to make up about 20% of all marine crustaceans caught, farmed, and consumed worldwide, amounting to 1½ million tonnes annually. Among these marine crabs, the portunid crabs or swimming crabs of the family Portunidae (Class: Crustacea; Order: Decapoda; Infraorder: Brachyura) assume greater significance in the marine industry owing to their delicate meat with nutritional qualities. These crabs live in a variety of habitats, from soft substrates like mudflats and sandflats to harder bottoms like coral reefs and rubble. Although several species of portunid crabs are edible and commercially important, only a few species of *Scylla* and *Portunus* have been widely cultivated. This is largely due to the lack of information on the biology and biodiversity of portunid crabs.

Keeping this in view, this volume is an attempt to present the biology and aquaculture of marine portunid crabs. The different chapters include (i) biology and ecology of commercially important portunid species; (ii) profile (habitat, distribution, morphology, food and feeding, reproduction and lifecycle, etc.) of about 280 species of the different subfamilies of the family Portunidae; (iii) aquaculture of portunid crabs: seed production and larval rearing of portunid crabs; monospecies and monosex culture of portunid crabs; polyculture of portunid crabs with fish, shrimps and seaweeds; fattening of portunid crabs; and soft-shell crab farming; (iv) nutritional values of portunid crabs; (v) pharmaceutical and industrial applications of portunid crab wastes; and (vi) diseases and parasites of portunid crabs.

It is hoped that the present publication will be of great use for students and researchers of disciplines such as fisheries science, marine biology, aquatic biology and fisheries, and zoology. The volume can also serve as a standard reference book for libraries.

I am highly indebted to my international friends for sharing some species photographs. I sincerely thank Dr. K. Venkataramanujam, former

Dean, Fisheries College and Research Institute, Tamilnadu Veterinary and Animal Sciences University, Thoothukudi, India, for his valuable suggestions. I am very grateful to all my international friends who were very kind enough to provide me with certain photographs necessary for this volume. My thanks are also due to Mrs. Albin Panimalar Ramesh for her help in secretarial assistance and photography.

*—Ramasamy Santhanam, PhD*

# CHAPTER 1

# INTRODUCTION

## CONTENTS

1.1 Portunid Crabs ............................................................... 2
1.2 Subfamilies and Genera of Portunid Crabs .................................. 3
1.3 Commercially Important Portunid Crabs .................................... 3
1.4 Role of Portunid Crabs in Aquaculture .................................... 4
1.5 Nutritional Values of Portunid Crabmeat .................................. 4
1.6 Potential Value of Portunid Crabs ........................................ 5
1.7 Indirect Economic Value of Portunid Crabs ............................... 5
1.8 Socio-Cultural Value of Portunid Crabs ................................... 5
1.9 Scientific Value of Portunid Crabs ....................................... 6
1.10 Environmental Problems ................................................... 6
1.11 Conservation of Portunid Crabs .......................................... 6
Keywords ...................................................................... 7

People living on the coast have depended on marine resources for many thousands of years. Millions earn their living in commercial fisheries, and many people still depend on seafood as their primary source of protein. The very high catch rates of modern industrialized fishing fleets have lead to widespread effects on fished species, non-target species, and marine habitats. Further, overfishing continues to be a major threat to marine ecosystems and vulnerable species throughout the world.

Although dwarfed by fisheries for finfish, crab fisheries are important in many developing countries. As early as prehistoric times, blue crabs were collected for subsistence.

## 1.1 PORTUNID CRABS

Brachyuran true crabs (Class: Crustacea; Order: Decapoda; Infraorder: Brachyura) comprise about 6700 known species worldwide (Melo, 1996) and, due to their great abundance, they are considered as one of the most relevant groups of the marine benthos, both in terms of biomass and community structure. Man consumes many species of crabs of the families—Lithodidae, Macidae, Caneridae, Portunidae, Xanthidae, Potamidae, Geryonidae, Gecarcinidae, and Ocypodidae (Muthiga, 1986). Among these crabs, portunid crabs or swimming crabs assume greater significance in the marine industry owing to their delicate meat with nutritional qualities. These crabs, which reach their greatest species diversity in the Pacific and Indian Oceans, are distinguished from the rest of the round-fronted crabs by the adaptation of some of their legs for swimming, for which these limbs are transformed into flattened paddles. Their strong, sharp claws, allow many species to be fast and aggressive predators. They also have a power of darting at high speed through the water. Corresponding to this mobility, they have a thin flattened form of body, and a lightness gained at the expense of the protective cuticle. These crabs are benthic to semipelagic with diverse habits. Their proper haunt is a space of quiet waters, such as the lagoon of a coral atoll, and these places are generally bottomed with white or greyish coral sand. Portunid crabs have rather diversified feeding habits, including algae, vascular plants, foraminiferous species, cnidarians, molluscs, polychaetes, other crustaceans, fish and other groups. They have ecological importance in the estuarine-marine ecosystem as they serve as a food resource for other aquatic organisms, such as fish and coastal birds. True crabs make up 20% of all crustaceans caught and farmed worldwide, with about 1.4 million tonnes being consumed annually. The horse crab, *Portunus trituberculatus* accounts for one quarter of that total. Other portunid species of great fishery value include *Scylla*

Introduction 3

*serrata, Portunus pelagicus, Portunus sanguinolentus,* and *Charybdis feriatus*—each of which provides more than 20,000 tonnes annually (https://en.wikipedia.org/wiki/Crab_fisheries).

## 1.2 SUBFAMILIES AND GENERA OF PORTUNID CRABS

- Subfamily 1. **Caphyrinae**: *Caphyra, Coelocarcinus, Lissocarcinus,* and **Mioxaiva**[†].
- Subfamily 2. **Carcininae**: *Carcinus,* **Cicarnus**[†], *Echinolatus, Portumnus,* and *Xaiva*.
- Subfamily 3. **Carupinae**: *Carupa, Catoptrus, Laleonectes, Libystes,* **Neptocarcinus**[†], **Rakosia**†, and *Richerellus*.
- Subfamily 4. **Macropipinae**: *Benthochascon,* **Boschettia**[†], *Brusinia, Liocarcinus, Macropipus, Ovalipes,* and *Parathranites*.
- Subfamily 5. **Podophthalminae**: *Euphylax, Podophthalmus,* **Psygmophthalmus**[†], **Sandomingia**[†], **Saratunus**[†], and **Viaophthalmus**[†].
- Subfamily 6. **Polybiinae**: *Bathynectes, Coenophthalmus,* **Falsiportunites**[†], **Geccheliicarcinus**[†], **Maeandricampus**[†], **Megokkos**[†], **Minohellenus**[†], *Necora, Nectocarcinus,* **Ophthalmoplax**, **Pleolobites**[†], *Polybius,* **Pororaria**[†], **Portufuria**[†], **Portunites**[†], **Proterocarcinus**[†], *Raymanninus,* and **Rhachiosoma**[†].
- Subfamily 7. **Portuninae**: **Acanthoportunus**[†], *Arenaeus, Atoportunus, Callinectes, Carupella,* **Colneptunus**[†], *Cronius,* **Euronectes**[†], *Lupella, Lupocyclus,* **Necronectes**[†], *Portunus,* **Pseudoachelous**[†], **Rathbunella**[†], *Sanquerus,* and *Scylla*.
- Subfamily 8. **Thalamitinae**: *Charybdis,* **Eocharybdis**[†], *Gonioinfradens, Thalamita,* and *Thalamitoides*.

*Note:* The gene marked with bold and dagger symbol are extinct.

## 1.3 COMMERCIALLY IMPORTANT PORTUNID CRABS

Though majority of portunids are edible, a small number of species have been reported to be commercially important with major or minor fisheries. They are: *Charybdis affinis, Charybdis anisodon, Charybdis annulatata, Charybdis bimaculata, Charybdis feriatus, Charybdis helleri, Charybdis hongkongensis, Charybdis japonica, Charybdis lucifera, Charybdis natator, Charybdis smithi, Charybdis vadorum, Charybdis variegate, Charybdis truncata, Podopthalmus vigil, Portunus gladiator, Portunus gracilimanus, Portunus haanii, Portunus hastatoides, Portunus pelagicus, Portunus sanguinolentus, Portunus trituberculatus, Scylla serrata,* and *Scylla olivacea.*

## 1.4 ROLE OF PORTUNID CRABS IN AQUACULTURE

Unlike shrimp culture, portunid crab culture provides sustainable marine food supply with lesser culture duration and free from unwanted diseases. Fisheries exploitation and aquaculture production of the world's crabs has increased sevenfold over the past thirty years. Several large of the swimming crabs are considered as candidate species for aquaculture. The most common tropical species of portunid crabs suitable for aquaculture include *Scylla serrata, Scylla olivacea, Scylla tranquebarica, Scylla paramamosain, Portunus pelagicus, Portunus sanguinolentus,* and *Charybdis feriatus.* One temperate species, *Portunus trituberculatus* is also considered as a suitable species for aquaculture especially in Japan. Widespread across Indo-Pacific including Southeast Asia, portunid crabs are among the valuable commodities across many countries, fetching higher market price compared to other locally consumed marine fish species. Increasing capture of small-sized crabs, berried females due to high market demand and decreasing recruitment and capture fishery landing have raised concern over the sustainability of current crab exploitation practices. Thus, the development of aquaculture for portunid crabs and resources management for sustainable fisheries should be the prime focus in ensuring the continuity of availability of portunid crabs.

## 1.5 NUTRITIONAL VALUES OF PORTUNID CRABMEAT

Crabmeat, like fish, has been reported as a nutritious food, low in calories and fat but high in protein and omega-3 fatty acids, which help prevent heart disease. These fatty acids are especially important for women who are pregnant or breastfeeding because they are needed for their child's

Introduction 5

developing nervous system. Omega-3s, which are higher in crabmeat than in shrimp, may also reduce inflammation, enhance immune function, and even lower the risk of certain types of cancer. Crabmeat is also rich in vitamins and minerals, with high levels of vitamin B-12, a vitamin critical for healthy nerve function.

## 1.6 POTENTIAL VALUE OF PORTUNID CRABS

The crabmeat is very tasty and nourishing. Crab curry is a reputed cure for asthma. Similarly, the soup made from the swimming crabs *Portunus sanquinolentus* and *Portunus pelagicus* are commonly used by the people just after recovery from malaria and typhoid. *Scylla serrata* serves as a cure for diarrhea and dysentery. All the edible crabmeats are rich in vitamins and are good for colds, asthma, eosinophilia, primary complex, wheezing, etc. They are also believed to stimulate brain cells. Crab shells are also used nowadays for preparing suture threads. Ointment prepared from crab shells is known to heal wounds quickly and avoids scars.

## 1.7 INDIRECT ECONOMIC VALUE OF PORTUNID CRABS

All the portunid crabs are indirectly important as they play a dominant role in the marine food web. The faeces of these crabs consist of carbon, nitrogen, phosphorus and trace metals which form a rich food for other consumers of the marine ecosystems. The crabs and their larvae are consumed by many predators and omnivorous fishes, and they play a vital role in the transfer of energy through the food chain. Thus they are also of immense help in recycling nutrients which enhance the richness of the soil by "ploughing".

## 1.8 SOCIO-CULTURAL VALUE OF PORTUNID CRABS

Portunid crabs have been associated with religion since first century. Romans worshiped the crab as Sea God and named the crab as Neptune. The crab *Charybdis feriatus* was the symbol of cross by Christians from coastal areas of India. They do not normally eat this crab and let it free to swim in the sea after fishing. *Portunus sanguinolentus*, which has three reddish round spots each encircled by whitish ring, is considered to be the manifestation of Lord Siva who has three eyes. The abnormal growth of body cells (due to cancer/tumor) resembles the nesting ground of *Cancer* spp. So the generic name is used to denote this killer disease.

## 1.9  SCIENTIFIC VALUE OF PORTUNID CRABS

Many highly active biocompounds have recently been isolated from portunid crabs associated with antimicrobial, antileukemic, anticoagulant and cardio active properties. A few species of portunid crabs have been labeled as 'Pests' as they are causing damage to coastal plants and culture ponds. However, the smaller intertidal crabs which are not of direct economic value have been used in the preparation of high-energy yielding, cheaper artificial pellet feeds for the aquaculture of edible varieties of seafood. Portunid crabs are always the chosen as test animals since their tolerance level is very wide. They have been used as tools in the elucidation of physiological mechanisms, fertilization, regeneration and cell association and mechanisms of drug action.

## 1.10  ENVIRONMENTAL PROBLEMS

Ocean acidification and warming seas, caused by carbon emissions could compound crab's troubles. Blue crabs and their catches have been projected to be detrimentally impacted by climate change and excess nutrients especially in the Chesapeake Bay. It is high time to know about the present status of the industry, share insights on relevant issues and identify the problem areas for further research and development on portunid crabs aquaculture and fisheries management for a sustainable portunid crabs industry.

## 1.11  CONSERVATION OF PORTUNID CRABS

At present, there is no ban on fishing immature and the berried crabs and the minimum size at capture is not implemented in many countries. As a conservation measure, the only possibility is to educate fishermen to release the juvenile, berried and soft crabs to the sea while they are alive. The governments should take necessary steps to implement ban during peak spawning seasons to prevent indiscriminate fishing. The best method to ensure a sustainable fishery throughout the year as well as to improve the quality of the yield is to ban fishing and marketing of undersized and berried crabs.

## KEYWORDS

- aquaculture
- commercial species
- conservation
- indirect economic value
- nutritional and health value
- socio-cultural and scientific value

# CHAPTER 2

# BIOLOGY AND ECOLOGY OF PORTUNID CRABS

## CONTENTS

2.1  Ecology of Portunid Crabs ................................................................. 9
2.2  Biology of Portunid Crabs ............................................................... 10
Keywords .............................................................................................. 22

## 2.1  ECOLOGY OF PORTUNID CRABS

### 2.1.1  HABITAT

The various portunid species live in a variety of habitats, from soft substrates like mudflats and sandflats to harder bottoms like coral reefs and rubble. Adult habitats range from near shore estuarine to offshore open water and deep benthic environments. Most portunid species inhabit inshore habitats as adults and move to more offshore areas such as mouths of inlets and open waters to spawn.

### 2.1.2  DISTRIBUTION

Portunid crabs are ubiquitous around the world and the majority of these crabs are present in the Pacific and Indian Oceans.

## 2.2 BIOLOGY OF PORTUNID CRABS

### 2.2.1 MORPHOLOGY OF PORTUNID CRABS

#### 2.2.1.1 External Anatomy

**Antenna/antennae:** These are long segmented appendages located behind the eyestalks. These allow the crab to interact with its environment by touch and chemoreception.

**Antennules:** These are shorter segmented appendages located between and below the eyestalks. These are sensory organs for chemoreception to "smell" and "taste".

**Apron:** This sis the abdomen of a crab, which is folded under the body; male's abdomen is shaped like an inverted Y. An immature female's abdomen is triangular (pyramid shaped) and mature female's abdomen is semi-circular, like the dome of the capitol building.

**Eyes:** These are visual organs mounted on the ends of eyestalks. The eyestalks contain cells that release hormones that inhibit molting.

**Lateral spines:** Paired points on the widest outside edges of the carapace.

**Mouth:** It is opening to the digestive system, and is located between the antennae. The mouth contains jaws that hold and push food into the esophagus.

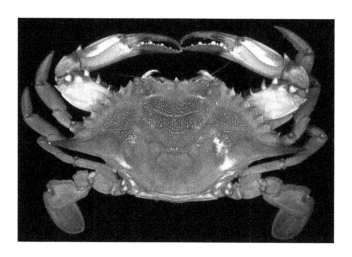

# Biology and Ecology of Portunid Crabs

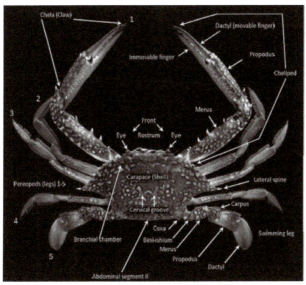

**External features of a portunid crab: dorsal view**

**Carapace:** It is the shell covering the body that provides rigidity and protective covering. It is made of chitin and is the part of exoskeleton (hard outer covering) that covers the head and thorax (center) of the crab. Carapace of portunid crabs may be hexagonal—transversely ovate to transversely hexagonal or almost circular. Dorsal surface is generally relatively flat, and is often ridged or granulose. Front is broad and its margin is usually multidentate. There are 5 to 9 teeth on each anterolateral margin. Posterolateral margins are usually distinctly converging. The antero-lateral margins of certain common genera of portunid crabs are given in the figure.

*Scylla*                  *Portunus*

*Thalamita*          *Charybdis*

**Carapace of certain portunid genera**

**Fluorescent patterns of carapace:** The carapace of fresh specimens of certain portunid species have been observed to have fluorescing patterns which are of great use as a diagnostic character in species identification. These fluorescent patterns vary inter- and intra-specifically and are consistent between males and females of each species, but not amongst juveniles (Ze-Lin et al., 2012).

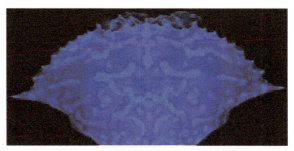

**Carapace of *Portunus pelagicus* under UV lamp**

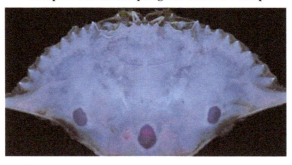

**Carapace of *Portunus sanguinolentus* under UV lamp**

Biology and Ecology of Portunid Crabs

**Appendages:** There are 10 legs (five pairs) including a claw-bearing pair with spines (chelipeds or pincers) used for feeding and defense, followed by three pairs of sharply pointed walking legs, and a pair modified as flat swimming paddles at the rear, swimming legs.

**Cheliped:** The first pair of legs in portunid crabs carries the large claw which is used for defense and obtaining food. These appendages are called as pincers (chelipeds), which are also armed with sharp spines. Often, one pincer is slightly larger than the other. Male's claws are blue tipped with red; female's are red.

**Swimmerets (pleopods):** These are paired abdominal append ages under the apron (abdomen) of the female crab on which the eggs are carried until they hatch.

**Walking legs:** These legs are mainly used for movement. The portunid crabs are capable of walking forward or diagonally, but usually they walk sideways.

### 2.2.1.2 Internal Anatomy

**Gills:** These are the places of respiration and filtration and are consisting of many plume-like filaments arranged around a central axis. There are eight gills on each side of a crab's body.

**Heart:** It is the pump of the circulatory system. It is broad in size and located in the lower center part of the body.

**Hepatopancreas (midgut gland):** It is an extremely large organ with several functions, including the secretion of digestive enzymes and absorption and storage of digested food. It fills most of the area around the stomach, depending on its contents of food reserves and water.

**Intestine:** It is the portion of the digestive system through which digested food passes.

**Stomach:** This is the organ of the digestive system that breaks down swallowed particles of food. It is lined with small hard plates and projections which aid digestion.

**Testes:** These are the parts of the male reproductive system, located on top of the hepatopancreas on either side of the stomach.

**Cartilage:** It encases muscles that help in movement of the legs. The muscles are the edible portions of the crab.

## 2.2.2 FOOD AND FEEDING

Portunids include filter feeders, sand cleansers, mud, plant, and carrion feeders, predators, commensals, and parasites. The diet of portunids varies between the species ranging from slow-moving prey such as snails and annelids to fast-moving ones such as fish and shrimps. The crab uses its mouthparts to chop the food into small pieces and then the gastric mill ossicles further reduce the food to fragments. They are all opportunistic omnivores with a preference for animal prey, but within that framework only rarely feed on more mobile prey such as fish and prawns. Crustaceans have been observed to constitute the most favored food item in *Portunus pelagicus* diet, followed by molluscs and fish. The presence of detritus (80%) in the stomachs of this species suggests that these crabs are also detritivorous, consuming both fresh and decaying flesh of all kinds of animals.

## 2.2.3 BEHAVIOR

Portunid crabs are particularly active at night, but are often also out and about during the day. Besides the large adults, small juvenile swimming crabs are also hidden among the seagrass and seaweed, and other nooks and crannies. These active crabs may possess all kinds of colors. Some can react fiercely by waving spiny pincers. If disturbed, swimming crabs often fearlessly wave their pincers menacingly. This is not an idle threat. When people come too close, this crab might just give a good nip that draws blood.

## 2.2.4 MOLTING AND GROWTH

One of the distinguishing characteristics of the phylum Arthropoda is the presence of an all- enveloping exoskeleton, which is composed of chitin. In all crabs, this chitinous exoskeleton is further strengthened by the deposition of calcium salts. For a crab to increase in size (growth), it must periodically shed its rigid outer covering and form a new one. T his process is called molting or ecdysis. Prior to molt, a new, soft exoskeleton is

Biology and Ecology of Portunid Crabs          15

formed inside, and the old shell becomes loosened. As the crab begins to molt, the carapace is lifted, revealing a gap between it and the abdomen. The shell splits on either side of the under surface, and the posterior part of the crab begins to protrude. Over a period of a few minutes, the crab completes the molting process, gradually backing out of the old shell. Immediately after molting, the shell (carapace) is thin, soft and wrinkled. Subsequently, the new cuticle becomes stretched, and the crab is considerably larger than before the molt. Growth in size of the crab results from the copious intake of water after molting. The number of times that a crab molts during its lifetime, and the length of time between molts, vary among species and are influenced by factors such as temperature and the amount of food available.

### 2.2.4.1  Age and Growth Studies

In studies relating to the age and growth of *Portunus sanguinolentus* and *Portunus pelagicus*, the size frequency analysis indicated that the growth rate was high and more or less uniform in juveniles, while the adults showed relatively low rate of growth and marked variation in males and females. In *Portunus sanguinolentus*, the mean monthly growth rates were found to be 10.3 m and 8.8 mm and attained a carapace width of 124.1 and 112.5 mm by males and females respectively on completion of one year. In *Portunus pelagicus*, the average growth rates were 11.0 and 9.6 mm and attained a size of 145.2 and 132.5 mm by males and females respectively, at the end of one year (Sukumaran and Neelakantan, 1997).

### 2.2.5   *AUTOTOMY AND REGENERATION OF LOST LEGS*

Crabs often lose one or more legs during a lifetime (unless they have reached a stage where they no longer molt) and are able to re place these with new limbs. Most crabs have the ability to drop an injured leg. This process is referred to as autotomy, which is the severance of an appendage at a preformed breakage point as the result of a reflex action. When a leg is severed at this breakage point, the resulting opening is closed by a valve, and there is very little loss of blood from the stump. It has also been shown that regeneration, or growth of a new limb, takes place more rapidly at this point. The process of autotomy is an adaptation that permits the crab to get rid of its injured limb quickly at a point where the possibilities for healing and regeneration are maximum. Autotomy may also be of value as a means of escape when an appendage is grasped by an enemy. Usually two or three molts are necessary to produce a claw or leg of normal size.

## 2.2.6 REPRODUCTION AND LIFECYCLE

### 2.2.6.1 Sexual Dimorphism and Sexual Characters

As in other crabs, in portunid crabs, sexes are separate and sexes can be distinguished from the shape of the abdomen. In males the abdomen is narrow, inverted 'T' shaped and in addition mature males have larger and broader chelae. The first and second abdominal appendages (pleopods) are highly modified to form an intromittent copulatory organ. Females, on the other hand, have a broad abdomen which is conical or oval in shape (according to the stage of maturity) and bear four pairs of pleopods. Many species of portunid crabs show sexual dimorphism, with males being larger, smaller, or possessing special or enlarged structures. In some species, the females are the larger. Most commonly, males have proportionately much larger chelipeds or chelae. Males always have only two pairs of gonopods (uniramous swimmerets or pleopods), which are specially modified for copulation (most crabs have internal fertilization). The first gonopod is a highly modified pleopod which has been folded or rolled longitudinally to form a cylindrical tube. The form of the first gonopod varies from broad to very slender, straight to sinuous, and even recurved.

**Ventral views of female and male crabs**

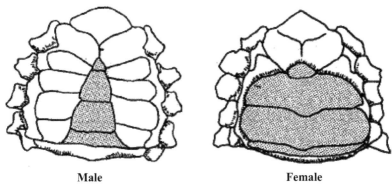

**Male**         **Female**
**Abdomen of *Portunus pelagicus***

## 2.2.6.2 Reproductive System

The male reproductive system of portunid crabs is bilaterally symmetrical creamy to whitish in color, composed of a pair of testes, a pair of vas differentia, and a pair of ejaculatory ducts internally, and a pair of pleopods externally as accessory reproductive organs, present on the inner side of the abdominal flab. The vas differentia has been divided into three distinct regions, viz. anterior, median and posterior vas deferens. The female reproductive system on the other hand is composed of a pair of ovaries, a pair of seminal receptacles (or) spermatheca, and a pair of oviducts which open to the exterior through the female genital opening situated on the left and right sternites of sixth thoracic segment. The ovaries, according to the size, color and external morphology are categorized into five stages viz. immature, early maturing, late maturing, ripe and spent. In general males mature earlier than females and the size at first maturity varies from species to species.

## 2.2.6.3 Mating and Spawning

Mating takes place as soon as the female crab molts with a hard male. The sperms are transferred and stored in the spermatheca of the female crab. After the spawning, the eggs are attached to the endopodites of the pleopods and females carry the 'berry' till hatching and release the planktonic larvae (zoeae). The embryonic development takes about 10 days. Hatching generally takes place during early morning hours. Spawning can occur year round, but each species spawns within certain months.

**Mating in *Charybdis feriatus***

**Mating behavior in *Charybdis feriatus*:** For convenient, the mating behavior of this species is divided into the following stages.

***Pre-molt guarding:*** When the hard-shelled male crab is with a female crab which is at the verge of premolt stage, the female tries to attract the male. Now the male crab contacts the female by moving towards her and extending his bigger chelate in front of her without showing any other attracting courtship display. The male after making contact with the female ultimately moves towards her and catches the female with the use of his chelae and embraces in a short while using his walking legs. This position is called as cradle-carrying position. During this position, both crabs are facing in the same direction. In general the male is active and aggressive but the female is defunct. Premolt guarding is last for 92 hours.

***Mating or copulation:*** Mating is initiated when the female's exoskeleton is soft. The male crab becomes very active and rotates the female by using his walking legs and by using chelate he turned over her. The female crab positions herself upside down beneath the male crab and extends her abdomen exposing gonophores allowing the male to insert his paired gonopods into her genital pores. By this time male and female crabs are facing in opposite direction. The female is in reverse position, her lower side is directed towards the males ventral side and the abdomen of both are flung backwards. During copulation, the male often walks around with the female, holding her with third and fourth walking legs. The copulation lasts for about 7 hrs.

***Post-copulatory guarding:*** After the completion of copulation, the male crab liberates the female from mating position and embraces her in a short while (for only few hrs.). The female is inactive until she attains normal hardness of her exoskeleton. Post-copulatory guarding lasts for about 12 hrs. The total mating sequence lasts for 119 hrs.

***Spawning:*** The male crabs deposit spermatophores in the female's spermatheca during mating. During the process of spawning or extrusion, the eggs are liberated from the ovaries passing through the seminal receptacles. In seminal receptacles, the stored sperms are liberated from the spermatophores to fertilize the eggs and fertilized eggs are extruded through the gonophores present in the sternites of the sixth thoracic segments of third pair of legs and these eggs become attached to the smooth setae present in the endopodites of the four pairs of pleopods in the abdominal flab. The egg mass segregated and carried on the abdominal flab is called as berry or sponge. The females carrying eggs is also called as berried crabs (ovigerous females). The freshly extruded eggs are initially orange in color and become black before hatched into zoeae. The spawning of the female takes about 17 days after copulation.

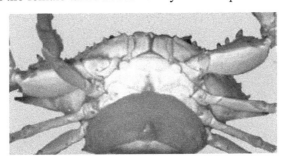

**Ovigerous female crab of *Portunus pelagicus***

**Ovigerous female crab of *Scylla serrata***

## 2.2.6.4 Fecundity

Fecundity is an index of reproductive capacity and is expressed in terms of the number of eggs produced by an animal. Among portunid crabs, the fecundity varies from species to species. Fecundity in ovigerous (berried) females is positively correlated with the size of the egg-bearing females in all species. The relationship between female size and egg number is usually described as an allometric function equivalent to that between size and weight. The increase in fecundity is here explained by positive allometric relationship (increase in egg number with the increase in total width). For portunid crabs, correlation is often high and body size is the prime factor in fecundity per brood. For example, in *Portunus pelagicus*, the fecundity has been reported to range between 60,000 and 19,76,398 eggs in crabs with carapace widths of 100 to 190 mm from Indian waters. In the same species from Malaysia, the fecundity has been reported to range from 1,48,897 to 8,35,401 eggs within a carapace width of 102–140 mm. The fecundity in *Portunus sanguinolentus* has been reported to range from 1,58,608 eggs to 22,50,000 eggs.

## 2.2.6.5 Life cycle

**Embryology:** In the Portunidae, there are generally seven zoeal stages and one postlarval or megalopal stage. Occasionally, an eighth zoeal stage is also observed.

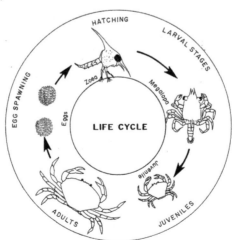

**Lifecycle of a portunid crab**

## Description of larval stages of portunid crabs

**Crab zoea:** Crab zoea has a strong and almost spherical carapace to which the legs it uses for swimming are attached. Legs are absent on the long and extended back. The often long and prominent rostrum always points downwards, but it may also be absent. Crab zoea larvae usually have several other outgrowths on their body, amongst them a very prominent vertical outgrowth on the back.

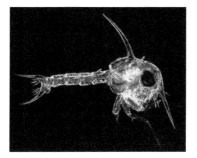

**Crab zoea**

**Crab megalopa:** The megalopa larval stage of different species of portunid crabs is distinguished by the following features; the segmentation and setation of the antennae; length, curvature, and thickness of the rostrum; the presence or absence and location of spines on the cheliped; the presence or absence of coxal spines on pereopods 2–5; length and shape sternal spines; the presence of a paddle like dactyl on the 5th (swimming) leg, as well as the number of stiff hooked setae on the dactyl; presence or absence of lateral spines on the 5th abdominal segment; shape of the telson; and the number of setae on the uropod.

**Crab megalopa**

Larval release usually occurs at high tide assuring larval abundance is at its peak during the ebbing tide. Crab larvae are advected offshore, and complete development in the coastal shelf waters. The typical time for development through the 7 zoeal stages is between 30 and 50 days before metamorphosis to the megalopal stage. The megalop is capable of postponing metamorphosis to first crab and may persist from 6 to 58 days. It is widely believed that the megalopal stage subsequently returns to the estuaries for settlement, and recruitment to adult populations.

## KEYWORDS

- ecology
- food and feeding
- molting and growth
- autotomy and regeneration
- reproduction
- life cycle

# CHAPTER 3

# PROFILE OF PORTUNID CRABS (FAMILY: PORTUNIDAE)

## CONTENTS

3.1 Characteristics of the Family Portunidae ....................................... 23
3.2 Subfamily Portuninae .................................................................. 24
3.3 Subfamily Thalamitinae ..............................................................115
3.4 Subfamily Podophthalminae ....................................................... 202
3.5  Subfamily Carcininae ................................................................ 207
3.6 Subfamily Polybiinae ................................................................. 215
3.7 Subfamily Macropipinae ............................................................ 235
3.8 Subfamily: Carupinae ................................................................ 253
3.9 Subfamily Caphyrinae ............................................................... 262
Keywords ............................................................................................ 271

## 3.1 CHARACTERISTICS OF THE FAMILY PORTUNIDAE

The crabs of this family are commonly called as swimming crabs. Carapace is usually round or oval. Dorsal surface is relatively flat to gently convex and is, usually ridged or granulose. Front is broad and margin is usually multidentate. There are usually 5 to 9 teeth on each anterolateral margin. Posterolateral margins are usually distinctly converging. Endopodite of second maxillipeds is with strongly developed lobe on inner margin. Legs are laterally flattened to varying degrees and the last two segments of last pair are paddle-like. Male abdominal segments (3 to 5) are completely fused and immovable.

## 3.2 SUBFAMILY PORTUNINAE

### 3.2.1 CHARACTERISTICS OF SUBFAMILY PORTUNINAE

Carapace of the crabs of this subfamily is subhexagonal; rarely subcircular; and is usually distinctly broader than long. Antero-lateral margin is divided into more than five teeth (5–9). Basal antennal segment is elongated and is lying obliquely. Chelipeds are longer than ambulatory legs and are bearing a set of spines on merus, carpus and palm. Last pair of legs is with paddle-shaped propodi and dactyli. Male first gonopod is without subterminal spines.

*Portunus argentatus* (Milne-Edwards, 1861)

**Common name:** Formosa crab
**Distribution:** Madagascar, Mozambique, Red Sea, Republic of Mauritius, South Africa, and Tanzania.
**Habitat:** 30–100 m depths.
**Description:** Carapace is moderately broad (breadth is 1.7–1.9 times length) and its surface is with widely spaced granular areas being separated by regions with a dense pubescence; and regions are well recogniz-

able. Front is with 4 rounded lobes, of which laterals are much broader than laterals. Antero-lateral borders are with 9 teeth, of which last one is much the longest and is directed outwards. Posterolateral junction is rounded. Antero-external angle of merus of third maxillipeds is markedly produced into a lobe. Chelipeds are stout; merus is with 2 spines at posterior border, anterior border is with 4 spines; carpus is with 2 spines of normal length; upper surface of palm is with 1 distal spine, and lower surface is with squamiform markings. Posterior border of merus of swimming leg is serrated and bearing a few spines; and is dactyl with an obvious red spot.

**Biology:** This species is omnivorous and it mainly preys on crustaceans and fish.

**Fisheries:** Not known.

### *Portunus anceps* (Saussure, 1858)

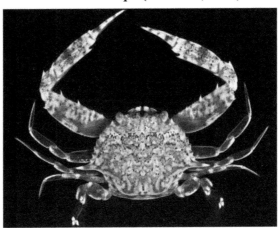

**Common name:** Delicate swimming crab

**Distribution:** Western Atlantic Ocean, Occidental Atlantic: North Carolina, Bermuda, Florida, Gulf of Mexico, West Indies, and Brazil (Amapá to Rio de Janeiro).

**Habitat:** Depth range (m): 2–57.

**Description:** Not known.

**Biology *Pigments:*** The pigmentary system of this species consists of white, black and yellow pigments. These pigments are present in separate chromatophores which are more numerous on the legs than elsewhere on the body. The white pigment bodies, which are the most numerous of the three, are distributed more or less uniformly over the dorsal surfaces of the appendages. In the central portion of the leg segments, especially the merus, the white pigment is absent or scarce, and in this region a cluster of black pigment bodies is situated. The yellow pigment, unlike the black, is uniformly distributed. The chromatophores are most distinct in the joints between the leg segments, especially between the merus and the ischium, and between the basis and the coxa. In addition to the pigments already mentioned, a blue color is observed on the legs. This blue pigment occurs as a haze in the vicinity of the black pigment (Abramowitz, 1935).

**Fisheries:** No information

### *Portunus binoculus* (Holthuis, 1969) (=*Achelous binoculus*)

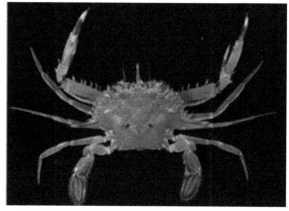

**Common name:** Red-spotted swimming crab

**Distribution:** W. North Atlantic, Caribbean Sea, and North Atlantic Ocean.

**Habitat:** Muddy with much terrestrial detritus, such as leaves, etc.—depth range 74 and 291 m.

**Description:** Carapace of this species is longer than, broad and arch formed by anterolateral margins and front is higher. Although cervical groove is distinct, other depressions are less conspicuous. Two transverse ridges over gastric area are mere rows of granules. Epibranchial region is more sunken. Front bears four very sharp triangular teeth, which are more sharply pointed and more markedly set off from frontal margin. Antero-lateral margin of carapace bears nine teeth, of which second, fourth, and sixth are somewhat smaller than others. Last tooth is extremely long, being almost four times as long as preceding tooth. Branchial ridge is distinct. There are two transverse gastric ridges, anterior being wider than posterior; and they are connected by a longitudinal ridge which extends over gastric region. No other ridges are present. Posterolateral margin of carapace is evenly rounded or somewhat angular. Distal lobe of merus of third maxilliped is distinctly larger; it is wider distally and reaches farther forward. Last three segments of this maxilliped are iridescent. Cheliped has merus which is more slender. Anterior margin of merus carries four or five spines; and its posterior margin ends in a distal spine. Upper margin of palm bears a strong subdistal tooth; and outer surface bears a sharp tooth near articulation with carpus. Lower margin of chela shows an elevated ridge. Inner surface of chela is flat and in male is heavily clothed with long red hairs. Fingers have cutting edges provided with many small sharp teeth of different size. Upper margin of dactylus bears four, and lower margin of fixed finger bears two carinae. Upper margin of dactylus bears a dense fringe of long red hairs. Carapace is uniformly greyish green and there are two distinct submedian red spots in middle of carapace. A dark median longitudinal line extends over cardiac and intestinal regions. Upper surface of merus of chelipeds is spotted over greater part of its length. Carapace length of females ranges from 14 to 24 mm and carapace breadth from 29 to 46 mm.

**Biology:** No information
**Fisheries:** Unknown.

### *Portunus brocki* (de Man, 1888)

**Common name:** Not designated.

**Distribution:** From Japan to eastern Indian Ocean including China, Singapore, Philippines, Indonesia and Australia.

**Habitat:** Sandy mud; 30 m depth.

**Description:** Carapace is moderately broad (breadth is slightly less than twice length); surface is with elevated areas, whose granulation is not always distinct and which are separated by areas of fine pubescence; regions are well recognizable; and front is truncate and very indistinctly quadrilobate. Antero-lateral borders are with 9 teeth, of which last one is much larger than the preceding ones and projecting slightly backwards; and posterolateral junction is rectangular. Antero-external angle of merus of third maxillipeds is markedly produced into a lobe. Chelipeds are moderately stout; granulated and tomentose; merus is with a spine and a lobulated tooth at postero-distal border; anterior border is with 2–3 spines; carpus is with 2 spines; upper surface of palm is with one distal spiniform lobule; and lower surface smooth. Posterior border of swimming leg is with the fringing hairs.

**Biology:** No information

**Fisheries:** Not known.

### *Portunus convexus* (De Haan, 1835)

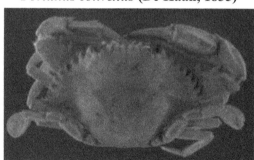

**Dorsal view**

**Frontal view**

**Common name:** Red-legged swimming crab

**Distribution:** Indo-West Pacific: Comprend Afrique orientale, Erithrée, Somalie, Europa, Madagascar, Seychelles, Réunion, Maurice, Indonésie. Includes East Africa, Eritrea, Somalia, Europa, Madagascar, Seychelles, Reunion, Mauritius, Indonesia and Guam.

**Habitat:** Soft bottom (sand or silt); mudflats and estuaries.

**Description:** Not known.

**Biology:** It feeds mainly on fish

**Fisheries:** Not known.

## *Portunus dayawanensis* (Chen, 1986)

Image not available

**Common name:** Not designated as it is a new species.

**Distribution:** Sanmen I., Daya Bay (= Dayawan), Guangdong, China (type locality), and Tolo Harbor, Hong Kong.

**Habitat:** Not known.

**Description:** It is a small species (maximum 3–4 cm in carapace width). Carapace is dorsoventrally flattened, dorsal surface is covered with small granules, slightly tomentose or naked; regions are moderately well defined; mesogastric, cardiac, protobranchial, mesobranchial, and lateral and median postcardiac regions are elevated and each is separated by grooves. Metagastric region is slightly depressed. Front is armed with 3 rounded lobes (excluding inner supraorbital teeth), of which median is slightly less pronounced than laterals. Anterolateral margin is armed with 9 distinct teeth, of which anterior 4 are slightly smaller than preceding 4. Epibranchial teeth are straight, longest, and are more prominent than preceding 8. Anterolateral border is fringed with plumose setae. Posterolateral and posterior borders are acute, and are armed with small but conspicuous spine. Chelipeds are slightly unequal; right is larger than left; anterior margin of merus is armed with 4 sharp teeth; posterior margin is with 2 teeth; anterior and posterior margins are fringed with short, plumose setae; carpus is armed with 2 teeth; and palm is slightly longer than fingers. All ambulatory legs are compressed, and are fringed with setae on ventral edge on all segments. Male abdomen has an inverted "T" shape and telson is ovate, narrow and longer than broad. 6th

abdominal somite is longer than broad and lateral margins are convex. 3$^{rd}$–5$^{th}$ somites are fused.
**Biology:** Not known.
**Fisheries:** No information

## *Achelous depressifrons* **(Stimpson, 1859)** *(= Portunus depressifrons)*

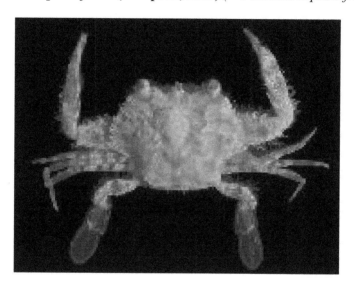

**Common name:** Flatface swimming crab; flat-browed swimming crab.

**Distribution:** Western Atlantic Ocean, Caribbean sea, Gulf of Mexico, Bermuda, North Carolina, and Florida.

**Habitat:** Reef-associated; sandy bottoms of coves and inlets of warmer coastal waters.

**Description:** Lateral spine IS very little longer than those preceding it. Upper half of outer surface of pam IS granulated between carina. A fringe of hair is seen on upper margin of movable finger. Posterodistal margin of merus of swimming legs is entire.

**Biology:** No information
**Fisheries:** No information

## *Portunus gibbesii* (Stimpson, 1859)

**Common name:** Iridescent swimming crab
**Distribution:** Occidental Atlantic: Massachusetts to Florida, Gulf of Mexico, Venezuela, Guiana's and Brazil (Bahia).
**Habitat:** Gulf and bay; Intertidal zone – 90 m.
**Description:** Carapace is broader than long and is compressed. Last pair of legs is flattened into paddle-shaped appendages. About 9 pairs of spines (including the outer orbital spines) are seen along lateral edges of carapace and most posterior pair (called lateral spines) is about 3–4 times as long as others. 8 low blunt teeth are seen between ocular orbits (eye sockets). About 4–7 spines are found on inner margins of merus (2nd segment from claws) and a small spine on inner and outer dorsal carpus (segment next to claw). Color of carapace is olive green to brown dorsally, yellowish to white underneath, white spot on each side of lower half of carapace, and iridescent patches at base of lateral spines. Legs are purple or brown with purple tint. Claws are long and slender. Palms are white. Fingers in males are blue, and in females hey are white with brown spot on movable finger. Maximum carapace width of this species is 7.5 cm.
**Biology**: Not known.
**Fisheries:** No information

## *Portunus gracilimanus* (Stimpson, 1858)

**Common name:** Hairy crab; real estate broker crab
**Distribution:** Indo-West Pacific: Singapore, China, and Taiwan.
**Habitat:** Water depths of 20 to 60 m; sand and mud bottom.
**Description:** A dense, short fine hair is present on surface of carapace. Posterolateral junction of carapace is rounded. Hand of cheliped is extremely slender, and is much less massive than arm. 4 to 6 spines are seen on anterior border of arm. Mesogastric region of carapace is with transverse granulated ridges. Lateral frontal teeth are almost the same width as medians. Teeth of the anterolateral margin are minute. An acute tooth or short spine at the posterior corner of carapace. Third pair of ambulatory legs is long. One spine is present near base of fingers on upper side of hand. This species is of a white color, punctate with dark-brown.
**Biology:** No information.
**Fisheries:** Minor commercial fisheries exist for this species in China.

## *Portunus granulatus* (Milne Edwards, 1834)

**Common name:** Not designated.
**Distribution:** Africa, Japan, Australia and Hawaii.
**Habitat:** Marine; near-shore; 30–120 m depths.
**Description:** Carapace is narrow (breadth about 1.5 times length); surface is with close-packed granular areas with a fine pubescence among the granules; regions are well recognizable on denuded specimens; front is with 2 broad, reasonably sharp lateral lobes, and a central confluent lobular area without clear lobes; anterolateral borders are with 9 teeth, of which last one is slightly most protruding; and posterolateral junction is rounded. Antero-external angle of merus of third maxillipeds is markedly produced into a lobe. Chelipeds are elongate, and propodus is nearly as long as merus; merus is with 2 spines at posterior border; anterior border is with spines set among spiniform granules; carpus is with 2 spines; upper surface of palm is with 1 distal spine; and lower surface is smooth. Posterior border of swimming leg is with usual fringing hairs.
**Biology:** Not known.
**Fisheries:** No information

## *Portunus haani* (Stimpson, 1858) *(= Portunus gladiator)*

**Common name:** Han's swimming crab; gladiator swimming crab.

**Distribution:** Indo-West Pacific: New Caledonia, Mozambique, Mauritius and Madagascar.

**Habitat:** Eelgrass beds and water bottoms; 10–345 m depths.

**Description:** Carapace moderately broad (breadth about 1.8 times length) and depressed; surface is with widely spaced granular areas being separated by regions with a dense pubescence; regions are well recognizable; front is with 4 lobes, of which medians are acute, and laterals are right-angled and much broader. Antero-lateral borders are with 9 teeth, of which last one is longest and directed outwards; and posterolateral junction is rounded. Antero-external angle of merus of third maxillipeds is markedly produced into a lobe. Chelipeds are stout. Cheliped merus is with 2 spines at posterior border and anterior border with 4 spines. Carpus is with 2 spines of normal length; upper surface of palm is with 1 distal spine; and lower surface is with squamiform markings. Posterior border of merus of swimming leg is serrated and is bearing a few spines; and dactyl is without a red spot. Penultimate segment of male abdomen is not convex on outer border. Distal end of the swimming leg is red. Carapace length of male is 47 mm and breadth is 85.7 mm.

**Biology:** It is a companion of the colorful blue crab.

**Fisheries:** Minor commercial fisheries exist for this species in China. Frozen cooked crab pincers of this species are of export value.

This crab is with high protein and minerals and low- fat. Hence it is an ideal seafood.

### *Portunus hastatoides* (Fabricius, 1798)

**Common name:** Spear-shaped swimming crab
**Distribution:** East Coast of Africa, Madagascar, India, Philippines to Australia and Japan.
**Habitat:** Bottoms of sand or mud; 30 to 100 m deep.
**Description:** Carapace is very broad (breadth 2–2.4 times length); surface is with conspicuous elevated granular areas; and areas are well delimited. Front is with 4 acute lobes, of which outer pair is much larger than inner ones; antero-lateral borders are with 9 teeth, of which acuteness increasing fore to aft; last one is very large and projecting straight out laterally; and posterolateral junctions bear a spine. Antero-external angle of merus of third maxillipeds is markedly produced into a lobe. Cheliped merus is with postero-distal end with 2 long sharp spines; its anterior border is with 4 spines; carpus is with 2 spines; and palm is with 1 spine on upper distal border with smooth, lower surface. Posterior border of merus and propodus of swimming leg are serrated and other joints are without spines or spinules.
**Biology:** An association between the bivalue, Amussium pleuronectes (Linne) and this species has been reported by Sethuramalingam et al. (1980).
**Fisheries:** Minor commercial fisheries exists for this species in China.

### *Portunus hastatus* (Linnaeus, 1767)

**Common name:** Lancer swimming crab; eyespot swimming crab.
**Distribution:** Eastern Atlantic and throughout the Mediterranean.
**Habitat:** Demersal; sandy and muddy sand bottoms; depth range 0–55 m.
**Description:** Carapace is dorsoventrally flattened, rhomboid and wider than long. Anterolateral edges have 8 spines and lasts about 4 times longer than others. 4 blunt spines are seen between eyes. Chelae are large with 2 spines on propodus. Brown bands are seen on chelae and grooves of dorsal surface of carapace. Maximum carapace length is 2.6 cm and maximum width is 5.6 cm.
**Biology:** No information
**Fisheries:** Minor commercial fisheries exist for this species.

### *Portunus innominatus* (Rathbun, 1909)

**Common name:** Not designated.

**Distribution:** Andaman Islands, Burma (Gulf of Martaban and Arakan Coast), Malay Peninsula, Gulf of Thailand, Java Sea, Sunda Strait, Ambon and New Caledonia.

**Habitat:** Sandy bottom; 6 m depth.

**Description:** Carapace is nearly twice as broad as long and convex. Four frontal lobes are somewhat protruding beyond inner supraorbital lobe; and acutely rounded medians are much narrower than broadly rounded laterals. Orbits are finely granulated; inner supraorbital lobes are truncated; and suborbital lobe is produced beyond frontal margin. Carapace is covered by very short dense tomentum and its granular ridges and areas are slightly elevated. Nine anterolateral teeth, of which 1st is relatively obtuse, reaching level of inner supraorbital lobe; following teeth (2–8) are smaller than 1st and are gradually increasing in size; first three of them are obtuse and rest are sharp. Last tooth is twice longer than longest of preceding teeth, sharp, and slightly curved forward. Chelipeds are 2.6–2.7 times longer than carapace. Abdomen somites 1 and 2 are completely fused; somites 3 and 4 are partly fused; three proximal somites are with usual transverse keels, and somite 4 is with keel (strong in larger female) in median third. Carapace length varies between 9 and 12.5 mm and carapace width is between 16 and 22 mm.

**Biology:** Unknown.

**Fisheries:** No information

### *Portunus iranjae* (Crosnier, 1962)

**Common name:** Iranjae swimming carb

**Distribution:** Madagascar, Japan, Taiwan, China, Sulu Archipelago.
**Habitat:** Shallow waters; intertidal; depth range 13–80 m.
**Description:** Carapace is very broad (breadth about 2 times length); surface is with elevated granular areas; regions are well recognizable; front is with 4 low set and rounded lobes, of which outer pair is very much broader and longer than inner ones; antero-lateral borders are with 6–9 (mostly 8–9) teeth, of which last one is very large and projecting obliquely backwards; and posterolateral junction is elevated and is bearing a spine. Antero-external angle of merus of third maxillipeds is rounded, and is not markedly produced into a lobe. Chelipeds are moderately long (not more than 2.5 times as long as carapace). Cheliped merus is with 1 spine at postero-distal border and anterior border is with 3–4 spines; and carpus is with 2 spines. Upper surface of palm is with 2 distal spines and its lower surface is smooth to naked eye. Posterior border of merus of swimming leg is very finely serrated, and remaining joints are smooth. Male and female measure 4.0 mm and 7.0 mm, respectively.
**Biology:** No information
**Fisheries:** No information

***Portunus longispinosus* (Dana, 1852)** *(=Amphitrite longispinosa)*

**Dorsal view**

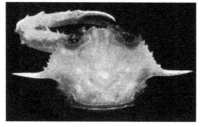

**Carapace**

Profile of Portunid Crabs (Family: Portunidae)

**Common name:** Long-spined swimming crab
**Distribution:** Indian and western Pacific Oceans, India, Maldives, Sri Lanka, Australia, Persian Gulf, Japan, and Hawaii.
**Habitat:** Sandy or shelly bottoms; depth range 1–305 m.
**Description:** Carapace is flattish and its length is from 2/3 to ¾ its breadth without spines. Its surface is subtomentose and is cut up into well-defined sub-regional elevations. Front is prominent and beyond epistome, cut into four usually acute, triangular teeth, of which middle two are small and receding and outer ones are very large and prominent. Supra-orbital margin is cut by 2 fissures. Antero-lateral borders are moderately oblique, and are armed with a variable number of small and inconspicuous teeth, and ending in a lateral epibranchial spine that is about half the breadth of carapace in length. Number of teeth, including outer orbital angle and lateral spine, varies from 6 in young to 9 in adult. Posterior border is nearly straight and makes a dentiform angles. Chelipeds are very long and slender and are more than 2.5 times as long as carapace.Merus is with 1 spine at postero-distal border and anterior border is with 4 spines.Carpus is with 2 spines. Upper surface of palm is with 2 distal spines.. First three pairs of legs are slender. Carapace width and length are 24.3 mm and 11.7 mm, respectively.
**Biology:** No information
**Fisheries:** No information

*Portunus macrophthalmus* **(Rathbun, 1906)**

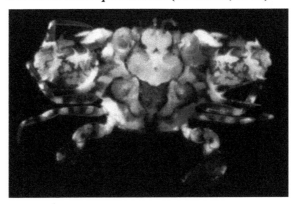

**Common name:** Big-eyed swimming crab
**Distribution:** Indo-Pacific, Seychelles, Cargados Carajos Islands, Mauritius, Japan, Philippines, Indonesia, and Hawaiian Islands.

**Habitat:** Uncommon in sandy rubble at scuba depths.

**Description:** Posterolateral junction of carapace is angular or spinous; front is distinctly 4-toothed or 4-lobed, of which median frontal teeth are very much shorter than, and less prominent than, laterals; and lateral frontal teeth are triangular, projecting, and typically sharp. Eyes are large. Carapace is moderately strongly embossed, with tuberculate but not spiniform elevations in cardiac and mesobranchial areas; metagastric areas are with 2 tubercles; posterior border of arm of cheliped is with single spine; and upper surface of palm of cheliped is with 2 distal spines, or 1 spine and 1 tubercle. Chelipeds are of moderate length and robustness, less than 2.5 times the carapace length. It attains a carapace width of only 2.5 cm.

**Biology:** Unknown.

**Fisheries:** No information

### *Portunus nipponensis* (Sakai, 1938)

**Common name:** Not designated.

**Distribution:** Réunion, Japan, Borneo, Indonesia, and Tuamotu Archipelago.

**Habitat: Depth range** 15–130 m.

**Description:** Carapace is broad (breadth is 2 times length); surface is naked, with distant granules on anterior and lateral surfaces; finer granules are present on posterior and posterolateral surfaces; regions are well recognizable; 4 transverse granular ridges are seen on gastric regions; and epibranchial ridges are long and distinct. Front is with 4 subequal teeth; antero-lateral borders are with 9 teeth, of which last one is very large and projecting sideways; posterolateral junction is elevated and with a spine; and pterygostomial region is with a long stridulating organ. Antero-external angle of merus of third maxillipeds is markedly produced into a lobe. Chelipeds are heavy and asymmetrical. Cheliped merus is with 1 spine at posterior border; anterior border of merus is with 3 spines; carpus is with 2 spines; and upper surface of palm is with 1 distal spine. Posterior border of swimming leg is with the usual fringing hairs.

**Biology:** Unknown.

**Fisheries:** No information

### *Portunus pankowskiorum* (new species)

Image not available.

**Common name:** Not designated.

**Distribution:** Southern California, USA.

**Habitat:** Not known.

**Description:** Carapace is ovate; it is wider than long and its length is about 65% carapace width. Frontal margin is with four spines to six spines including inner orbital spine; and inner two are closely spaced. Anterolateral margin is arcuate; it is with at least eight anterolateral spines which are becoming slightly larger posteriorly; each spine is with straight anterior margin and curved posterior margin; first seven are directed anterolaterally; and last spine is longest and stoutest, directed laterally, marking widest part of carapace. Posterolateral margin is arcuate and concave; and posterior margin is straight. Chelipeds are weakly heterochelous. Major chelipeds are with short ischium, possibly with spines on upper surface. Cheliped merus is much longer than wide; outer surface is keeled; and upper and lower surfaces are with spines. Carpus is very short. Manus is much longer than high; outer surface is with 2 keels; and upper surface is with keel. Fixed finger is long, slender, with molariform denticles with black tips on occlusal surface; and movable finger is arcuate, slender, with denticles on occlusal surface. Pereiopods 2–4 are with more flattened proximal elements than chelipeds; and pereiopod 5 is with flattened, short, ovate carpus, manus, and dactylus. Maximum width of carapace is 54.0 mm and maximum length is 34.9 mm.

**Biology:** Not known.

**Fisheries:** No information

## *Portunus pelagicus* (Linnaeus, 1758)

**Common name:** Flower crab, blue crab, blue swimmer crab, blue manna crab or sand crab.

**Distribution:** Indo-Pacific Ocean, Red Sea to Tahiti.

**Habitat:** Sublittoral sediments; adults occur on sand, sandy-mud, from intertidal to 65 m, near reefs, mangroves, seagrass and algal beds; juveniles commonly occur in the intertidal areas and in estuaries.

**Description:** Carapace is very broad; surface is coarsely granulated, and with a short but dense pubescence between granules; mesogastric, epibranchial, and indistinct metagastric ridges, are distinct; and cardiac and mesobranchial ridges are with low granular eminences. Front is with 4 acute teeth, of which outer pair is larger and more prominent than inner ones; antero-lateral borders are with 9 teeth, of which last one is very large and projecting straight out laterally; and posterolateral junction is rounded. Cheliped merus is with postero-distal border spinous; anterior border is with 3–4 (usually 4) sharp spines; carpus is with inner and outer spines; and lower surface of palm is smooth. Posterior border of swimming leg is without spines or spinules. Maximum carapace width is 20 cm (males). Common size in male: 140 mm CW (64 mm carapace length).

### Biology

*Behavior:* A marine nocturnal crab and is an active swimmer, but during inactive periods it buries in sediment.

*Food and feeding:* Blue swimmer crabs are opportunistic, bottom-feeding carnivores and scavengers. They are most active in foraging and feeding at sunset. Their diet chiefly consists of a variety of sessile and slow-moving invertebrates, including bivalve molluscs, crustaceans, polychaete worms and brittle stars and seagrasses and algae.

*Growth:* Blue swimmer crabs molt one or more times. Just before molting, the underlying skin secretes substances that sever the connections between the skin and the old shell. A thin layer of new shell is excreted just below the old shell, which begins to split. Once the old shell is discarded, the crab takes up water to stretch the new shell. The crab absorbs calcium from the seawater and these calcium sources are used for the recalcification of the shell. Crabs then ingest large amounts of calcareous material, such as shell fragments, to complete calcification. Under laboratory conditions, juvenile crabs of 11–25 mm CW attained a size of 140–145 mm CW at the 12th molt after a period of 14 months.

Age and growth studies were also made in this species by Sukumaran and Neelakantan (1997). The size frequency analysis indicated that the growth rate was high and more or less uniform in juveniles, while the adults showed relatively low rate of growth and marked variation in males and females. The average growth rates were 11.0 and 9.6 mm and attained a size of 145.2 and 132.5 mm by males and females respectively, at the end of one year.

*Longevity:* Longevity of this species determined has been reported to be 2.5 years (152 mm CW).

*Reproduction:*

*Seasonal variation:* In tropical regions this species has been reported to breed throughout the year, whereas in temperate regions, reproduction is found restricted to the warmer months.

*Size at maturity:* Blue swimmer crabs mature at about one year of age. The size at which maturity occurs can vary with latitude or location and within individuals at any location.

*Fecundity:* The number of eggs produced by females varies with the size of the individual as well as between individuals of a similar size. Generally, larger females produce more eggs than smaller females. Fecundity may vary from 900,000 to 1,600,000 eggs per batch. A female has the ability to produce three batches of eggs between molts and therefore it is possible to produce as many as 3,600,000 eggs in six weeks.

## Larval Stages in Portunus pelagicus:

The larval development of Portunus pelagicus includes four zoeal stages and a megalopa. The megalopa stage metamorphoses into the crab stage. Each zoea has a long rostrum, a dorsal spine, and a pair of short lateral spines on the carapace.

*First zoea:* Carapace length is 0.44–0.54 mm; abdomen-telson length is 1.07–1.23 mm. Eyes are sessile. Abdomen is five-segmented plus the telson. Telson is forked, with each fork bearing one inner and one dorsal spine. Inner margin of each fork bears three long, serrated setae.

*Second zoea:* Carapace length is 0.72–0.77 mm; abdomen-telson length is 1.46–1.54 mm. Eyes are stalked. Abdomen as in previous stage, except for pair of medium-sized setae on dorsal surface of first somite. Telson has a pair of short, plumose setae on median margin of cleft part. Other structures as in previous stage

*Third zoea:* Carapace length is 0.79–0.87 mm, abdomen-telson length is 2.02–2.21 mm. Rudimentary buds of thoracic appendages are seen behind second maxilliped. Abdomen is 6-segmented. Paired pleopod buds are seen at ventral posterior end of somites 2–5. Telson is similar to that of previous stage.

*Fourth zoea:* Carapace length is 0.98–1.06 mm; abdomen-telson length is 2.61–3.03 mm. Pleopod buds are well developed. Telson is similar to that of zoea III, except for additional short seta on inner margin.

*Megalopa:* It is very similar to that of other portunids. Rostral spine is present. Eyes project as far as lateral margin of carapace. Carapace length (including rostrum) is 1.69–1.81 mm, and carapace width is 1.16–1.31 mm. Abdomen sis ix-segmented, with dorso-ventrally flattened telson. Total length including rostrum is 3.00–3.2 mm. Antenna is elongated and 11-segmented. Proximal segments are comparatively larger and bearing simple setae. Eighth segment bears 4 setae on distal margin. Distal segment is with 4 setae. Mandible has a simple cutting edge and bears 3-segmented palp. First pereiopod (cheliped) is 5-segmented and is well developed. All segments bear a few short setae, mostly on propodus. Ischium and carpus each bear one short, stout spine. Second to fifth pereiopods are with well-developed endopods, each of 5 segments.

*Crab stage:* The megalopa metamorphoses to crab instar I, which resembles the adult crab. Carapace width is 2.0–2.5 mm. Margin of carapace is serrated, with 9 anterolateral spines. Pereiopods are well developed, with setae, especially on propodus and dactylus of fifth pair.

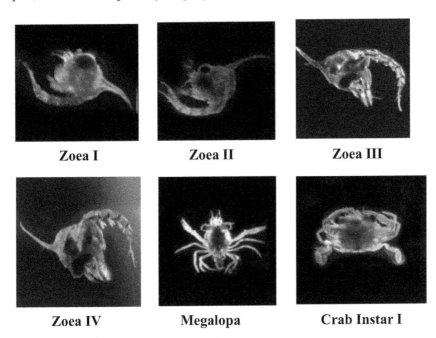

*Larval Stages of Portunus pelagicus*

*Predation:* The smooth stingray, *Dasyatis brevicaudata,* southern fiddler, *Trygonorhina fasciata guanerius*, and gummy shark, *Mustelus antarcticus* are known predators of adult crabs. Generally these crabs are most vulnerable to predation immediately after molting.

**Fisheries:** Commercially important in near shore fisheries.

### *Portunus petreus* (Alcock, 1899)

**Common name:** Not designated.

**Distribution:** Madagascar, Providence Islands, Cargados Carajos Islands, India, Laccadive Islands, Japan.

**Habitat:** Coral reefs; shallow sandy bottoms; 3–30 m deep.

**Description:** Carapace is moderately broad (breadth is 1.5–1.6 times length); surface is with widely spaced granular areas which are being separated by regions with a dense pubescence; and regions are well recognizable. Front is with 4 lobes; of which medians are much lower than laterals; antero-lateral borders are with 9 teeth, of which last one is longest and directed outwards; and posterolateral junction is rounded. Antero-external angle of merus of third maxillipeds is markedly produced into a lobe. Chelipeds are stout; cheliped merus is with 2 spines at posterior border; anterior border is with 4 spines; and carpus is with 2 spines of which inner is enormously long, two-thirds palm length. Upper surface of palm is with 1 distal spine and lower surface is with squamiform markings. Posterior border of merus of swimming leg is serrated and is bearing a few spines; dactyl is without a red spot. Penultimate segment of male abdomen is not markedly convex on outer border.

**Biology:** Not known.

**Fisheries:** No information

### *Portunus pseudohastatoides* (Yang & Tang, 2006)

Image not available.

**Common name:** Not designated as it is a new species

**Distribution:** South China, Hong Kong, west coast (Taiwan Strait) and northeastern Taiwan.

**Habitat:** Not known.

**Description:** Carapace is extremely dorsoventrally flattened; dorsal surface is naked or very slightly tomentose; regions are well defined; and mesogastric, cardiac, median, and lateral post cardiac regions are elevated and separated by deep grooves. Front is armed with 4 teeth (excluding inner supraorbital teeth), of which medians are acute, and narrower yet extending beyond rounded; and, broader laterals. Anterolateral border is with 9 distinct teeth, of which 2nd to 4th are smaller and successive ones are increasing in size. Chelipeds are elongated and subequal; anterior and posterior borders of cheliped merus is fringed with plumose setae; 4 or unusually 5 acute teeth are present on anterior border; 2 are small on basal 1/4 portion; 1 or 2 in middle, and 1 on near distal end; and 2 stouter teeth on distal portion of posterior border. Carpus is with 2 spines, of which 1 is on outer surface and 1 is in inner angle; and 3 granular ridges are seen. All ambulatory legs are compressed, and are fringed with setae on ventral edge on all segments; posterodistal edge of merus of last ambulatory leg is fringed with row of small teeth; and propodus and dactylus are flattened; fringed with plumose seta; and unarmed. Male abdomen has an inverted "T" shape. Telson is longer than broad and its basal border is convex and triangular, with angles rounded. 6th somite is longer than broad, and its lateral borders convex.

**Biology:** Not known.
**Fisheries:** No information

### *Portunus pseudotenuipes* (Spiridonov, 1999)

**Common name:** Not designated as it is a new species.
**Distribution:** Ambon, Philippines, and probably Malaya.
**Habitat:** Sandy to muddy bottom; 8–28 m depths.
**Description:** Carapace is much broader than long (average width/length ratio, 2.41). Three frontal lobes are slightly producing beyond inner supraorbital lobe; and median lobes are sharply rounded, and are much narrower than evenly rounded laterals. Anterolateral margin is hirsute. Nine antero-lateral teeth, of which 1st is low and truncated; teeth 2–7 are subequal, smaller than 1st, truncated or rounded; tooth 8 is smallest, almost evenly rounded; and last tooth is largest, slightly directed forward.

Chelipeds are long (40 mm). Abdomen is smooth. Penultimate somite is slightly broader than long (length 3.5 mm, width 4.0 mm) with outer edges which are slightly convex and converging distally. Ultimate somite is broader than long (length 1.5 mm, width 2.0 mm), and is acutely rounded. Genital opening in females is bulb-shaped, and is located nearly in middle of proximal portion of sternite III. Carapace of freshly preserved crabs is brownish grey with characteristic dark hatched bounding carapace areas. Size of males is: 6.5–22 mm; and females, 8.5–20 mm.

**Biology:** This species is sometimes associated with its closely related species viz. *Portunus tenuipes*

**Fisheries:** No information

### *Portunus pubescens* (Dana, 1852)

**Common name:** Hairy swimming crab
**Distribution:** Hawaii and Indo-Pacific; depth range 20–30 m.
**Habitat:** Coastal species; uncommon in sand and rubble at night.
**Description:** Carapace is moderately broad (breadth is 1.63 times length); and surface is covered with a short dense pubescence. Continuous bow-shaped mesogastric, epibranchial, and metagastric ridges; are seen. Front is with 4 bluntly rounded lobes, of which outer pair is slightly larger than inner one. Antero-lateral borders of carapace are with 9 teeth, of which last one is distinctly longer and stouter than the preceding ones, projecting

slightly forward; and posterolateral junction is rounded. Cheliped merus is with postero-distal border which are smooth, and anterior border is with 3 spines; carpus is with 2 outer spines; upper border of palm is with 2 distal spines; and lower surface is granular. Posterior border of swimming leg is without spines or spinules. It attains carapace width of 50 mm.
**Biology:** Unknown.
**Fisheries:** No information

### *Portunus pulchricristatus* (Gordon, 1931)

**Common name:** Li-wen crab
**Distribution:** Indo-Pacific: Muscat, Madras, Andaman, Burma, Malaysian area, Philippines, China, north Queensland and Pakistan (northern Arabian Sea).
**Habitat:** Sand and mud bottom of 10–830 m.
**Description:** It is a rare species. Carapace is convex and is 2.3 times as broad as long. Regions are distinct, and each one is composed of patches of coarse granules. Front is cut into 4 lobes, of which inner one is narrow and small and outer one is broader and rounded. Dorsal margin of orbit is serrated, with 2 fissures. Antero-lateral margin is armed with 9 teeth. First tooth is somewhat acute and triangular; second and third are obtuse; and sequent 5 teeth are nearly triangular. Last one is long and large and its length is 1–1/3–1 1/2 times breadth of carapace. Posterior margin is straight, microscopically serrated, and its outer angle is acutely triangular. Merus of third maxilliped is narrower, with outer-distal angle rounded. Merus of cheliped is armed with 3 spines on its anterior margin and 2 on the posterior margin. Dorsal surface of carpus is with ridges, and is armed

with 2 spines. Palm is armed with a sharp spine Fingers are slender, bearing ridges. Merus of last ambulatory leg is with a spine at distal end of posterior margin. Sixth segment of male abdomen is 1.5 times as long as basal breadth. Seventh segment is somewhat acutely triangular. Female abdomen is oval. Carapace length of male is 11.2 mm and breadth is 26 mm. Corresponding values for females are 8.1 mm and 19.5 mm, respectively.

**Biology:** Unknown.

**Fisheries:** No information

### *Portunus rubromarginatus* (Lanchester, 1900)

**Dorsal view**

**Frontal view**

**Common name:** Red swimming crab
**Distribution:** Indo-West Pacific
**Habitat:** Depth range 23–107 m.
**Description:** Front of carapace is 4 lobed. Last anterolateral spine is distinctly widest and most protruding and directed slightly forwards. Posterolateral junction is rounded. Sharp posterolateral spine is seen between last anterolateral tooth and posterolateral junction. Cheliped merus is with 1 spine on outer border. Carapace width is 54 mm.
**Biology:** This species has three zoeal and the megalopa stages. The megalopae differ from other related species in having a large spine on the ischium of the first pereiopod of the first pereiopod and in having relatively very small sternal cornuae (Greenwood and Fielder, 1979).
**Fisheries:** No information

### *Portunus sanguinolentus* (Herbst, 1783)

**Common name:** Blood-spotted swimming crab; three-spot swimming crab.
**Distribution:** Indo-Pacific; Tropical to temperate.
**Habitat:** Intertidal zone (especially juveniles) to depths of 30 m; sandy to sandy-muddy substrates; also in Brackish water.
**Description:** Carapace is very broad (breadth is 2.0–2.5 times length). Surface is finely granulated anteriorly and is smooth posteriorly with recognizable mesogastric, epibranchial, and metagastric ridges. Front is with 4 triangular teeth, of which outer pair is broader and very slightly more prominent than inner ones. Antero-lateral borders are with 9 teeth; of which first is clearly longer and much more pointed than following 7, and

last one is very large and projecting straight out laterally. Posterolateral junction is rounded. Cheliped merus is with smooth postero-distal border and anterior border is with 3–4 sharp spines; carpus is with inner and outer spines; and lower surface of palm is smooth. Posterior border of swimming leg is without spines or spinules. Color of carapace is olive to dark green, with 3 prominent maroon to red spots on posterior 1/3 of carapace. It is easily distinguished from other species by its very distinct color markings. Maximum carapace width of males is 20 cm.

**Biology**

***Food and feeding:*** This species is omnivorous and it mainly preys on crustaceans and fish. It is a predator of sessile and slow-moving benthic macroinvertebrates. It may also be a scavenger.

***Age and growth:*** This species has been reported to have a rapid growth and short lifespan of about 2–5 years. Studies on these aspects were made in this species by Sukumaran and Neelakantan (1997). The size frequency analysis indicated that the growth rate was high and more or less uniform in juveniles, while the adults showed relatively low rate of growth and marked variation in males and females. The mean monthly growth rates were 10.3 m and 8.8 mm and attained a carapace width of 124.1 and 112.5 mm by males and females respectively on completion of one year.

***Size at sexual maturity:*** This species has been reported to attain sexual maturity at a size of 64–70 mm short carapace width (= 83–89 mm long carapace width) in male and 63–71 mm SCW (= 81–93 mm LCW) in female crabs. Males mature at 60.8 mm SCW while females mature at 63.5 mm SCW. The smallest female crab having eggs on its abdominal pleopod measures 63 mm SCW. Berried females were found in almost throughout the year. Minimum and maximum number of eggs were found to be 272,000 and 1,395,000 (Rasheed and Mustaquim, 2010).

***Mating behavior and broodstock development:*** Normally the female crab was found to avoid the small crab and accepted the bigger crab for its aggressive interaction and eventually successful mating happened. In another experiment in which two same sized males (100 g) in which one male had one chelate and another one with two chelae were stocked with a soft shell female. In this case, the female always readily accepted to mate with two chelate male than single chelate male.

***Reproductive Biology:*** The male reproductive system of this species is bilaterally symmetrical, and is creamy to whitish in color. It is composed of a pair of testes, a pair of vas differentia, and a pair of ejaculatory ducts internally, and a pair of pleopods externally as accessory reproductive organs, present on the inner side of the abdominal flab. Male measuring below 8.5 cm carapace width (CW) is immature stage, above 10.5 cm CW is mature stage, and CW in between 14.5 and 10.5 cm is maturing stage. The female reproductive system composed of a pair of ovaries, a pair of seminal receptacles (or) spermatheca, and a pair of oviducts. The ovaries of *P. sanguinolentus* are categorized into five stages, according to their size, color and external morphology. They are immature, early maturing, late maturing, ripe and spent. Female crabs are known to attain sexual maturity when they reach the size of 9.1–9.5 cm CW and the male crabs attain sexual maturity when they reach the size of 9.5–10.0 cm CW. Soundarapandian et al. (2013) reported that the smallest berried female attain sexual maturity when they reach a size of 8.6 cm CW. Based on the change in color, increase in the size and change in the shape of the berry eggs, three different stages of egg development have been observed. Stage-I has pale yellow to deep yellow colored egg mass (260.16 μm). Stage-II has yellow to gray colored egg mass (290.2 μm). Stage-III has deep gray to black colored egg mass (340.32 μm).

***Microbial activity of serum:*** A naturally occurring haemagglutinin (HA), with activity against bacteria and yeast cells was detected in the serum of this species (Meena et al., 2011).

**Fisheries:** Major commercial fisheries exist for this species.

### *Portunus sayi* (Gibbes, 1850)

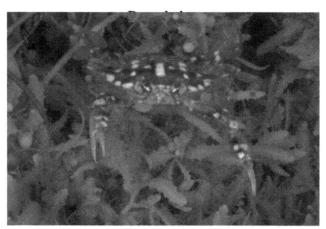

**Crab with seaweeds**

**Common name:** *Sargassum* swimming crab; gulf-weed crab.

**Distribution:** Western Atlantic Ocean and Caribbean Sea; from Nova Scotia and Gulf of Maine to southern Florida, Bahamas and Gulf of Mexico; at depths down to about 18 m.

**Habitat:** Reef-associated; Gulf and the bays; lives among floating sargassum; depths down to 18 m.

**Description:** Its smooth, shiny, compressed carapace is nearly twice as broad as it is long and has six frontal teeth on each side. Last pair of legs is flattened into paddle-shaped appendages. 9 pairs of spines (including the outer orbital spines) are seen along the lateral edges of the carapace, most posterior pair (called the lateral spines) is about 3 times as long as the others. 6 low blunt spines are found between ocular orbits (eye sockets), including inner orbit spines. Spines are present on inner margins of cheliped merus and a small spine on inner and outer dorsal carpus. Chelipeds are larger in males than in females and their spines have orange margins. Carapace, claws and legs are usually pale brown with large white or flesh-colored blotches but brown color may have a purplish or greenish tinge. This crab, like other members of its family, is specially adapted for swimming rather than walking as their fourth pair of legs is modified with flattened, paddle-like surfaces. It swims sideways rather than forwards and can move surprisingly swiftly. It grows to a length of about 5 to 7.5 cm.

## Biology

**Food and feeding:** When these become large they behave as formidable predators and are quick to consume any smaller animal that comes within reach. Fish, other crustaceans, and even smaller members of their own species are not safe from this hungry crab.

***Behavior:*** The coloring of this crab provides camouflage among the fronds of seaweed through which it hunts or lies in wait to ambush small items of prey. It has been shown that the crab chooses *Sargassum* spp. over similar seaweeds such as *Thalassia testudinum* by detecting chemical cues in the water.

This crab responded to chemical odors from *Sargassum* spp. Crabs visually located habitats but did not visually distinguish between different habitats. In habitat selection trials, crabs selected *Sargassum* spp. over artificial *Sargassum* and *T. testudinum*. These results suggest that crabs isolated from *Sargassum* likely use chemoreception from longer distances; and within visual proximity of a potential patch, they use both chemical and visual information (West, 2012).

**Association:** This crab has been found as an epibiont of loggerhead turtles along with the goose barnacles Lepas anatifera and Conchoderma virgatum, various other crabs, sea spiders, tunicates and hydroids.

**Fisheries:** No information

### *Portunus sebae* (Milne-Edwards, 1834)

**Dorsal view**

**Frontal view**

**Mating behavior of ocellater swimming crab**

**Common name:** Ocellate swimming crab
**Distribution:** Western Atlantic Ocean; Caribbean, Bahamas and South Florida.
**Habitat:** Reefs and seagrass beds; depth range 5–30 m.
**Description:** Last walking legs have last segment enlarged to paddles, which are used for swimming. At two-thirds of carapace is one spine on both sides. On the rear of carapace are two large dark spots ringed with white. Claws are long and slender. Size of carapace is up to 10 cm wide.
**Biology:** Unknown.
**Fisheries:** No information

## ***Portunus segnis*** **(Forskal, 1775)**

**Common name:** Not designated.

**Distribution:** Eastern Mediterranean to east Africa in the Indian Ocean, and to Pakistan, Red Sea and Persian Gulf.

**Habitat:** Wide range of inshore and continental shelf areas, including sandy, muddy or algal and sea grass habitats, from the intertidal zone to at least 50 m depth; shallow bays with sandy bottom.

**Description:** A prominent inner spine is seen on cheliped carpus. It has bidentate front, with nine teeth which are present along antero-lateral margin, of which last one is most prominent. Male has a triangular abdomen. First pereiopod is considerably longer than the carapace width. A mottled pattern with yellow-whitish spots and lines is seen on carapace and legs. Swimming pereiopods are proximally mottled and a vivid blue coloration is found at distal end. Fingers on second, third and fourth pereiopods have a reddish fringe, whilst its surface is blue. Carapace width ranges from 7.5 to 17.5 cm for males and form 7.0 to 16.5 cm for females. Carapace length is from 4 to 8 cm for males and from 3.5 to 8 cm for females.

### Biology

***Sex ratio:*** Sex ratio (M:F) of this species has been found to be 0.92:1.

***Food and feeding:*** The main food items of this species include crustaceans, mollusca and fish. The crustaceans include shrimp, barnacle and crab. The dominant molluscs are the bivalve Cardita bicolor and gastropod Cerithium erythraeonense.

***Gonad development and ovigerous female:*** In this species, ovarian development has been classified by size and color of the ovary as follows: Stage 1 – Gonad immature, ovary very thin and transparent (colorless).

Stage 2 – Early gonad maturing, ovary changed color to creamy, but not extending into hepatic region. Stage 3 – Gonad maturing, ovary became enlarge and changed color to yellow, extending some 1/3–1/4 of the hepatic region. Stage 4 – Gonad mature, the ovary covered most part of hepatic region, and turned orange or reddish orange. Stage 5 – Females carrying eggs (Ovigerous female). All the five stages of ovarian development of this species have been observed throughout the year.

***Size at first sexual maturity:*** In this species the carapace width of mature females ranges from 92 to 165 mm, with average size of mature females at 123.6 mm carapace width. 50 percent of female crabs are found in stages 4 and 5, in which mean carapace width (CW) at sexual maturity is 113 mm.

***Fecundity:*** In this species each female with carapace width ranging between 103 and 155 mm can produce 521,027 to 6,656,599 eggs. Mean fecundity of this species is 2397–967 eggs.

***Reproductive biology:*** All the five stages of ovarian development of this species have been observed throughout the year. The size of ovigerous crabs varied from 103 to 155 mm carapace width. This crab is found to spawn all year round with a spawning peak in mid-winter to early of spring season. The fecundity of ovigerous crabs ranges from 521,027 to 6,656,599 eggs, with average fecundity of 2,397,967 eggs. The minimum CW of female crabs that reach sexual maturity is 92–138 mm and the length at which 50% of all ovigerous females is 113 mm carapace width (Safaie et al., 2013).

**Fisheries:** This species supports substantial commercial fishery in the Persian Gulf and is an important component of many recreational fisheries in Iran and other parts of the world.

### *Portunus sinuosodactylus* (Stephenson, 1967)

Profile of Portunid Crabs (Family: Portunidae) 59

**Common name:** Not designated as it is a new species
**Distribution:** Not available
**Habitat:** Not known.
**Description:** Front has four distinct lobes, of which medians are slightly protruding and only slightly narrower than laterals. Among anterolateral teeth, first is stout and broad; and last is stout with tip curving forwards. There is slight alternation between larger and smaller teeth, but fifth and seventh are relatively large. Ninth anterolateral tooth is approximately twice as long as eighth. Carapace is moderately convex, hirsute, and is bearing granular ridges and some patches. Postlateral junction is rounded and moderately long (B/L 1.65). Frontal and postfrontal patches are moderately well developed. Chelipeds are elongated and very slender; propodus is much less massive than merus. Merus-posterodistal border is with two sharp spines. Anterior border is with four spines, of which the three (proximal spines) are well separated from subterminal distal. Undersurface is with terminal boss bearing spine. Carpus is elongated; inner and outer spines are well developed and acute; Y-shaped carina is seen on upper surface, and three carinae are on outer surface. Propodus is long, thin, narrow and strongly carinated. Spine at carpus articulation is long and sharp. Upper surface is with two conspicuous carinae, and inner with very large curved subterminal spine. Outer surface is with single well-developed central carina; under surface is with granules approximating to squamiform arrangement and inner surface is with central carina. Index and dactylus are extremely long, sharp and sinuously curved. Fifth leg merus is 1.6 times as long as broad with conspicuous spine on posterodistal border.
**Biology:** Not known.
**Fisheries:** No information

*Portunus speciosa* (**Dana, 1852**) *(= Amphitrite speciosa)*
Image not available
**Common name:** Not designated.
**Distribution:** Japan
**Habitat:** Not known.
**Description:** Longitudinal impressed line is present on merus-joint of chelopoda Color in living specimens is dark grey, mottled with white.
**Biology:** Not known.
**Fisheries**: No information

## *Portunus tenuipes* (de Haan, 1833)

**Common name:** Not available

**Distribution:** Red Sea, Andaman Islands, Japan, China, Singapore, Philippines, Indonesia, Australia.

**Habitat:** Depth range, 15–35 m.

**Description:** Carapace is moderately broad (breadth twice or just less twice length); surface is with well defined granular elevations, between which is a fine pile of hairs is present; all regions are discernible; front is with 3 teeth, of which outer pair is much broader than median; antero-lateral borders are with 9 teeth, of which last one is very large and directed slightly backwards; posterolateral junction is elevated to form a short ridge bearing an almost right-angled border. Antero-external angle of merus of third maxillipeds is rounded. Cheliped merus is with postero-distal border bearing 1 spine of very variable shape; anterior border is with 3–4 spines; carpus is with 2 spines; lower surface of palm is granular or with squamiform markings. Posterior border of merus of swimming leg finely is serrated.

### Biology

**Food and feeding:** It feeds on bivalves, small benthic animals and detritus (it is the chief food source for squids and fish).

**Fisheries:** No information

### *Portunus trituberculatus* (Miers, 1876)

**Common name**: Gazami crab, Japanese blue crab or horse crab.

**Distribution:** Indian and West Pacific Oceans: Southeast and East Asia (from Japan, Korea, China and Formosa and the Bay of Bengal), to the West, North and East of Australia.

**Habitat:** Sandy and sand-muddy depths in shallow waters to 50 m depth.

**Description:** Carapace is very broad (breadth is just over 2–2 1/3 times length); surface is finely granulated; it is usually with recognizable mesogastric, epibranchial, and indistinct metagastric ridges, cardiac and mesobranchial ridges are with low granular eminences; front is with 2 acute teeth; antero-lateral borders are with 9 teeth, of which last one is very large and projecting straight out laterally; and posterolateral junction is rounded. Cheliped merus is with postero-distal border which is spinous; anterior border is with 3–4 (usually 4) sharp spines; carpus is with inner and outer spines; and lower surface of palm is smooth. Posterior border of swimming leg is without spines or spinules. Carapace color is dull green to brown. Maximum carapace width is 15 cm (males). A male of 149 mm CW has a carapace length of 70 mm.

### Biology

*Fecundity:* In this species, oocyte number was found to increase with increasing female body size and estimates ranged from 0.8 to 4.5 million for CW of 130–240 mm. Size of first ranged between 1.22 and 1.62 mm and it decreased with advancing hatching date, e.g., increasing hatching temperatures. Number of first zoeas increased with body size of females and there was no fluctuation throughout the breeding season in females of the same size. Predicted numbers of first zoeas ranged from 0.7 to 3.8 million for CW of 130–240 mm. Number of first zoeas was found to be less than the oocyte number for females of the same size and it is largely due to the egg loss which occurred from oviposition to hatching (Hamasakia et al., 2006).

**Fisheries:** It is an important commercial species in Japan, where it is a common edible crab, and it is collected in large numbers. Minor commercial fisheries exist for this species in China also.

## *Portunus tuberculosus* (Milne-Edwards, 1861)

**Common name:** Blood-spot swimming crab
**Distribution:** Northwest Pacific: Taiwan, Persian Gulf, Hawaii, Madagascar and Tanzania.
**Habitat:** No information
**Description:** Not available
**Biology:** Not known.
**Fisheries:** No information

## *Portunus tweediei* (Shen, 1937)

**Common name:** Not designated.
**Distribution:** Pacific Ocean: China and Hong Kong.
**Habitat:** No information
**Description:** Median frontal lobe is smaller and less protruding. Male abdomens vary in breadth but some are distinctly narrower. Distinction between parallel-sided proximal portion and converging distal portion in abdomen is still ·evident, but is less obvious. Mesogastric and cardiac granulated elevations of the carapace are tubercular eminences.
**Biology:** Unknown.
**Fisheries:** No information

## *Portunus ventralis* (Milne-Edwards, 1879)

**Common name:** Not designated.
**Distribution:** Gulf of Mexico

**Habitat:** Depth range, 0.37 m.

**Description:** In this species there is no spine at extremity of outer margin of arm. A spine is found at distal third of upper margin on manus. Frontal margin is four-toothed, equally developed. Carapace and chela are light tan; and swimming feet are brown. It is brown all over with a deeper shade on chela. Carapace is light pink and chela is deep red. Length of carapace is 10 mm and width of carapace, 21 mm.

**Biology:** No information

**Fisheries:** Unknown.

### *Portunus vossi* (Lemaitre, 1992)

Image not available.

**Common name:** Not designated.

**Distribution:** Gulf of Mexico: West coast of Florida.

**Habitat:** Not known.

**Description:** Carapace is 1.6–1.8 times as long as broad (including last anterolateral tooth). Anterolateral margins are forming an arc of a circle with center near middle of cardiac region. Regions of carapace are distinct, and are more or less elevated. Dorsal surface is covered with very short pubescence. Frontal lobes, dorsal and ventral surface of anterolateral teeth, and protogastric, mesogastric, metagastric, cardiac and lateral postcardiac regions are unevenly covered with granules. Front is with four (two submesial and two lateral) rounded teeth, which are separated by U-shaped sinuses. First anterolateral tooth (outer orbital) is blunt; anterolateral teeth 2–9 are sharp, with dark-colored tips; and last anterolateral tooth is twice or more as long as preceding tooth, weakly curved anteriorly. Chelipeds are slender, with granules slightly larger than those on carapace; and are with granulated ridges and spines with dark-colored tips. Merus of cheliped is with 4–5 strong spines on anterior margin, and sharp granules between spines; and posterior margin is with granules and distal spine. Carpus is with strong inner spine, outer ventrodistal spine, and two longitudinal granulated ridges on dorsal and dorsolateral surface. Palm is with two spines, one on dorsal margin near articulation of dactyl, and one on outer proximal surface adjacent to carpus; and outer surface is with three longitudinal granulated ridges and is lacking iridescence.

Fingers are straight except for curved, crossed tips, and are longitudinally ribbed; cutting edges are with strong, subtriangular or rounded unequal teeth; and movable finger is with scattered setae. Pereopods 2–4 (walking legs) are unarmed; propodus and dactyl are with fringe of setae on ventral and dorsal margins. Pereopod 5 (swimming leg) is covered with short, dense setation on dorsal surface of merus, carpus, propodus, and median region of dactyl. Merus of pereopod 5 is longer than broad; dorsodistal margin is armed with row of small spines; and posterodistal margin is with strong spine (often bifid). Male abdomen is with penultimate segment which has weakly curved lateral margins. Male first pleopod is sinuous in posterior view, and is not reaching level of suture between sixth and seventh thoracic stemites; and distal portion is with minute, microscopic spinules. Female abdomen is broadly triangular. Female gonopores are narrow, transversely ovoid. Male measures $11.0 \times 20.0$ to $18.3 \times 31.0$ mm and females, $12.2 \times 19.8$ to $19.8 \times 34.0$ mm.

**Biology:** Unknown.

**Fisheries:** No information

### *Portunus (Lupocycloporus) wilsoni* (Moosa, 1981)

Image not available.

**Common name:** Not designated as it is a new species.

**Distribution:** Northern territory, Australia.

**Habitat:** No information

**Description:** Front carapace is with 4 lobes. Internal lobes are markedly more prominent that laterals. Lateral lobes are polygonal and are only slightly wider than internals which are triangular with rounded tip. Anterolateral border is with 9 teeth. Ninth is the largest. Fourth and sixth are smallest. Eight is smaller that second, third, fifth and seventh but larger than fourth and sixth. Carapace is moderately broad. Breadth is less than 1.5 times length. Carapace is covered with short relatively dense hairs. Frontal and post-frontal regions are with granular patch. Carapace measures $12.9 \times 19.2$ mm.

**Biology:** Unknown.

**Fisheries:** No information

## *Portunus xantusii* (Simpson, 1860)

**Common name:** Xantus swimming crab
**Distribution:** Central California to Sea of Cortez.
**Habitat:** Sandy and muddy areas; eelgrass beds; from low intertidal to as deep as 181 m.
**Description:** Carapace is broad, flat, and finely pubescent. Each lateral angle of carapace is armed with a long spine. Last pair of legs is paddle-like and each merus is armed with spinules and not spines. Males measure from 4.9 × 8.2 to 35.5 × 70.7 mm; females from 6.2 × 11.8 to 36.2 × 73.1 mm; ovigerous females from 15.2 × 29.8 to 31.0 × 55.9 mm; and young from 2.8 × 3.5 mm body width (including spine) to 70.7 mm in males and 73.1 mm in females.
**Biology:** A quick, predatory crab that has been observed preying on sand crabs, Emerita, by thrusting a claw into the sand to grab one, then running in a circle to twist it up out of the sediment.
**Fisheries:** It is not sufficiently numerous to be exploited commercially.

## *Portunus yoronensis* (Sakai, 1974)

Image not available.
**Common name:** Not designated as it is a new species.
**Distribution:** Eastern Indian Ocean to Philippines and French Polynesia.
**Habitat:** Not known.

**Description:** Carapace is narrow (breadth is 1–1/3 times length); surface is with close-packed granular areas with a fine pubescence among granules; regions are well recognizable on denuded specimens; front is with 4 sharp teeth, of which lateral ones are more protruding than medians; antero-lateral borders are with 9 subequal teeth, of which last one is not protruding more than preceding; posterolateral junction is rounded. Antero-external angle of merus of third maxillipeds is markedly produced into a lobe. Chelipeds are elongated; merus is with 1 spine at posterior border, and anterior border is with 3 acuminate teeth; carpus is with 2 spines, of which inner is one very long; upper surface of palm is with 2 subdistal spine and, lower surface is smooth. Posterior border of swimming leg is with usual fringing hairs and there are no spines on propodus and dactylus.

**Biology:** No information

**Fisheries:** Unknown.

*Lupocycloporus aburatsubo* **(Balss, 1922)** *(=Portunus aburatsubo)*
Image not available.

**Common name:** Not designated.

**Distribution:** Japan

**Habitat:** No information

**Description:** Carapace is moderately broad (breadth is 1.7 times length); and its surface is with elevated areas, whose granulation is distinct and which are separated by the furrows betwee n the regions. Front is quadridentate and median teeth are smaller than lateral ones. Antero-lateral borders are with 9 teeth of which last one is much larger than preceding ones and projecting sideways. Posterolateral junction is rounded. Antero-external angle of merus of third maxillipeds is markedly produced into a lobe. Cheliped hand is extremely slender and is much less massive than arm. Cheliped merus is with 2 spines on posterior border and its anterior border is with 7 spines; carpus is with 2 spines and upper surface of palm is with 2 distal spines. Posterior border of swimming leg is with usual fringing hairs.

**Biology:** No information

**Fisheries:** No information

## *Scylla olivacea* (Herbst, 1796)

**Dorsal view**

**Frontal view of adult male**

**Common name:** Orange mud crab; brown mud crab.

**Distribution:** Pakistan, southern China, Kupang, Arafura Sea, Australia, Indonesia, India.

**Habitat:** Estuaries and mangrove areas; shallow areas below the low tide mark.

**Description:** Frontal lobe spines are low. Antero-lateral carapace spines are broad with outer convex. Carpus of chelipeds is usually with one small blunt prominence (may be spinous in juveniles) ventro-medially on outer margin; and reduced second spine may be present dorso-distally in juveniles and young adults. Palm of cheliped is usually with a pair of blunt prominences on dorsal margin behind insertion of the dactyl. Chelipeds, legs and abdomen are all without obvious polygonal patterning for both sexes. Color varies from red through brown to brownish/black depending on habitat. The brown mud crab is smaller and it grows up to 1.5 kg and 150 mm carapace width.

## Biology

*Food and feeding:* These crabs emerge at night to forage for food; they eat almost anything. However, they mainly eat slow-moving or stationary bottom-dwelling animals such as molluscs, smaller crabs and worms. They also eat plant materials.

*Behavior:* These crabs bury themselves in the mud during the day.

*Mating behavior:* In this species, the mating process which lasts for 82.0 h is divided into four phases: precopulation, molting, copulation, and postcopulation. Courtship displays and fighting are shown by mature males while they are courting females. Precopulatory position lasts for 55.2 h before the pairs disengage for the female to molt. The molting process lasts for 4.6 h. Copulation (mean duration is 6.6 h) occurs while the females exoskeleton is still soft. Postcopulatory guarding lasts for 13.6 h. Separation of the mating pairs indicates the end of postcopulation phase. Mating success percentage is found unaffected by sex ratio, but is inversely affected by stocking density (Waiho et al., 2015).

*Predators:* Larvae of these crabs are eaten by small fish. It is presumed that the post-larvae and juveniles are preyed on by fish species that live in mangrove areas. Predators of adult crabs include turtles, rays, crocodiles and large fish, such as barramundi and sharks.

**Fisheries:** It has a major commercial fisheries. This species is the most common species in many markets of Southeast Asia. It is usually marketed alive.

## *Scylla paramamosain* (Estampador, 1949)

**Dorsal view**

**Frontal view of adult male**

**Common name:** Green mud crab

**Distribution:** Indo-West Pacific: northern parts of South China Sea and parts of Java.

**Habitat:** Rock areas; near reef; mangroves.

**Description:** Carpus of chelipeds is without two obvious spines on distal half of outer margin; palm of cheliped is with a pair of distinct spines on dorsal margin behind insertion of the dactyl, followed by ridges running posteriorly. Frontal lobe spines are high and are triangular with straight margins and angular interspaces. Chelipeds and legs are with weak polygonal patterning for both sexes. Carapace is usually green to light green; palm is green to greenish blue with lower surface and base of fingers is usually pale yellow to yellowish orange. Max length of carapace is 20.0 cm CW.

**Biology**

*Gonad development and maturity:* The reproductive characteristics and size at sexual maturity of the male mud crab Scylla paramamosain were investigated. Gonad development was classified into three stages: (i) Immature (Spermatogonia), (ii) Maturing (Spermatocytes), and (iii) Mature (Spermatids and Spermatozoa). Among the sample population, the highest 72% was under gonad development stage I, whereas mature stage III was only 12%. The mean size at first sexual maturity and 50% maturation of males were 96 mm and 109 mm internal carapace width (ICW), which revealed that 88% individuals were immature. The present result suggested that the minimum legal size of male S. paramamosain capture may be >110 mm ICW (Islam and Kurokura, 2013).

*Immune-associated components:* This species is one of the most important marine breeding crabs and often suffers from pathogen infection with high mortality. However, few effective immune methods are being utilized for controlling or minimizing the mortality of these crabs in the farms. In this regard, immune-associated components could play a very important role in protecting these crabs from infection. Several new immune-associated components including a membrane lipid rafts related gene *Sp*FLT-1, a hemichannel-associated transmembrane protein named as Sp-inx2, two antimicrobial peptides named as Sphistin and SpHyastatin, have been separately identified in different tissues of this species (Wang et al. http://digitalcommons.library.umaine.edu/cgi/viewcontent.cgi?article=1168&context=isfsi).

***Hemocyanin content:*** Four subunits of hemocyanin (an important respiratory protein) viz. cDNAs (SpHc1, SpHc2, SpHc3, and SpHc4) have been isolated from this species. These findings suggest that hemocyanin may potentially be involved in the crab immune response, and that the role of the four subunits may differ in different tissues and during various developmental stages (Wang et al., 2015).

**Fisheries:** No information

*Scylla serrata* **(Forskål, 1775)**

**Dorsal view**

**Frontal view of adult female**

**Common name:** Giant mud crab; mangrove crab.

**Distribution:** Indo-West Pacific: from East and South Africa to southeast and east Asia (from SE of China and Sri Lanka), Northeast Australia, Marianas, Fidji and Samoa Islands.

**Habitat:** River mouth or estuary; not far from shore.

**Description:** Carapace is smooth and glabrous with exception of granular lines on gastric regions and an epibranchial line starting from tip of last antero-lateral tooth and reaching to branchial regions; front is with 4 subequal and are equally spaced teeth with acute to rounded tips; antero-lateral borders are with 9 very acute and subequal teeths, of which last one is smallest. Basal antennal joint is short and broad, with a lobule at its antero-external angle. Chelipeds are heterochelous; merus is with 3 spines on anterior border and 2 spines on posterior; carpus is with a strong spine on inner corner and another on outer face; and propodus is with 2 acute spines at distal end of upper face and a strong knob on inner face at base of fixed finger. Swimming leg is without spines on posterior border of either of the joints. Maximum carapace width in males is between 25 and 28 cm and maximum weight is between 2 and 3 kg.

**Biology**

*Behavior:* In soft muddy bottoms where it digs deep burrows. Migrations offshore (up to 50 km) to spawn. Diet based on molluscs (bivalves: Mytilidae, gastropods) and crustaceans (grapsid crabs), rarely on plant material and fish. Adults remain buried at day, emerging at sunset and night to feed.

*Food and feeding:* These crabs emerge at night to forage for food; they eat almost anything. However, they mainly eat slow-moving or stationary bottom-dwelling animals such as molluscs, smaller crabs and worms. They also eat plant materials.

*Predators:* Larvae of these crabs may be eaten by small fish. The post-larvae and juvenile stages are often preyed on by fish species that live in mangrove areas. Adult mud crabs have fewer predators. However, turtles, rays, crocodiles and large fish, such as barramundi and sharks, prey on adult mud crabs.

**Reproduction and larval development**

**Mating:** In the warmer months, mature females release a 'pheromone' (chemical attractant) into the water to attract males. Once paired, the male climbs on top of the female, clasps her with his hind legs, picks her up and carries her around for up to 4 days. The male crab releases the female crab when she begins to molt, and after she has shed her shell, he turns her upside down to mate. The male crab deposits a capsule of sperm inside her reproductive opening, where it is stored for months until her developing 'ova' (eggs) are ready to be fertilized. After mating, the male flips the female upright and holds her under him for a few more days while her shell hardens.

**Spawning and hatching**: The female crab migrates offshore to spawn. The fertilized eggs are released in batches of 2 to 5 million. After digging a hole in the sand or mud with her abdominal flap, the female releases her eggs into it. Using her swimming legs, she gathers the eggs up to carry them under her abdomen. The eggs hatch in two to four weeks. The life-cycle then begins again.

**Berried female:** It commonly carries between 2 and 5 million eggs. During the incubation period of 10–14 days, the eggs change color from orange to black.

**12-day-old eggs:** These are attached to setae, and are showing pigmentation of eyes and body. Fertilization rates are typically greater than 90%.

The different *Scylla* spp. pass through 5 zoeal stages and a megalopa stage before it molts to the crab stage, taking 21–25 days for the entire cycle.

**Newly hatched first zoeal stage (Zoea 1):** Zoea 1 hatches as a non-motile pre-zoea which molts within 15 minutes in the free swimming and is actively feeding.

**Fifth zoeal stage (Z5):** Z5 stage appears between days 11 and 14 in the larval development. Pleopods which have developed since the Z1 stage are seen on the abdomen.

**Megalopa:** During the fifth molt, it transforms into a 'megalopa,' which has functional claws. After a week, it moves inshore and settles to the seabed. After a few days, it molts into a stage one juvenile crab.

**Juvenile crab**: A miniature version of the adult, it is about 4 mm wide. About a month after hatching, when 10–20 mm wide, it moves to an estuary and settles in a sheltered area.

**Young adult**: The crab reaches sexual maturity at 18 to 24 months.

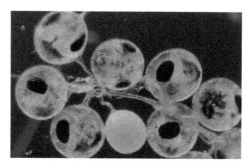
**12-day-old eggs attached to setae**

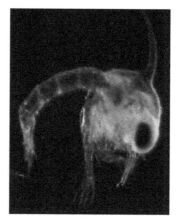
**Newly hatched first zoeal stage**

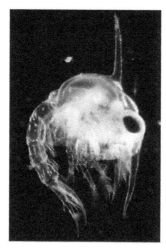
**Fifth zoeal stage**

Profile of Portunid Crabs (Family: Portunidae) 75

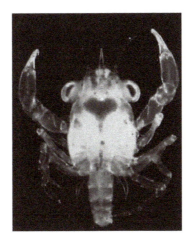

**Megalopa**

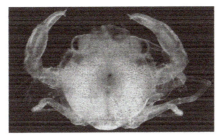

**First crab stage**

**Fisheries:** Major commercial fisheries exist for this species especially in Western Central Pacific; Indonesia and Thailand.

***Scylla tranquebarica* (Fabricius, 1798)** *(= Scylla oceanica)*

**Dorsal view**

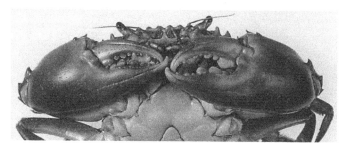

**Frontal view of adult male**

**Common name:** Purple mud crab

**Distribution:** Eastern Indian Ocean; Northwest Pacific; Western Central Pacific: Indo-West Pacific including shelf waters.

**Habitat:** Coastal waters; mangrove areas; do not live in burrows.

**Description:** Carpus of chelipeds is with two obvious spines on distal half of outer margin. Frontal lobe spines are of moderate height (mean height c. 0.04 times frontal width measured between medial orbital sutures), and are blunted with rounded interspaces. Antero-lateral carapace spines are broad, and outer margin of carapace is convex. Polygonal patterning is weak on chelipeds and first two pairs of legs. Last two pairs of legs are with stronger patterning for both sexes; patterning is variable on abdomen of female, and is absent on male. Color is variable and is similar to Scylla serrata. Max length: 20.0 cm CW male/unsexed; Max published weight: 2.0 kg.

**Biology**

***Food and feeding:*** Diet of mud crab *S. tranquebarica* consists of crustaceans, molluscs, fish remains, detritus, debris and undigested flesh.

***Mating between Scylla olivacea and Scylla tranquebarica:*** As these two species live associated in the wild, investigation was done on the mating success between these species by determination of the duration and recording the process of mating activity. Mating success of control trials of *S. olivacea* (T1) and *S. tranquebarica* (T2) was 60 and 50% respectively. Mating success of hybrid trials for male *S. olivacea* with female *S. tranquebarica* (T3) was 40% and male *S. tranquebarica* with female S. olivacea (T4) was 30%. The highest mean duration of pre-copulatory guarding is T3 (12,240 min) while the lowest is T2 (8,064 min). As for copulation, the highest mean duration is T4 (59.3 min) while the lowest is T1 (59.3 min). Meanwhile, the highest mean duration of post-copulatory

Profile of Portunid Crabs (Family: Portunidae)

guarding is T2 (312.0 min) and the lowest is T3 (82.5 min). Longer duration of pre-copulatory guarding and copulation were observed on hybrid trials (T3, T4) compare to the controls (T1, T2). It has been observed that all successful mating trials in the present study show general mating activity, involving pre-copulatory guarding, copulation and post-copulatory guarding. Result of the present study shows that hybridization can occur in captivity and there are also possibilities of hybridization between the two mud crab species to occur in the wild (Baiduri et al., 2014).

***Embryonic development:*** Embryonic development and morphological changes in the embryos, egg size and development, incubation period of mud crab, *Scylla tranquebarica* were studied. The incubation period has been reported to be 6–7 days. Five embryonic stages, viz., blastula, gastrula, eye placode, pigment and heartbeat were identified. The size of the developing egg increased at every stage and yolk mass decreased in size. The color of the egg was initially yellow changed to orange, brown and then black colors (Sonawane et al., 2015).

***Larval development in laboratory:*** Larval development of this species has been studied under hatchery conditions from first zoea to first crab stage. The crabs were fed daily with mussel (meretrix) at 20% of body weight and water quality was maintained by flow-through system of seawater. Newly hatched zoeae were stocked at density of 50 ind/ml and fed with rotifer, *Brachionus sp.* (50 ind/ml), *copepod, Acartia sp.* (60 ind/ml) and brine shrimp nauplii (1 to 5 ind/ml) as the larvae developed to megalopa. The larvae were reared at temperature of 28.7°C, salinity of 33.1 ppt, dissolve oxygen above 6.7 ppm and pH of 8.4. The results showed that the fecundity estimated ranged from 1.46 to 4.68 million eggs per female per batch. The color of the egg was initially yellowish and it gradually changed to orange, brown, grey and dark grey before hatching. The diameter of the freshly laid egg was 308.29 µm and 381.57 µm just prior to hatch. Hatching rate was 17.47 to 47.26 %. The duration (mean) of each zoea stage was 3.6 d and 3.4 d for first and second zoea, respectively, while for both third and fourth zoea was 3.2 d and 2.4 d for fifth zoea. However, megalopa took 7.0 d while crab instar took 3.4 d, and reached the first crab stage in 27 to 28 d from hatching (Zohri, 2011).

**Fisheries:** It is commercially important species

## *Callinectes amnicola* (Rochebrune, 1883) *(= Callinectes latimanus)*

**Common name:** Big-fisted swim crab
**Distribution:** Eastern Atlantic: Cape Verde Islands and Mauritania to Angola
**Habitat:** Demersal; freshwater; brackish and marine
**Description:** Carapace is bearing four frontal teeth is with variably rounded tips. Anterolateral margins are arched. Surface of carapace is coarsely granulate, but granules are more widely spaced or absent near margins, on epibranchial surfaces, and along regional sulci. However, there are most closely crowded granules on mesogastric, cardiac, and mesobranchial areas. Epibranchial line is prominent and almost uninterrupted. Chelipeds with propodus and carpus are moderately ridged. Chelae of large specimens are very strong and major one is very broad with fingers heavily toothed. Telson is lanceolate and is much longer than broad. Sixth segment of abdomen is broadened distally. Mature female's abdomen and telson are reaching about midlength of thoracic sternite IV. Telson is elongate triangular with inflated sides. Carapace length ranges from 2.2 cm to 16.8 cm with weight of 3.5 g to 277.1 g. Color of carapace is uniform greenish brown with articulations and internal face of chela and dactyl are bluish. Ventral aspect is yellowish white.

**Biology**
*Food items:* The food items of this species include mollusc shells, fish parts, shrimps and crab appendages and occasionally higher plant materials (Lawal-Are, 2003).

*Size at Maturity:* Maturity occurs in this species between 6.2 and 16.5 cm CW for the females and between 7.3 and 15.3 cm CW for the males. Fifty percent maturity was observed sat 10.8 cm CW in males and 11.0 cm CW in females (Lawal-Are, 2010).

***Fecundity:*** The fecundity ranges from 478,400 to 4,480,500 eggs. The average fecundity has been observed as 780,480 eggs. The diameter of the eggs ranges from 0.25 mm to 0.35 mm with a mean of 0.29 mm (Lawal-Are, 2010).

***Survival and growth:*** The effect of salinity on the survival and growth of this species was monitored in the laboratory for 22 weeks. This species was found to be euryhaline and tolerate a salinity range of 5 to 25 per thousand and had 90% survival at 15 and 20 per thousand. The highest gain in weight (173.0%) and carapace width (56.1%) was obtained at salinity of 15 per thousand. The highest specific growth rate (1.98) was obtained at 15 per thousand, while the lowest specific growth rate (–0.28) was recorded at 35 per thousand. Complete molting was obtained in this species at a salinity of 15 per thousand in the 12th week of the experiment. The crab with carapace width of 6.8 cm increased to 8.1 cm (19.1%) after molting (Lawal-Are and Kusemiju, 2010).

***Accumulation of toxic metals:*** This species has been observed to accumulate toxic metals. The concentrations of metals observed however, were found lower than the WHO permissible level of 2.00 µg/g for Lead (Pb) and Cadmium (Cd) in foods. The mean values obtained ranged in Fe (0.041–0.219 µg/g); Zn (0.647–1.774 µg/g); Cd (0.093–0.635 µg/g); Ni (0.261–0.825 µg/g); Pb (0.160–0.261 µg/g). Although, the levels were lower than the permissible levels, it was observed that this species had higher level of metals than their immediate environment. This is an indication of bioaccumulation of the persistent toxic metals in the organisms (Oyebisi et al., 2013).

**Fisheries:** Commercial fisheries exists for this species.

### *Callinectes arcuatus* (Ordway, 1863)

**Common name:** Arched swimming crab; cuata swim crab; blue swim crab.

**Distribution:** Tropical: Eastern Pacific: Mexico and Peru.

**Habitat:** Demersal; sand and mud substrates; estuaries, mangrove swamps and tidal flats; intertidal to 50 m deep.

**Description:** Carapace of this species is inflated with four triangular frontal teeth. Anterolateral margins are arched. Surface of carapace with granulation which is fairly uniform and is most crowded on gastric, mesobranchial, and cardiac regions; more scattered near anterolateral margins, and smooth along frontoorbital, posterolateral, and posterior borders. Chelipeds are with sharply granulate ridges on propodus, basal portion of dactyl, and exposed surfaces of carpus. Male abdomen and telson are reaching beyond suture between thoracic sternites IV and V and telson is triangular, longer than broad; Mature female abdomen and telson are reaching as far forward as in male. Telson is triangular with slightly inflated sides. Gonopores of females are elliptical with long axis in transverse plane. Carapace length of the largest male is 54 mm, and that of the largest female is 55 mm. In the case of males, the carapace is dull olive gray-green; chelipeds are olive green dorsally and whitish ventrally; and swimming legs are olive green. In females, the carapace is generally blue, central portion is blue violet and anterolateral portions are deep purplish-vinaceous. Chelipeds are with base of merus are olive; inner portion of hands is blue-violet and remainder is purplish but varied. Fingers are barred with purple, and propodal fingers are usually white tipped. Remaining legs are Italian blue. Hairs are olive. Swimming legs are with articulations and margins are narrowly violet. Abdomen is violet. Joints and sternum are white.

### Biology

***Food and feeding:*** Its diet comprises fish, molluscs, shrimp and plant matter. Molluscs were found to be the most important item in the diet of this species and they constituted nearly 28 % of the total dry weight of the foregut contents. Bivalves occurred nearly 5 times as frequently as gastropods and accounted for more than 25 times as much of the dry weight of food, with *Tagelus affinis* being the dominant species in the former group, and *Cerithidea mazatlanica* in the latter. Next most abundant in the foreguts were crustaceans, particularly crabs, which together formed over 22 % of the total dry weight of food.

As other members of the order Decapoda, this species is also mostly gonochoric. Precopulatory courtship ritual is common (through olfactory and tactile cues) and usually there is indirect sperm transfer in this species.

**Fisheries:** Commercial fisheries exist for this species. A small fishery for this species is carried out in Costa Rica,

### *Callinectes bellicosus* (Stimpson, 1859)

**Common name:** Warrior swimming crab, Cortez swimming crab, green crab, hard shell blue crab, or Gumbo crab.

**Distribution:** East Pacific, native to Mexico.

**Habitat:** Warm waters and shorelines.

**Description:** Carapace is with two slender frontal teeth separated by a space often bearing a rudimentary submesial pair of teeth. Anterolateral margins are broadly arched. Lateral spines are short, about twice length of preceding tooth. Surface is finely granulated and remarkably smooth except on anterolateral sulci and lines of granules are more prominent on young than on adults. Chelipeds are provided with prominent and sharply tuberculate or spiniform ridge on outer surface of propodus. Other ridges are lower and nearly smooth. Male abdomen and telson are reaching a bit beyond suture between thoracic sternites IV and V. Telson is triangular and longer than broad. Mature female abdomen and telson are reaching about same level as male and telson is with inflated sides. Gonopores of female are asymmetrically ovate in outline. Carapace length of largest male is 76 mm and that of female is 89 mm. Carapace is mottled greenish yellow to brownish green, sometimes with dark spot on center of orbit and dark green areas roughly outlining epibranchial ridge. Arms are generally greenish yellow to greenish brown. Wrist articulations are purple red. Hand is with blotch at level of finger articulation.

**Biology**

***Behavior:*** It is a cousin of the true blue crab (*C. sapidus*). It is known to be more aggressive than the true blue crab. It develops a hard shell and strong flavor due to the salinity of the water.

***Maturity:*** Studies on size (carapace width, CW) at maturity of this species have been made in the coastal lagoon in the southern Gulf of California by Rodríguez-Domínguez et al. (2012). Segregation according to the size at maturity was observed. The size at maturity was greater in the inside lagoon (114.08 mm CW) than at the mouth of the lagoon (103.73 mm CW). The pooled size at maturity was found to be 107.78 mm CW. These findings are important not only for fisheries management but also for the general biological knowledge of Callinectes species. This study offers an improved approach for evaluating fisheries management.

***Fisheries:*** This species is wild-captured in the Sea of Cortez (also called the Gulf of California); and it is prepared and eaten in the same manner as blue crabs. Its large size and flavor mean that it can also be grilled, steamed, boiled, or fried.

***Callinectes bocourti*** **(Milne-Edwards, 1879)** *(= Callinectes maracaiboensis, Callinectes diacanthus)*

**Common name:** Dead-tooth-swimming crab; blunt-tooth swimming crab; bocourt swimming crab; red-blue crab; stump tooth swimming crab.

**Distribution:** Native range extends from Jamaica and Belize south to Brazil; Atlantic Coast from Virginia to Florida.

**Habitat:** Marine, estuaries, bays, intertidal zones; coastal waters and mangroves.

**Description:** Carapace is bearing four triangular frontal teeth with tips reaching a nearly common level. Anterolateral margins are moderately arcuate. Surface of carapace is dorsally smooth and glistening around perimeter (when wet) and on epibranchial surfaces. Central portion is granulate and coarsest granules are seen over mesobranchial and rear half of cardiac areas and lateral half of branchial lobes. Chelipeds are remarkably smooth except for usual spines and obsolescent granules on ridges. Fingers of major chela are heavily toothed and lower margin of propodal finger is often decurved near base in adults. Male abdomen and telson are long, extending nearly to juncture between thoracic sternites III and IV. Telson is lanceolate and is much longer than broad. Mature female abdomen and telson are reaching as far forward as in male. First gonopods of male are very long, often exceeding telson and crossed near tips. Gonopores of female are asymmetrically ovate in outline with apex on long axis directed anteromesad. Carapace length of the largest male is 76 mm and that of female is 70 mm. Overall cast is olive green with prominent reddish markings. Carapace is olive, grayish green, greenish chestnut, or forest green with variable purplish to red markings.

**Biology**

**Behavior:** This species may compete for food and refuge with *Callinectes sapidus* in estuaries. It is a threatened and an endangered species.

**Fisheries:** Small-scale fishery exists for this species.

### *Callinectes danae* (Smith, 1869)

84 Biology and Culture of Portunid Crabs of World Seas

**Common name:** Dana swimming crab

**Distribution:** Occidental Atlantic: Bermuda, Florida, Gulf of Mexico, The West Indies, Colombia, Venezuela and Brazil (Paraíba to Rio Grande do Sul).

**Habitat:** Wide range of habitats from muddy estuaries in mangroves and algae-covered broken shell bottom, to beaches and open ocean depths down to 75 m.

**Description:** Carapace is bearing four frontal teeth. Metagastric area of adults is with anterior width about 2–2.5 times length, and posterior width is about 1.5 times length. Anterolateral margin is with 9 teeth. Surface of carapace is rather evenly and smoothly granulate, except granules more widely spaced on epibranchial region and near anterolateral border; most crowded on gastric, mesobranchial, and cardiac regions; and nearly smooth along frontoorbital, posterolateral, and posterior borders. Chelipeds are with granulate ridges and upper surface of carpus is bearing slightly developed interrupted ridges. Male abdomen and telson are reaching beyond suture between thoracic sternites IV and V. Telson is triangular and is longer than broad with somewhat inflated sides. Sixth segment of abdomen is with sides which are nearly straight and are diverging proximal. Carapace is olive, becoming indigo on edges of lateral spines and outer anterolateral teeth in some individuals and more uniformly olive in others. Teeth and spines on chelae are white tipped. A white patch is seen in deepest part of depression above third walking leg. Cheliped with upper surface of palm, dactyl, part of carpus, and spined edge of merus is indigo to purple; same color in splashes on inside of fingers and distally on merus and laterally on carpus. Largest male measures a length of 58 mm and width at base of lateral spines is 104. Largest female measures length and carapace width 48mm and 108 mm respectively.

**Biology**

***Food and feeding:*** This species is tolerant of salinities ranging from hyposaline (3.5 ppt) to hypersaline. It feeds on molluscs, other bottom invertebrates, and some fishes, carrion and detritus.

***Reproduction:*** Reproductive features have been studied in this species. Females with developed gonads/ovigerous females were found in greater abundance than females with rudimentary/developing gonads, mainly in deeper transects. Although the areas sampled have different environmental characteristics, the reproductive pattern of this species did not change, showing continuous reproduction throughout, with more abundance of reproductive females on spring and summer. Males reached maturity at larger sizes than females in all three bays where sampling was done (Andrade et al., 2015).

**Fisheries:** Commercial fisheries exists for this species.

## *Callinectes exasperatus* (Gerstaecker, 1856)

**Common name:** Rugose swimming crab

**Distribution:** Eastern Pacific and Western Atlantic: from South Carolina to Florida and Texas, to Mexico, Belize, Guatemala, Honduras, Nicaragua, Costa Rica, Panama (Miraflores Locks), including the West Indies, to Colombia, Venezuela, the Guianas and Brazil (from Para to Santa Catarina).

**Habitat:** Estuaries and shallow oceanic littoral zones, especially in association with mangroves and near river mouths, down to 8 m.

**Description:** Carapace of this species is less than twice as broad as long. There are 9 stout teeth on strongly arched anterolateral margin; and all except outer orbital tooth and short lateral spine are usually swept forward. Front is bearing 4 well-developed teeth (excluding inner orbital angles). Coarse scattered and transverse lines of granules are seen on convex dorsal surface. Pincers are robust and ridges and crests are coarsely granulated. Fifth legs are as flattened form of paddles. Male with T-shaped abdomen is reaching posterior quarter of thoracic sternite 4. First pleopods are reaching slightly beyond suture between thoracic sternites 6 and 7, sinuously curved, overlapping proximally, diverging distally to tips curved abruptly inward, and armed distally with scattered minute spinules. Adult male is dorsally purplish red, more accented on proto-, meso-, and metagastric areas and at base of lateral spines and anterolateral teeth. Branchial region and anterolateral teeth are obscure maroon. Dorsal surface of all legs is purplish red with intense orange red on articulations. Inferior portions of merus, carpus, and fingers of chelipeds are intense violet. Internal and external portion of chelae as well as remaining ventral aspect of animal are white with tints of soft purple.

**Biology**

*Food and feeding:* This species feeds on molluscs, other bottom invertebrates, fishes, carrion and detritus. It digs the clam *D. denticulatus* out of the sand with one of its claws then handles and cracks its shell.

*Adaptation:* This species has been reported to tolerate extreme hyposaline conditions of 3.5 ppt

*Reproduction:* As other members of the order Decapoda this species is mostly gonochoric.

*Mating behavior:* Precopulatory courtship ritual is common (through olfactory and tactile cues); usually indirect sperm transfer.

**Fisheries:** No commercial fisheries for this species.

*Callinectes gladiator* **(Benedict, 1893)** *(= Portunus haani, Portunus gladiator, Neptunus gladiator)*

**Common name:** Gladiator swimming crab; blue crab.
**Distribution:** Indo-Pacific, Japan, Hong Kong to Durban in South Africa.

**Habitat:** Brackish waters of estuaries and lagoons.

**Description:** Carapace is depressed and is moderately broad (breadth about 1.8 times length). Surface of carapace is with widely spaced granular areas being separated by regions with a dense pubescence. Regions are well recognizable. Front is with 4 lobes, of which medians are acute and laterals are right-angled and much broader. Antero-lateral borders are with 9 teeth, of which last one is longest, and is directed outwards. Posterolateral junction is rounded. Antero-external angle of merus of third maxillipeds is markedly produced into a lobe. Chelipeds are stout; their merus is with 2 spines at posterior border; anterior border is with 4 spines; carpus is with 2 spines of normal length; upper surface of palm is with 1 distal spine; and lower surface is with squamiform markings. Posterior border of merus of swimming leg is serrated and is bearing a few spines and dactyl is without a red spot. Penultimate segment of male abdomen is not convex on outer border. Body coloration is uniform gray-green or gray-blue with spot of blue on palm and proximal internal part of fingers of chela. Due to its beautiful mottled carapace with bright blue legs, it is called as the marine or deep-sea blue swimming crab. Largest male has a length of 48 mm and width of 92 mm. Largest female has a length of 60 cm and width of 108 mm.

**Biology:** Unknown.

**Fisheries:** Not known.

### *Callinectes marginatus* (A. Milne-Edwards, 1861)

**Common name:** Sharp-tooth swim crab; and marbled swimming crab.

**Distribution:** Atlantic Ocean: from southeastern Florida to Brazil rarely in Bermuda and North Carolina, also in the Cape Verde Islands and along the African coast from Mauritania to Angola.

**Habitat:** Shallow littoral environments; from intertidal pools to 3 m; shallow salt water on sandy or sandy mud bottom; Brackish waters of estuaries; on algae and grass flats, sandy beaches, rocky pools, eroded coral bases, oyster bars, and edge of mangroves; and depth range, 1–421 m.

**Description:** Carapace is bearing four frontal teeth. Central trapezoidal (metagastric) area is short. Anterior width is about 2.4 times length and posterior width is about 1.5 times length. Anterolateral margins are arched slightly; anterolateral teeth exclusive of outer orbital and lateral spine are without shoulders; and last two teeth are piniform. Lateral spine is moderately long and slender. Carapace surface is coarsely granulate. Posterior and posterolateral margins are smooth. Chelipeds are with smoothly granulate prominent ridges on propodi and reduced ones on carpi; fingers are compressed but broadened dorsoventrally producing a pointed spatulate shape; and major chela are with usual enlarged proximal tooth on dactyl. Propodus is often with decurved lower margin. Male abdomen and telson are narrow, reaching slightly beyond suture between sternites IV and V. Telson is about 1.8 times longer than wide. Sixth segment is nearly parallel sided but is somewhat broadened proximally. Sixth segment of abdomen is longer than fifth. Carapace is brown with areas of bluish black. Chelae are brown above and fingers are dark on external face except for tips and proximal portion. Internal face of fingers is dark in distal two-thirds. Carapace length of largest male is 67 mm and width of 118 mm. Largest female measures a length of 49 mm and width of 95 mm.

**Biology:** Not studied.

**Fisheries:** Commercial fisheries exists for this species.

### *Callinectes ornatus* (Ordway, 1863)

**Common name:** Ornate blue crab

**Distribution:** Western Atlantic Ocean and Caribbean coastlines.

**Habitat:** Depths of up to 75 meters on sand and mud bottoms; river mouths; offshore; seagrass beds in the lagoon; juveniles in shallower habitats.

**Description:** Carapace is with prominent lateral pair of frontal teeth. Submesial pair is small and is often almost completely rudimentary. Metagastric area of adults is not deeply sculptured. Anterior width is 2.8–2.9 times length and posterior width is 1.75 times length. Anterolateral margins are broadly arched. Of anterolateral teeth, first five teeth are with posterior margins which are longer than anterior margins; last two teethwith margins which are approximately equal in length. Surface of carapace is with granulations which are most prominent on anterior half and on mesobranchial regions. Granulations are smaller and more closely crowded on meso-metagastric and cardiac regions. Granulations are nearly smooth along posterolateral and posterior borders. Chelipeds are with smoothly granulated ridges on chelae. Carpus is almost smooth dorsally. Major chela is usually with strong basal tooth on dactyl. Male abdomen and telson are reaching beyond suture between thoracic sternites IV and V. Telson is slightly longer than broad with somewhat inflated sides. Sixth segment of abdomen is relatively narrow and its sides are slightly constricted. Males are usually slightly larger than the females measuring up to 60 mm while the female carapace can measure up to 58 mm. Color of this species varies considerably. Carapace of adult males is uniformly olive to dark brown with a large orange spot posteriorly that can appear round to blotchy. There are distinct ivory white tips on all the spines on the front of the carapace. Juveniles are not brilliantly colored and are appearing olive-yellow to greenish.

### Biology

*Adaptation:* This species is usually found in tropical seas in temperatures ranging from 18–31°C. Though specimens of this species have been trapped in fresh water but most collections of *C. ornatus* come from waters with higher salinities. Laboratory experiments have shown that this species has lower tolerance to hyposaline conditions (reductions in salinity). Individuals of *C. ornatus* were shown to begin dying off when seawater was diluted by 25% (8.7 ppt).

***Food and feeding:*** This species is both a saprophagous species, feeding on decaying matter, as well as an active predator. *C. ornatus* will dig into the substratum in search of food, feeding on molluscs, especially bivalves, and other crustaceans, including brachyura (true crabs), other species of Callinectes, algae, polychaetes, echinoderms, and foraminifera. Its competitors of the ornate blue crab are other crustaceans, in particular *C. sapidus* and *C. similis*.

***Reproduction:*** This species has been reported to spawn year-round. Adult females move offshore to find temperature and/or salinity conditions that are best for spawning. One female has been reported to mate with several males during the same reproductive period.

**Fisheries:** As this species occupies the same habitats as the commercial species, *C. sapidus*, it is impacted by the crab fishery along western Atlantic coast. In the Ubatuba region of Brazil it is a major fishery (Smithsonian Marine Station at Fort Pierce, http://www.sms.si.edu/irlspec/Callinectes_ornatus.htm).

### *Callinectes pallidus* (Rochebrune, 1883)

**Common name:** Gladiator swim crab
**Distribution:** Eastern Atlantic: from Mauritania to Angola.
**Habitat**: Brackish and saltwater at depths of less than 30 m on bottoms of mud and sand.
**Description:** Carapace of this species is rhombic with broad anterior and posterior edges and a rough surface. Sides of carapace are elongated to form a lateral spine and antero-lateral teeth are present. Fourth pair of

walking legs is broad and flattened to form a paddle, which helps in swimming. Carapace length is 5.9 cm.

**Biology**

*Food and feeing:* Laboratory experiments have been made on the food and feeding habits of this species. The crabs fed on a variety of food items which included fishes, molluscs, crustaceans, higher plant materials, algae and diatom. Fishes were found to be the most important food item constituting 75.6%. This was closely followed by mollusc accounting for 71.3%. The least item consumed was diatom with 11.6%. Fishes still remained the most important feed item relative to size and sex. However, there was slight variation in food and feeding habit relative to size. The smaller crabs (carapace width <3.99 cm) fed more on algae and diatom than medium size (4.00, 4.99 cm carapace width) and large crabs (carapace width >5.00 cm). There was also a slight variation in food and feeding habit relative to sex with males consuming more fishes (82.4 %) than females (70.9%). Further, females consumed higher plant materials (53.4%) than the males (38.2%). This study revealed that Callinectes pallidus is an omnivore and an opportunistic feeder (Jimoh et al., 2014).

*Callinectes rathbunae* **(Contreras, 1930)**

**Common name:** Sharptooth swimming crab
**Distribution:** Western Central Atlantic: Gulf of Mexico.
**Habitat:** Estuarine waters of ditches, lagoons, and river mouths.
**Description:** Carapace is bearing four acuminate frontal teeth with tips reaching a nearly common level. Submesial teeth are narrower and slightly shorter than laterals. Metagastric area length and posterior width are about equal. Anterior width is 2 times length. Anterolateral margins

are slightly arcuate. Anterolateral teeth exclusive of outerorbital and lateral spine are all acuminate with edges which are variably granulate. Anterior margins of teeth are slightly shorter than posterior margins. Chelipeds are with sharply granulate ridges and usual spines. Fingers of major chela are heavily toothed but not gaping. Male abdomen and telson are long, extending nearly to juncture between thoracic sternites III and IV. Telson is lanceolate and is much longer than broad. Sixth segment of abdomen is broadened distally. Telson is elongate and triangular with inflated sides. Length and width of largest male are 61 mm and 107 mm, respectively. Corresponding values for largest females are 66 mm and 116 mm, respectively. Color of carapace as obscure or dark green and underparts are white.

**Biology**

***Food and feeding:*** This species feeds on a variety of materials including mollusks, other bottom invertebrates, and some fishes, carrion and detritus.

***Growth:*** Effects of salinity on growth and molting of this species have been studied. When megalopae of this species were exposed to salinities of 5, 15, and 25 at 25.0°C (through crab stage 16), their survival decreased significantly at higher salinities. It had significantly higher survival in salinities of 5 and 15. There was no difference in survival of *C. rathbunae* by gender; and survival of both genders was generally lowest at a salinity of 25. Additionally, males had significantly greater survival than females in salinities of 15 and 25. Females of this species grew faster than males at all salinities and both genders grew fastest in a salinity of 15. There was no significant difference in intermolt duration between genders of this species. However, intermolt duration among salinity treatments for this species differed significantly in crabs ≥ stage 7, with the longest duration in the highest salinity. Results of this study suggest that *C. rathbunae* is more tolerant of low salinity habitats (Cházaro-Olvera and Peterson, 2004).

***Aquaculture:*** This species is suitable for cultivating in floating cages.

**Fisheries:** A minor commercial fisheries exists for this species.

## *Callinectes sapidus* (Rathbun, 1896)

**A: Dorsal view**

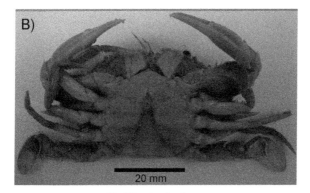

**B: Ventral view**

**Common name:** Blue crab

**Distribution:** Western Atlantic Ocean from Nova Scotia to Argentina; introduced, accidentally or deliberately, into both Asia and Europe.

**Habitat:** Brackish coastal lagoons and estuaries.

**Description:** Carapace of this species is bearing two broad either obtuse or acuminate, triangular frontal teeth with mesial slopes (incorporating a pair of rudimentary submesial teeth) longer than lateral slopes. Metagastric area is with posterior width approximately 1.2 times length and anterior width is about 2 times length. Anterolateral margins are slightly arched. Anterolateral teeth exclusive of outer orbital and lateral spine obtuse to acuminate and directed outward more than forward. Much of surface is smooth, with scattered granules. But granules are found concentrated on mesobranchial, posterior slope of cardiac and anterior portion of mesogastric area. Lateral spines are varying from rather stout, blunt, and forward trending to slender, elongate, and slightly backward trending. Propodus and carpus of chelipeds are with moderate finely granulate ridges. Male abdomen and telson are reaching about midlength of thoracic sternite IV. Telson is lanceolate and is much longer than broad. Sixth segment of abdomen is broadened distally. Mature female abdomen and telson are reaching about midlength of thoracic sternite IV. Telson is with inflated sides almost equilaterally triangular and fifth and sixth abdominal segments are equal in length. Color of carapace and chelipeds is grayish, bluish, or brownish green of varying shades and tints. Mature females are with orange fingers on chelae tipped with purple. Underparts are off-white with tints of yellow and pink. Values of carapace in length and width of largest male are 91 mm. and 168 mm, respectively. Corresponding values for females are 75 mm and 143 mm, respectively.

**Biology**

***Digestive system:*** The digestive system of the blue crab is a simple tube system with a foregut, midgut, and hindgut. The hindgut gland, or hepatopancreas, produces digestive enzymes, absorbs nutrients, and stores lipids. An important organ system in the blue crab often targeted by diseases, parasites, and symbionts is the respiratory system, specifically the gills. The blue crab has eight gills that function in respiration and ion regulation and excretion. Blue crabs have an open circulatory system with a combination of hemal sinuses and blood vessels that transport hemolymph.

***Food and feeding:*** Predation by blue crabs helps in regulation of marine bivalve populations in shallow, unvegetated soft and hard bottom communities. Blue crabs prefer molluscs such as oysters (*Crassostrea virginica*) and hard clams (*Mercenaria mercenaria*) as their primary food sources, though older juveniles and adults sometimes incorporate some plant material such as Ulva, eelgrass, and Spartina into their diet. Foraging behavior is preceded by an increases in gill bailing, antennule movement and flexion of the mouthparts. The dactyls of the anterior walking legs are used to probe the substrate for buried bivalves, and to manipulate them after they are located.

*Locomotion:* Blue crabs are highly mobile, but are more active during daylight hours than in the evenings. They are able to move from 0–140 m per hour, with an average of 15.5 m/hr. The total distance traversed per day for these crabs is 215 m.

*Environmental parameters and growth:* Growth occurs at temperatures from 15–30°C, but is prevented at temperatures below 10°C. A hibernative state is induced in blue crabs at temperatures below 5°C. The larvae of this species have a salinity requirement of at least 20 ppt. However, as these crabs grow, they are increasingly euryhaline. Both juvenile and adult blue crabs are able to inhabit fresh water areas, as well as highly saline habitats. Thus, salinity is not a major limiting factor in growth, molting or reproduction. This species has been reported to tolerate extreme hyposaline conditions of 3.5 ppt.

*Age, size and lifespan:* This species may attain 25 cm carapace length, with carapace width being approximately twice the length. New recruits enter estuaries. Maturity is reached by the second year at carapace lengths of 120–170 mm. Low temperatures (<10°C) have been reported to prevent molting and decrease growth rates in blue crabs. Generally, growth occurs in this species at temperatures over 15°C, and is mostly unaffected by salinity conditions. Under laboratory conditions these crabs are able to molt and grow regularly at temperatures between 15 and 30°C, and salinities as low a 3 ppt. Growth observed in blue crabs has been estimated to be between 12 and 35% per molt.

*Effects of salinity on growth and molting:* When megalopae of this species were exposed to salinities of 5, 15, and 25 at 25.0°C (through crab stage 16), there was no salinity effect on the survival of this species. There was no difference in survival for females among salinities, but males had lowest survival at the lowest salinity. Additionally, males had significantly greater survival than females in salinities of 15 and 25. Males and females of this species grew at the same rate, with the lowest growth rate for both genders at the lowest salinities. This species also had a significantly higher growth rate and shorter intermolt duration at the highest salinity. However, there was no significant difference in intermolt duration between males and females of this species. Intermolt duration was found to be shortest in a salinity of 25. Results of this study therefore suggest that *C. sapidus* is not tolerant of low salinity (Cházaro-Olvera and Peterson, 2004).

***Mortality:*** Mortality of adult blue crabs has been reported to occur under laboratory conditions when dissolved oxygen concentrations fall below 0.6 mg/l for over 24 hours at a temperature of 24° C.

***Predators:*** Predators on blue crabs include fish as well as other blue crabs. The major fish predators include the Black Drum (*Pogonias cromis*), Red Drum (*Scianops ocellata*), the American Eel (*Anguilla rostrata*), and the American Croaker (*Micropogonias undulatus*).

***Competitors:*** Competitors of blue crabs are generally other crustaceans such as *Callinectes similis* and *C. ornatus*. *Panopeus herbstii, Menippe mercenaria,* and *Carcinus maenas* also compete for resources with blue crabs.

***Reproduction:*** Two spawning periods are common in this species. Unlike males, female blue crabs mate only one time in their lives, following the terminal, or pubertal molt. When approaching this molt, females release a pheromone in their urine which attracts males. Male crabs vie for females and will protect them until molting occurs. It is at this time that mating occurs. Mating may last as long as 5–12 hours. Blue crabs are highly fecund, with females producing from 2 to 8 million eggs per spawn. Eggs are brooded 14–17 days, during which time females migrate to the mouths of estuaries so that larvae may be released into high salinity waters. Larvae of this species have a salinity requirement of at least 20 ppt, and show poor survival below this threshold.

***Larval stages* in Callinectes sapidus*:*** There are usually 7 zoeal stages and 1 postlarval, or megalopal, stage in this species. Occasionally an eighth zoeal stage is also observed. Larval release is often timed to occur at the peak of high tide and larval abundance is greatest when the tide begins to ebb. Blue crab larvae are advected offshore, and complete development in coastal shelf waters. Typical time for development through the 7 zoeal stages is between 30 and 50 days before metamorphosis to the megalopal stage. The megalopa then persists between 6 and 58 days. It is widely believed that it is the megalopal stage that subsequently return to estuaries for settlement, and eventual recruitment to adult populations.

Profile of Portunid Crabs (Family: Portunidae)

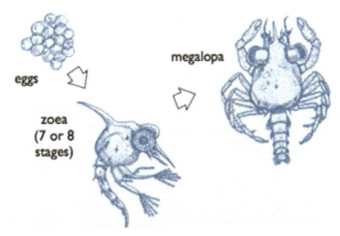

### Larval stages in *C. sapidus*

**First zoea:** Abdomen has five segments plus a telson. Eyes are not stalked. Second segment of abdomen bears a short lateral knob and third segment has a short hook on each side.

**Second zoea:** Eyes are stalked. On third segment of endopodite of first maxilliped, one spine is added. Exopodite bears 6 plumose swimming setae. Endopodite of second maxilliped has one additional subterminal spine

**Third zoea:** Setation of antennule and antenna are unchanged from previous stage. Mandible) has several small teeth in addition to broad cutting surface. Exopodites of both maxillipeds terminate in 8 swimming setae. A sixth segment has been added to abdomen.

**Fourth zoea:** A slight swelling is seen in basal region of antenna indicates beginning of the endopodite bud. A small, unsegmented palp appears with mandible. Exopodites of both first and second maxillipeds bear 9 swimming setae of unequal length.

**Fifth zoea:** Developing endopodite bud of antenna is larger than in previous stage. Maxillule remains as in previous stage but setation of maxilla is increased to 8 spines on coxal endite. Buds of third maxilliped, chela, and pereiopods are visible beneath carapace. Number of setae projecting from edge of carapace has increased.

**Sixth zoea:** Hairs appear on the small, unsegmented palp of the mandible. A plumose spine is added to basal segment of endopodite of the maxillule and coxal endite bears a total of 9 spines.

**Seventh zoea:** Basal portion of antennule is swollen and there is a slight indentation in distal half. Swimming setae have increased to 14 on first maxilliped and to 13 on second maxilliped. Developing thoracic appendages have increased in size and project below the margin of carapace.

**Eighth zoea:** A second spine is added below endopodite. Spines of basal and coxal endites of maxilla have increased to 15 and 10, respectively. Swimming setae on first maxilliped have decreased to 12 and 14 setae are found on the second maxilliped. On first maxilliped an epipodite, partially developed, bears short setae and soft, non-plumose hairs. Exopodite of third maxilliped bears two short terminal spines and the epipodite terminates in one small, non-plumose spine. Chela and pereiopods are larger and project well beyond border of carapace. Pleopod buds bear short non-plumose hairs. Spines on inner margin of telson total 10. Four small hairs project dorsally from posterior margin of first abdominal segment.

**Megalopa:** Rostrum is pointed, and is longer than antennules but shorter than antennae; and eyes are stalked. Appendages, eyes, and margins of carapace are provided with small hairs. Antenna is composed of 11 segments, some of which bear setae. Mandible has a palp of two segments with 11 bristles on distal segment. Endopodite of maxillule has 4 spines on terminal segment and 6 spines on first segment. Number of spines on coxal and basal endites has increased to 17 and 25, respectively. Endopodite of maxilla is reduced in size and is bearing only three spines. First, maxilliped is considerably modified from swimming appendage of zoeal stages. Endopodite broader with 8 non-plumose setae on distal border. Exopodite of two segments, is with 6 terminal setae on second segment. Epipodite is well developed and fringed with long, non-plumose hairs. Second, maxilliped has endopodite of 4 segments with stout spines on terminal segment. Exopodite is two-segmented with 6 terminal hairs. Epipodite is small. Third, maxilliped is with large endopodite bearing numerous spines on all segments; exopodite is unsegmented and bearing 6 terminal setae; and epipodite is fringed at distal portion by soft, non-plumose hairs. Fourth, abdominal segment retains lateral spines, projecting caudally past the smaller sixth abdominal segment.

**Fisheries:** *Callinectes sapidus* is beneficial in terms of its value as a commercial and recreational fishery species. Commercial fisheries for *C.*

*sapidus* exist along much of the Atlantic coast of the United States, and in the Gulf of Mexico.

## *Callinectes similis* (Williams, 1966)

**Common name:** Lesser blue crab

**Distribution:** Western Atlantic Ocean, Caribbean Sea and Gulf of Mexico from the United States to Colombia.

**Habitat:** Gulf and bay; oceanic littoral zone in salinities above 15 at a depth of 100 m; juveniles in estuaries

**Description:** Carapace is with four frontal teeth and submesial pair is small. Central trapezoidal (metagastric) area is short and wide. Anterior width is 2.75 times length and posterior width is 1.6–1.7 times length. Anterolateral margins broadly are arched. Anterolateral teeth exclusive of outer orbital and lateral spine are short and broad; tips of first five are nearly rectangular; and sixth and seventh are acuminate. First five teeth are with anterior margins which are shorter than posterior and are separated by narrow based rounded notches. Lateral spine is strong, slender, and curved forward. Surface of carapace is even and uniformly granulate except smooth along posterolateral and posterior slopes, and nearly smooth along anterolateral and anterior margins. Chelipeds are with very fine granulations on ridges; carpus is bearing two obsolescent granulate ridges; chelae are Strong and are not greatly dissimilar in size. Male telson is longer than wide. Sixth segment of abdomen is slightly sinuous sided but broader at all levels than telson. Mature female telson is slightly wider than long. Adult male's carapace is green dorsally. Chelipeds and portions of legs are similar in color or more tannish green dorsally, with iridescent areas on outer and upper edges of carpus and hands. Chelae are white on outer face and blue to fuchsia on inner surface. Carapace length of largest male measures 55 mm and width to 97 mm. Corresponding values for females are 45 mm and 76 mm, respectively.

**Biology**

*Food and feeding:* it is an opportunistic predator feeding chiefly on bivalves and other benthic macro-invertebrates. Other diet items include fish, crustaceans, squid, detritus, and plants.

*Growth:* Male crab grows to a maximum width of 122 mm and the female grows to a maximum width of 95 mm.

*Reproduction:* It has spring and fall spawning seasons. The egg carrying females have been reported to migrate to near-shore waters with higher salinities to release their larvae.

*Embryology:* There are generally 8 zoeal stages and 1 postlarval, or megalopal stage. Larval release occurs normally at high tide. Crab larvae are advected offshore, and complete development in coastal shelf waters. Typical time for development through the 7 zoeal stages is between 30 and 50 days before metamorphosis to megalopal stage. The megalop persists between 6 and 58 days. It is widely believed that the megalopal stage subsequently returns to the estuaries for settlement, and recruitment to adult populations.

*Behavior:* These crabs are very aggressive species and use both visual and chemical cues to evaluate danger and potential mates from a distance. They use color as a means of determining which mate they will try to attract then they display to the female.

*Effects of salinity:* Adults are not usually encountered in estuaries where salinities are below 15 ppt. It is a hyperosmoregulator and is able to respond to changes in salinity by maintaining a high hemolymph osmolarity by active ion pumping in the gills. Juveniles exhibit a higher tolerance for salinity change than adults allowing them to inhabit estuaries with salinities as low as 10 ppt. In laboratory experiments, juveniles survived in near freshwater (as low as 5 ppt) to hypersaline conditions, but were reported to be best adapted to 35 ppt.

*Dissolved Oxygen:* Juveniles tolerate high levels of hypoxia but will die under anoxic conditions. In laboratory studies, there was a noticeable decrease observed in feeding, growth and molting rates below normal oxygen concentration.

*Associated Species:* This species is usually associated with *Callinectes sapidus* where it competes for food and other resources.
**Fisheries:** This species has a commercial fisheries.

### *Callinectes toxotes* (Ordway, 1863)

**Common name:** Giant swim crab
**Distribution:** Southeast Pacific: Peru.
**Habitat:** Demersal
**Description:** Carapace is bearing four large, rounded frontal teeth. Metagastric area is with length approximately equal to posterior width. Anterior width is about 2 times length. Anterolateral margins are moderately arched. Surface of carapace is coarsely granulate and uneven; nearly smooth around margins and along regional sulci; more granulate over branchial and gastric areas; and most closely crowded granules on cardiac, mesobranchial, and anterior half of mesogastric regions. Epibranchialline is prominent and nearly uninterrupted. Propodus and carpus of chelipeds are with sharply and rather coarsely granulated ridges. Dactyl of major chela is with basal teeth (often a single strong tooth). Male abdomen and telson are reaching beyond midlength of thoracic sternite IV. Telson is much longer than broad, triangular with inflated sides. Sixth segment of abdomen is narrowest in proximal third. Mature female abdomen and telson are reaching no more than midlength of thoracic sternite IV. Telson is elongate triangular with inflated sides. Sixth segment is longer than fifth. Carapace length and width of larger male are 88 mm and 156 mm, respectively. Corresponding values for large female are 74 mm and 133 mm, respectively.

**Biology**

*Food and feeding:* Food and feeding: Important food items of this species include bivalve molluscs and fish. Feeding occurs mainly at night, especially around dark.

**Fisheries:** A commercial fisheries exists for this species

*Lupocyclus inaequalis* **(Walker, 1887)**

Image not available.

**Common name:** Not designated.

**Distribution:** Seychelles, Saya de Malha Bank, Japan, Singapore, South China Sea, Philippines, Indonesia, Australia, and Hawaii.

**Habitat:** Marine; depth range 18–94 m.

**Description:** Dorsal surface of carapace is with ridges. Two pairs of frontal teeth are protruding well forward. 5 to 6 spines are seen on anterior border of arm of cheliped. 5 main antero-lateral teeth are present. Epibranchial ridges of carapace are distinct and are almost joining metagastric ridges.

**Biology:** No information

**Fisheries:** There is no commercial fisheries for this species.

*Lupocyclus philippinensis* **(Semper, 1880)** *(=Lupocyclus sexspinosus)*

**Common name:** Philippine Island spike crab; Scissor swimming crab.

**Distribution:** Indo-Pacific, Madagascar, Seychelles, Pakistan, Laccadive Islands, India, Andaman Sea, West Malay Peninsula, Japan, Taiwan, China, South China Sea, Thailand, Philippines, Indonesia, Australia, New Caledonia, Vanuatu, Wallis and Futuna Islands.

**Habitat:** Depth range 13–305 m; bottoms of sandy mud or broken shells.

**Description:** Carapace of this species is subcircular; its length is bout 0.9 times breadth; it is covered with a fine dense pile; and it has hardly elevated granular patches on frontal, protogastric, mesogastric, metagastric, cardiac, and mesobranchial regions. Front is protruding with 4 teeth; all are very elongated and acute; and medians are more protruding than the laterals. Antero-lateral margin is with 6 teeth, of which, first is largest. Chelipeds are very elongate and are more than 3 times as long as carapace; merus is with 6–7 spines on anterior border and 2 on posterior border; carpus is with 2 spines; and palm is with 3 spines. Posterior border of merus of swimming legs is armed with one subdistal spine. Posterior border of propodus is smooth.

**Biology:** No information

**Fisheries:** There are no commercial fisheries for this species.

### *Lupocyclus quinquedentatus* (Rathbun, 1906)

**Common name:** Not designated.

**Distribution:** Cargados Carajos Islands, Seychelles, Japan, Indonesia, Hawaii, Marquesas Islands, Society, Tubuai Islands.

**Habitat:** Marine; depth range 80–110 m.

**Description:** Carapace of this species is 0.75 times as long as broad and is covered with a fine dense pile, except on transverse granulated ridges. Epibranchials are interrupted medially. Clusters of granules are seen on frontal region and near to antero-lateral teeth. Front is protruding, with 6 triangular and acute teeth, of which median ones more advanced than laterals. Antero-lateral borders are with 5 sharp subequal teeth, of which first is stoutest. Chelipeds are about 2.5 times as long as carapace in male, and 2.2 times in

female; merus is with 3–4 teeth on anterior border and posterior with a distal spine; and carpus is with an inner and 2 outer subdistal spines. Posterior border of merus of swimming legs is armed with 2 spines and greater part of posterior margin of propodus is armed with small stout denticles.
**Biology:** Unknown.
**Fisheries:** Unknown.

## *Lupocyclus rotundatus* (Adams & White, 1849)

**Common name:** Round spike crab
**Distribution:** Sri Lanka, Andaman Sea; Malay Peninsula, Taiwan, South China Sea, Philippines, Indonesia, and Australia.
**Habitat:** Habitat within 30 to 100 meters of seabed mud.
**Description:** Dorsal surface of carapace is with ridges. 2 pairs of frontal teeth are protruding well forward. 5 to 6 spines are seen on anterior border of arm of cheliped. 5 main antero-lateral teeth are present. Epibranchial ridges of carapace are distinct. A broad gap is seen between epibranchial and meta- gastric ridges of carapace.
**Biology:** Not known.
**Fisheries:** Not known.

## *Lupocyclus tugelae* (Barnard, 1950)
Image not available.
**Common name:** Not designated.

**Distribution:** Red Sea, Gulf of Aden, South Africa, Kenya, Madagascar, Japan, South China Sea, Philippines, Sulu Archipelago, Indonesia, Western Australia, New Caledonia.

**Habitat:** Shallow waters; 30–320 m depth

**Description:** Carapace is broader than long; its length is 0.8 times breadth; and is covered with a fine dense pile through which the main granular patches are visible. Distinctly elevated granular patches are seen on frontal, protogastric, mesogastric, metagastric, cardiac, and mesobranchial regions. Front is protruding with 4 teeth, of which medians are triangular and laterals are obtuse. Antero-lateral border is with 5 large teeth which are alternating with 4 smaller ones. Chelipeds are moderately elongated and is 2–2.5 times as long as carapace; merus is with 5 spines on anterior border and 2 on posterior; carpus is with 2 spines; and palm is with 3 spines. Swimming legs with posterior border of merus is armed with one subdistal spine and posterior border of propodus is smooth.

**Biology:** Unknown.
**Fisheries:** Unknown.

### *Achelous ordwayi* (Stimpson, 1860) *(=Portunus ordwayi)*

**Common name:** Redhair swimming crab; red crab
**Distribution:** Caribbean, Bahamas, Florida, Gulf of Mexico, and Bermuda.
**Habitat:** Depth range 0–15 m.
**Description:** Pronounced fringe of hair is seen on upper margin of chelipeds. Outer, upper surface of manus is smooth and iridescent. Posterodistal margin of merus of swimming legs is spinulous. Lateral spine of carapace is three times as long as preceding tooth. Length of carapace is mm and width is 11 mm.

**Biology**
*Pigments:* In this species, there are red, white, yellow and blackish-brown pigments, which are supplemented by a golden-green iridescence on certain regions of the appendages. A blue coloration is also noticed in the proximity of the red and dark pigments. In this species, the carapace has the ability to assume different hues (Abramowitz, 1935).
**Fisheries:** No information

*Achelous spinicarpus* **(Stimpson, 1871)** *(=Portunus spinicarpus)*

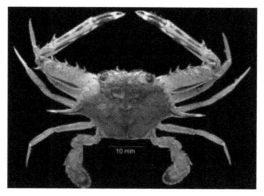

**Common name:** Longspine swimming crab
**Distribution:** South Atlantic Ocean, Western Atlantic Ocean, North Carolina to Florida and Gulf of Mexico, and the West Indies to Uruguay.
**Habitat:** Benthic; deep waters; soft bottoms; sand and mud bottoms; broken shells and corals; depth range 2–550 m (usually 5–55 m).
**Description:** Eyestalk has a conspicuous red spot on its anterior surface Pereiopods are greyish green with pink dactyli. Fifth legs are greyish green with a very distinct red spot in distal part of dactylus. Carapace length of males ranges from 15 to 34 mm; females from 13 to 32 mm and ovigerous females from 9 to 32 mm. For these three categories, total carapace breadth (the lateral spines included) is 35 to 64 mm, 30 to 58 mm, and 18 to 58 mm, respectively. Third maxilliped has last three segments which are iridescent. Upper surface of merus of cheliped shows a dark band. Carpus has an inconspicuous streak in basal part of inner margin. Long carpal spine is red, with some white spots in distal part; margin of spine facing chela has a fringe of long, red hairs. Carapace of living animal is greyish green, and is marbled all over with short, curved, reddish-brown lines and streaks. Maximum size of male is 4.3 cm CW.

## Biology

***Distribution:*** Abundance of individuals of this species has been reported to be proportional to the depth and inversely proportional to temperature.

***Food and feeding:*** It is a carnivorous species feeding on polychaetes.

***Growth:*** Males are usually larger than females. Among the collected individuals the smallest individuals (<28 mm carapace width) were only females, and the only two specimens larger than 60 mm were males. The size of the smallest specimen (13 mm) presumably indicates that individuals recruit to the adult population around this size.

***Maturity and fecundity:*** The size at maturity estimated for this species was found to be 37mm CW. Fecundity of this species, however, has a positive correlation between the number of eggs and the size of the female.

**Fisheries:** No information.

## *Achelous spinimanus* **(Latreille, 1819)** *(= Portunus spinimanus)*

**Common name:** Blotched swimming crab
**Distribution:** Gulf of Mexico and North Atlantic Ocean
**Habitat:** Gulf and deeper channels of bays.
**Description:** Carapace is broader than long and is compressed. Color of carapace is dark mottled brown dorsally and is white underneath. Claws are long and fingers are mottled brown, and some are orange. Last pair of legs is flattened into paddle-shaped appendages. Nine pair of spines (including the outer orbital spines) are seen along lateral edges of carapace; most posterior pair (called lateral spines) is not much longer than other spines (about twice as long); and tips of spines are lighter in color.

Eight low, blunt teeth are seen between ocular orbits (eye sockets), including inner orbital teeth; and outer 2 pairs and inner pair are more bi-lobed. Spines are seen on inner margins of merus (2 segments behind claws) and inner and outer dorsal carpus. Carapace width of this species is 11 cm.

**Biology**

*Fecundity:* The fecundity of this species has been reported to vary between 188.065 and 682.992 eggs for the range in female size between 56.2 mm and 86.6 mm, and the average fecundity of the population was 429.676 eggs. The diameter of the eggs analyzed varied between 103.0 and 159.7 μm. At the initial stage, the average diameter was 117.3 μm; at the intermediate stage, it was 127.9 μm; and at the final stage, it was 151.8 μm. The differences among sizes in the classes of egg's development were significant ($p < 0.05$).

**Fisheries:** No information

***Arenaeus cribarius* (Lamarck, 1818)** *(=Portunus cribarius)*

**Common name:** Speckled swimming crab, surf crab

**Distribution:** Argentina, Atlantic Ocean from Vineyard Sound, Massachusetts in United States, La Paloma in Uruguay, Brazil

**Habitat:** Shallow gulfs, common in the surf and prefers sandy or muddy bottoms.

**Description:** Carapace is broader than long and is compressed; color is brownish with white spots dorsally, yellowish to white underneath, and sometimes with white irregular area in center of carapace. Claws are long and their color is brown with white fingers. Last pair of legs is flattened

# Profile of Portunid Crabs (Family: Portunidae)

into paddle-shaped appendages. Nine pair of spines (including the outer orbital spines) are seen along lateral edges of carapace; and most posterior pair (lateral spines) is longest. Eight low blunt spines are found between ocular orbits (including the inner orbit spines). Inner orbit spines are bicuspid. Spines are seen on inner margins of merus (arms of claws) and a small spine is present on inner and outer dorsal carpus (joint of claw). Carapace width shows a range of 1.8 to 11.9 cm for males, and 2.9 to 9.3 cm for females. This species did not show a significant difference between the sexes. Males reach size and weight, on average, higher than females.

**Biology**

*Adaptation:* Hair-like setae line the anterior-lateral margin of the crab's carapace and dense tufts of similar setae are present on either side of the mouthparts are creating a respiration channel as the crab holds its chelipeds close to the body when buried. Such morphological adaptations allow this species to maintain an existence in the rather strenuous breaker zone of the near shore environment, where crabs must adapt to life in waves and shifting sand.

*Food and feeding:* It is usually found in bottom burrows at daytime and emerges at night, at temperatures of 27.5 to 28.6°C and salinities of 27.5 to 35 ppt. It is known to feed on a variety of items including molluscs, other bottom invertebrates, and some fishes, carrion and detritus

*Growth:* The age estimated for the first juvenile stage of this species was found to be 6.1 and 8.3 days for males and females, respectively. The maximum age determined was 1.8 years for males and 2 years for females, which correspond to a maximum size of 115.8 and 96.7 mm, respectively. The maximum size (carapace width, CW) estimated was 114.5 mm for males and 95.8 mm for females. Males have a precocious sexual maturity (5 months) when compared to females (6.8 months). The growth rate and size of *A. cribrarius* are higher than other portunid species, with great interest for aquaculture (Pinheiro and Hattoria, 2006).

*Reproduction:* The average size of first sexual maturation for this species is between 7.5 and 8.0 cm. Reproduction and recruitment only occurred during the warmer months of the year, and size at sexual maturity was smaller (50.1 mm CW) in males than in females (56.3 mm CW) (Andrade et al., 2015).

Reproductive behavior exhibited by this species in captivity has been studied. In the presence of premolt females, intermolt males exhibited a courtship display that became intensified when the potential mate was visually perceived. After mate selection, the male carried the female under itself (precopulatory position) for 29.8 d until the female molted. Afterwards, the male manipulated the recently molted female, and inverted her position under itself as to penetrate her with his first pair of pleopods (copulation), a process that took 17.1 h. After copulation the male continued to carry his soft-shelled mate for 29.7 d (postcopulatory position). The time elapsed between copulation and spawning was 57.8 d, and the time interval between successive spawns was 33.8 d. Total embryonic development took 13.5 d in temperature conditions of 25.0°C. During the last 4.7 d, embryos' eyes were already visible (Pinheiro and Fransozo, 1999).

*Larval development*

Eight zoeal stages and a megalopa have been identified in the larval development of this species.

*Characteristics of zoea I to zoea VIII:*

*Zoea I:* Carapace is bearing elongate dorsal, rostrum, and lateral spines. Minute spine-like setae are present on each side of dorsal spine base. Postero-ventral margin is finely serrated. Eyes are sessile.

*Zoea II:* Posterolateral margin of carapace is smooth and each margin is bearing I seta. Anterior margin is with medial seta, otherwise carapace as in zoea I. Eyes is stalked.

*Zoea III:* Posterolateral margins of carapace, each bearing 3 setae, otherwise, carapace as in zoea II.

Antennule is as in zoea II except second long seta rarely occurs.

*Zoea IV:* Posterolateral margins of carapace, each bearing 5 setae, otherwise, carapace as in zoea III. Antennule is as in Zoea III.

*Zoea V:* Posterolateral margins of carapace, each bearing 7–9 setae; rostral spine is slightly longer than antennal protopodite.

Zoea VI: Posterolateral margins of carapace, each bearing 9–11 setae; rostral spine is 1.25 times length of antennal protopodite.

*Zoea VII:* Posterolateral margins of carapace, each bearing 14–15 setae; rostral spine is about 1.5 times length of antennal protopodite.

*Zoea VIII:* Posterolateral margins of carapace, each bearing 17–19 setae; rostral spine is 1.8 times length of antennal protopodite.

**Megalopa:** Carapace is rectangular, without dorsal and lateral spines; rostrum is 0.8 times length of antenna; strong pair of spines is extending from 7th sternal segment posteriorly to anterior margin of 3rd abdominal segment. Eyestalks each is with a distinct pigment spot on dorsal surface. Pereopod I is chelate, 5-segmented and cutting edge of propodus is slightly longer than dactylus with bearing 2 large and 2 small teeth; carpus is with weak dorsal and strong medial spine; ischium is with hooked spine; pereopod 2 is with spine on ventral side of basis, otherwise, pereopods 2–4 are similar; and dactylus of pereopod 5 is paddle like.

**Fisheries:** It has a minor commercial fisheries.

### *Atoportunus gustavi* (Ng & Takeda, 2003)

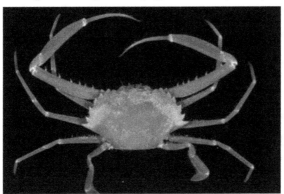

**Common name:** Not designated as it is a new species.

**Distribution:** Guam, Japan, Hawaii and Christmas Island (Indian Ocean).

**Habitat:** Not known.

**Description:** Carapace front is fourlobed. Median lobes are less than half as broad as laterals and laterals are rightangled with rounded tips. Among the anterolateral teeth, first is blunt; second to fourth are shorter than fifth to eighth; and ninth tooth is distinctly the largest (after the first). Carapace is moderately narrow (breadth ca. 1.4 times length), with almost rounded (suborbicular) outline. All granular areas distinct, giving embossed surface; granular patches are separated by moderately broad areas bearing a fine pile of hairs; granules in patches are relatively large and widely spaced. Chelipeds are elongate, granular, hirsute except on under surface which is polished and pitted. Two sharp spines are seen on posterior border of arm. Anterior border of arm is with 5 spines .Wrist is with 2 usual spines. Hand is strongly carinated . Fifth leg is of usual form and posterior border of merus is with rounded granules.

**Biology:** Not studied.
**Fisheries**: Not a commercial species.

### Atoportunus pluto (Ng & Takeda, 2003)

**Common name:** Not designated.
**Distribution:** Guam, Japan, Hawaii and Christmas Island (Indian Ocean).
**Habitat:** Not known.
**Description:** Anterolateral margin is with first four teeth which are lobiform and not spine-tipped; and next three are sharp and spine-tipped. Epibranchial ridge is lined with series of granules, many of which have sharp tips. Mesobranchial ridge is lined with distinct granules. Merus of cheliped is slender and its inner margin is with 4–6 prominent spines. Merus, carpus and propodus of ambulatory legs are slender. Dorsal surface of carapace is orangish and ventral surface is mostly dirty white.
**Biology:** Not studied.
**Fisheries:** Not a commercial species.

Profile of Portunid Crabs (Family: Portunidae) 113

*Cycloachelous orbicularis* **(Richters, 1880)** *(= Portunus orbicularis)*

Photo courtesy of Tadafumi Maenosono

**Common name:** Not designated.
**Distribution:** Western Pacific: Taiwan and the Philippines, Indian Ocean, Andaman Islands
**Habitat:** Coastal sandy sediment.
**Description:** This species resembles *P. granulatus* in size, form, and color of carapace. A distinctive feature, however, is that nine teeth of anterolateral border diminish in size from front to back. Adult of this species is $8 \times 10$ mm, and it can reach $18 \times 24$ mm.
**Biology:** No information
**Fisheries:** No information

*Cycloachelous orbitosinus* **(Rathbun, 1911)** *(= Portunus orbitosinus)*

**Common name:** Not designated.
**Distribution:** Japan, Australia, India, the western Pacific over from Madagascar of Japan, and French Polynesia
**Habitat:** Sandy bottom; shell bottom; coral gravel bottom; depth range 20–320 m.

**Description:** Carapace is narrow (breadth is 1.5 times length); surface is with well separated granular areas with a fine pubescence among the granules; and regions are well recognizable. Front is with 2 broad, reasonably sharp lateral lobes, and a central confluent lobular area without clear lobes; antero-lateral borders are with 9 teeth, of which last one is slightly most protruding; and posterolateral junction is rounded. Antero-external angle of merus of third maxillipeds is markedly produced into a lobe. Chelipeds are elongated; propodus is nearly as long as merus; merus is with 2 spines at posterior border and anterior border is with 4 spines; carpus is with 2 spines; upper surface of palm is with 1 distal spine; and lower surface is with squamiform markings. Posterior border of merus of swimming leg is with small serrations. Carapace attains a width of 4.1.cm.
**Biology:** Not known.
**Fisheries:** No information

*Cycloachelous suborbicularis* **(Stephenson, 1975)** [= *Portunus (Achelous) suborbicularis*]

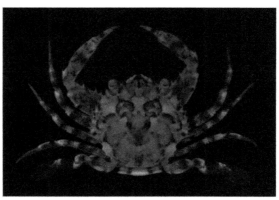

Photo courtesy of Tadafumi Maenosono

**Common name:** Albatross blue crab
**Distribution:** Indo-Pacific, Mauritius, Japan, Marshall Islands, Midway Island, Hawaiian, and Society Islands.
**Habitat:** Sea/Bay/Gulf; depth range 4–56 m.
**Description:** Carapace front is fourlobed. Median lobes are less than half as broad as laterals and laterals are rightangled with rounded tips. Among the anterolateral teeth, first is blunt; second to fourth are shorter than fifth to eighth; and ninth tooth is distinctly the largest

(after the first). Carapace is moderately narrow (breadth ca. 1.4 times length), with almost rounded (suborbicular) outline. All granular areas distinct, giving embossed surface; granular patches are separated by moderately broad areas bearing a fine pile of hairs; granules in patches are relatively large and widely spaced. Chelipeds are elongate, granular, hirsute except on under surface which is polished and pitted. Two sharp spines are seen on posterior border of arm. Anterior border of arm is with 5 spines .Wrist is with 2 usual spines. Hand is strongly carinated . Fifth leg is of usual form and posterior border of merus is with rounded granules.

**Biology:** Not available

**Fisheries:** A minor fishery of this species exists in Hong Kong.

### 3.3  SUBFAMILY THALAMITINAE

#### 3.3.1  CHARACTERISTICS OF SUBFAMILY THALAMITINAE

Carapace is subhexagonal or subtrapezoidal and is usually markedly broader than long. Antero-lateral border is divided into 4–7 teeth. Basal antennal segment is transversely broadened, usually filling orbital hiatus, with ridges, granules or spines. Antennal peduncle and flagellum are usually excluded from orbit. Chelipeds are longer than ambulatory legs, bearing a set of spines on merus, carpus and palm. Last pair of legs is with paddle shaped propodi and dactyli. Male first gonopod is with subterminal spines.

*Charybdis acuta* (Milne-Edwards, 1869)

**Dorsal view**

**Ventral view**

**Common name:** Indo-Pacific swimming crab, red-legged swimming crab, sharptooth Charybdis

**Distribution:** Japan, Korea, Taiwan and China.

**Habitat:** Marine; 10–50 m depth; rocks or sand and mud bottom

**Description:** Carapace is covered with a dense short tomentum. Transverse granular lines are seen on frontal (very indistinct), protogastric and mesogastric regions. Epibranchial line is interrupted at the cervical groove and across midline. Front has 6 acute triangular teeth with dark tips and medians are the most prominent. Antero-lateral borders are with 6 teeth. Among them, first tooth is smallest with slightly concave lateral border and acute at tip. Third, is largest and sixth is acute. Posterolateral junctions are rounded. Antennal flagellum is excluded from orbit. Chelipeds are quite hairy. Cheloped merus is with 3 strong spines on anterior border and posterior border is smooth. Carpus is with a strong internal spine and its outer border is with 3 spinules. Palm has 5 spines on upper border. Merus of swimming leg is with a subdistal posterior spine and propodus is with a row of spinules on posterior border. Size: 54.50 × 33.11 mm.

**Biology:** Not studied.

**Fisheries:** A minor fishery of this species exists in Hong Kong.

### *Charybdis acutidens* (Türkay, 1986)

**Common name:** Not designated.
**Distribution:** Tanzania, Malaysia, East Sumatra, Bay of Batavia, Timor, Moluccas and Solomon Islands.
**Habitat:** Bottoms of rock or coral reefs at 10–30 m in depth.
**Description**: Not known.
**Biology:** Not known.
**Fisheries:** No information.

### *Charybdis acutifrons* (De Man, 1879)

**Common name:** Straight-faced swimming crab
**Distribution:** Madagascar; Japan; Taiwan and Indonesia.

**Habitat:** Perched on the reef from 5 to 30 meters deep-water coral reefs or region.

**Description:** Carapace is densely pilose. All carapace ridges are present and are faintly granular. There are six frontal teeth which are subequal, sharp and with curving sides. Median teeth are separated from submedians by V-shaped notch and laterals which are projecting beyond rest are separated from submedians by deeper notch. Inner supraorbital lobe is acutely triangular and inner infraorbital lobe is denticulate and acute. There are seven anterolateral teeth. While second and fourth teeth are rudimentary, last is smallest and narrowest. Chelipeds are finely pilose. Cheliped merus is with 3 strong spines on its anterior border. Carpus is with strong spine on inner angle and three spinules at outer angle. Penultimate segment of male abdomen is with lateral borders which are parallel and then converging distally. Carapace is dark olive green with a large round red blot on the branchial regions. Chelipeds are pale flesh color. Fingers at proximal half are purple red and distal half and teeth are black. Spines reddish at base and legs are reddish, covered with red dots. Males measure 21.0 by 60.0 mm.

**Biology:** Not studied.

**Fisheries:** A minor fishery of this deep water species exists in the Digha coast of West Bengal in India.

### *Charybdis affinis* (Dana, 1852)

**Common name:** Smooth shelled swimming crab, Japanese swimming crab.

**Distribution:** India, China, Hong Kong, Taiwan, Thailand, Malaysia, Singapore and Indonesia.

**Habitat:** Sandy to muddy beaches; 10 to 30 m depths

**Description:** Carapace is flat and pilose. Frontal, cardiac and mesobranchial ridges are absent. There are six frontal teeth which are all triangular. Median teeth are most prominent and are separated by wide V-shaped notch. There are six anterolateral teeth. Among them, first tooth is notched; second to fifth are gradually increasing in size and last is elongate and spiniform, projecting laterally beyond preceding tooth. Chelipeds are swollen and their surface is finely pubescent. Anterior border of carapace merus is with three spines and carpus has a strong spine on inner angle and three spinules at outer angle. Propodus of natatory leg is smooth on posterior border. Penultimate segment of male abdomen is broader than long. This species is heterochelous, with pronounced right-handedness in both sexes. There is a marked sexual difference in relative growth of cheliped. Female abdomen increases in size at puberty, which is also accompanied by a reduced algometric growth rate. Dorsal surface of carapace is dark brownish-green. Maximum size of males is 37 by 54 mm.

**Biology:** Analysis of gonad development of this species gave an estimate of 50% gonadal maturity at 42 and 36 mm carapace width for females and males, respectively. Hepatosomatic index of female crabs is inversely related to the gonadosomatic index indicating the role of the hepatopancreas in storage of nutrients necessary for ovarian development in the reproductive season (Chu, 1999).

**Fisheries:** Minor commercial fisheries exists for this species in China.

### *Charybdis amboinensis* (Leene, 1938)

**Common name:** No common name.

**Distribution:** Japan, Taiwan, Philippine, Sulu Archipelago and Indonesia.

**Habitat:** Marine, intertidal to 14 deep; rocks and stones and among live corals.

**Description:** Carapace is covered with a dense short tomentum. Transverse granular lines are seen on frontal, protogastric and mesogastric regions. Epibranchial line is interrupted at the cervical groove and across midline. Front is with 6 teeth. Among these teeth, medians are most prominent and are with rounded tips. Antero-lateral borders are with 6 acute teeth. Sixth is acute and is directed antero-laterally. Posterolateral junctions are rounded. Antennal flagellum is excluded from orbit. Chelipeds are heterochelous. Cheliped merus is with 3 strong spines on anterior border and posterior border is smooth. Carpus is with a strong internal spine and its outer border is with 3 spinules. Palm is with 5 spines on upper border. While upper border is granulated, lower border is with squamiform markings. Merus of swimming leg is with a subdistal posterior spine and its propodus is with a row of spinules on posterior border.

**Biology:** Not known.

**Fisheries:** A minor fishery of this species exists in the Parangipettai coast in Tamilnadu in India.

*Charybdis anisodon* **(De Haan, 1850)**

**Common name:** Two-spined arm swimming crab
**Distribution:** Indo-West-Pacific, Madagascar, Red Sea, China, Taiwan, Hong Kong, Japan, Philippines, Thailand, Malaysia, Singapore, Indonesia, Sulawesi, Australia and New Caledonia.
**Habitat:** Muddy substrate at depths of up to 15 m; sandy beaches.
**Description:** Carapace is smooth and is marked by faintly granular transverse carapace ridges. Frontal, cardiac and mesobranchial ridges are absent. There are six frontal lobes. Among these lobes, medians are truncate; laterals are bluntly triangular and are separated from submedians by deep V-shaped notch; inner supraorbital lobe is triangular, bearing a smooth ridge and inner infraorbital lobe is obtuse. There are six anterolateral teeth. Among these teeth, first is obtusely triangular; second is smaller than third and last one is elongate and large, projecting laterally beyond preceding tooth. Antennal segment bearing fine granular ridge. Chelipeds are smooth and slightly unequal. Anterior border of cheliped merus is with two spines and its posterior border is granulated. Carpus is with strong spine on inner angle and three spinules at outer angle. Propodus of natatory leg is smooth on posterior border in larger specimens. Penultimate segment of male abdomen is with lateral borders. Abdominal surface is with a row of spinules on the lip and its outer surface is with longer row of bristles starting from tip and terminating lower than inner row as small spinules. Carapace is pale green on dorsal surfaces and whitish on the ventral surfaces. Large male measures 42.5 by 75.8 mm.
**Biology:** Not studied.
**Fisheries:** Minor commercial fisheries exists for this species in China.

*Charybdis annulata* **(Fabricius, 1798)**

**Common name:** Banded-legged swimming crab; stone crab.

**Distribution:** Indo-West Pacific Ocean: Tanzania, Madagascar, Pakistan, India, Sri Lanka, China, Taiwan, Japan, Thailand, Malaysia, Singapore, Indonesia, and Tahiti.

**Habitat:** Perched between the coastal intertidal reef to 10 m deep in shallow water; and depth within 50 m of rock and sediment near bottom.

**Description:**Carapace is slightly hexagonal,smooth and convex. Protogastric and mesogastric regions are with transverse granular lines. Epibranchial line is interrupted at the cervical groove and across midline. . There are six frontal teeth, separated by V-shaped notches. Medians and submedians are broadly triangular and former is projecting beyond latter. Lateral teeth are narrowest, and are acutely triangular, lying on a lower plane. Inner supraorbital lobe is bluntly triangular and inner infraorbital lobe is with a stout tooth. There are six anterolateral teeth; first and second are smaller and third which is largest is gradually decreasing in size to spiniform sixth. Chelipeds are stout and unequal. Anterior border of cheliped merus is with three spines and a spinule at distal end. Carpus is with strong spine on inner angle and three spinules at outer angle. Propodus of natatory leg is serrated on posterior border. Carapace is bluish grey and legs and chelipeds have circular purple and creamy bands. Large female measures 48 by 70 mm.

**Biology:** Two color morphs viz. orange morph and brown morph have been reported in this species. While orange morph shows specific micro-habitat preference, brown morph utilizes wide range of micro-habitats.

### *Charybdis beauforti* (Leene & Buitendijk, 1949)

Image not available.

**Common name:** Not designated.

**Distribution:** Not known.

**Habitat:** Depth range (m): 20–53.

**Description:** Céphalothorax is covered by a short rather dense pile and its regions are distinctly indicated and are crossed by six granular transverse ridges. Front is cut into six teeth. Its median teeth are slightly prominent beyond submedians; they are bluntly triangular and their outer margins slightly granular. Antero-lateral border is cut into six teeth. First tooth, which is rather blunt, has a nearly straight granular border. Its convex outer border is also granular. Second, up to the fifth teeth are of about the same shape, with slightly concave more or less granular anterior borders, and convex granular posterior borders and sharp tips. Posterior tooth is spine-like. Posterolateral borders converge rather strongly and form a curve with the finely granular slightly curved posterior border. Third, abdominal segment is about as broad at its anterior border as it is long. Its lateral margins are distinctly divergent over two thirds of their length and then they converge. Left cheliped is slightly more slender than the right. Upper surface of cheliped merus is granular. Palm is hairy and coarsely granular. Both fingers are channeled and movable one is longer than palm. Carapace length and breadth measure 30.5 mm and 45.2 mm, respectively.

**Biology:** Not studied.

### *Charybdis bimaculata* (Miers, 1886)

**Common name:** Two-spot swimming crab
**Distribution:** Indo-West Pacific: China, Taiwan, Korea Rep and Japan.
**Habitat:** Subtidal; depth range 0–439 m.
**Description:** Carapace is covered with a dense short tomentum; patches of granula are seen on cardiac and branchial regions; transverse granular lines are present on frontal, protogastric and mesogastric regions; and epibranchial line is interrupted at cervical groove but not across midline. Front is with 6 low, blunt and broadly triangular teeth, of which medians are little more prominent than the submedians; antero-lateral borders are with 6 teeth, of which first is truncate, third is largest and sixth is acute and directed antero-laterally. Posterolateral junctions are angular. Anten-

nal flagellum is not excluded from orbit. Chelipeds are slightly heterochelous; merus is with 3–4 strong spines on anterior border and posterior border with a spinule at its distal end; carpus is with a strong internal spine and outer border is with 3 spinules; palm is with 3 spines on upper border and, granulated on upper, and outer faces. Merus of swimming leg is with a subdistal posterior spine and a smaller, additional at the distal end and propodus is unarmed on posterior border. Carapace width is 41.3 mm.

**Biology**

*Reproduction:* The abundance, sex ratio, growth, sexual maturity, morphological sex dimorphism, and reproduction of this species have been studied. In Japan its relative abundance [inferred from catch-per-unit-effort (CPUE)] was higher and lower during summer and winter months, respectively. The proportion of males varied between 0.2 and 0.5 throughout the year and the overall sex ratio was greatly biased toward females. The population had a unimodal size frequency distribution and the life-span was estimated to be one year. Growth was not depressed in winter. The puberty molt in males brought about changes in the relative size of the chelipeds and gonopods and was estimated to occur in a carapace length (CL) range of 12.50–16.00 mm. Puberty in females was evidenced by a sharp increase in the pleon width and was estimated to occur at CLs between 11.34 and 16.74 mm. Morphological sexual dimorphism was observed in all features which showed secondary sexual development (cheliped dimensions, carapace and pleon width, and body weight). Ovarian development was uniformly arrested only during winter but began to develop in early spring, and mating (insemination) increased in frequency between spring and summer. Ovigerous females were found in all seasons except for winter, but were more common during the warmer months. Batch fecundity ranged from 8,300 to 38,400 eggs per female and was positively correlated with body size. This species was found to display the typical reproductive features and spawning pattern of temperate portunid crabs in spite of its short life-span (Doi et al., 2008).

**Fisheries:** Minor commercial fisheries exists for this species in China.

### *Charybdis brevispinosa* (Leene, 1937)

Image not available.

**Common name:** Not designated.

Profile of Portunid Crabs (Family: Portunidae)

**Distribution:** China, Indonesia, Malaysia and Singapore.
**Habitat:** Muddy, sandy bottoms; 30 m depth.
**Description:** Carapace is pilose. All anterior carapace ridges are present and granular. Two pairs of mesobranchial ridges are seen. There are six frontal teeth; medians are bluntly round, projecting beyond laterally directed submedians and laterals are acute and narrowest, separated from submedians by deeper notch. There are six anterolateral teeth. Among them, first is slightly notched and last is spiniform and directed laterally, projecting beyond preceding tooth. Chelipeds are finely pubescent and its lower surface is with squamiform markings. Anterior border of cheliped merus is granular, with three spines. Carpus is with strong spine on inner angle and two spinules at outer angle. Propodus of natatory leg is with one to two spinules on posterior border with spine on posterior border. Penultimate segment of male abdomen is with lateral borders which are strongly convex. Large males measure 23.5 by 38.0 mm. Color of crab is not known.
**Biology:** Not known.

### *Charybdis callianassa* (Herbst, 1789)

**Common name:** Not designated.
**Distribution:** India, Karachi, China, Thailand, Malaysia, Singapore and Australia.
**Habitat:** Sandy to muddy and shelly bottoms of 5–11 m depths.
**Description:** Carapace is convex and its surface is shortly pilose. Frontal and mesobranchial ridges are absent. Cardiac ridge is faintly granular. There are six frontal teeth. Among these teeth, medians are elliptical,

projecting beyond laterally directed submedians and laterals are narrowest, separated from submedians by deep notch. Inner supraorbital lobe is triangular. It has six anterolateral teeth with serrated borders. First tooth is notched; second to fifth are increasing in size and last is spiniforrn, projecting laterally beyond preceding tooth. Chelipeds are swollen and slightly unequal; surface is finely pubescent; anterior border of merus is with two spines and posterior border is finely granular. Carpus is with strong spine on inner angle and three spinules at outer angle. Propodus of natatory leg is smooth on posterior border. Penultimate segment of male abdomen is with lateral borders which are evenly convex. Abdomen's second to fourth segments are keeled. Color of carapace is dirty white or light grey throughout. Large male measures 25 by 38 mm.

**Biology:** Not known.

### *Charybdis curtidentata* (Stephenson, 1967)

Image not available.

**Common name:** No common name

**Distribution:** East Africa and Indian Ocean islands (Seychelles).

**Habitat:** Not known.

**Description:** Front of this species is six lobed. Medians are broad, with oblique outer borders, and approximately twice width of submedians. Deep and moderately wide fissure is seen between submedians and triangular laterals. There are 6 anterolateral teeth. First tooth is blunt, square-cut or truncate; second is distinctly smaller than first with blunt tip pointing forwards; third and fourth are subequal, broad, somewhat square-cut, with stout forwardly directed sharp tips; fifth is small (about as broad as second), with sharp forwardly directed tip; and sixth is narrow, protruding slightly more than remainder. Outer margins of all teeth except last are coarsely granular. Carapace is relatively narrow. General surfaces is microscopically granular and bearing fine hairs, with more coarsely granulated, elevated areas and some granular ridges. Postlateral junction is angular. Bases of median and submedian frontal teeth are coarsely granular but not elevated. Left cheliped is slightly larger than right, with very conspicuous coarsely granular carinae. Anterior border of cheliped merus is armed with three stout spines which are increasing in size distally. Cheliped carpus is coarsely granular, bearing three coarsely granular carinae. Fifth leg merus is short and broad with well-developed posterodistal spine. Its propodus is with posterior border bearing 4–7 spinules.

**Biology:** Not known.

### *Charybdis demani* (Leene, 1937)

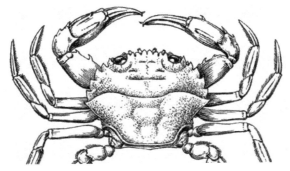

**Common name:** Not designated.
**Distribution:** Not known.
**Habitat:** Not known.
**Description:** Carapace is not hairy on the whole surface, but only the anterior part (especially the epibranchial regions) is covered with a dense pile, which is more developed in the female than in the male. Regions are fairly distinct. Whole surface is granular. Front is cut into six teeth. Median teeth are prominent beyond others and are elliptical. Submedian teeth are on a somewhat higher plane than medians. Lateral teeth are nearly as long as submedians; narrower than others and triangular. Antero-lateral borders are cut into six teeth. First tooth is notched. Second tooth is nearly as broad as first on the left side and somewhat smaller on right. Third tooth is much broader. Fourth tooth is nearly as broad as third. Fifth tooth is narrower than fourth. Second to fifth teeth have nearly same shape and are triangular. Sixth tooth is more spine. Posterolateral borders converge strongly posteriorly. The sternum is smooth and bare. Abdomen of male has carinae on second and third terga. Third, fourth and fifth terga are fused. On fourth tergum, there is a transverse keel. Sixth tergum has gradually converging sides and it is broader than long. Boundary between sixth tergum and telson is somewhat curved anteriorly and between fifth and sixth terga, it is straight.

**Biology:** Not studied.

## *Charybdis edwardsi* (Leene & Buitendijk, 1949)

**Common name:** Not known.
**Distribution:** Western Indian Ocean; East Africa; Gulf of Aden; India.
**Habitat:** Offshore beyond continental shelf.
**Description:** Céphalothorax is only slightly convex and its regions are fairly well defined. It is absolutely smooth. Front is cut into six small and sharp (spiniform) teeth. Median teeth, which are slightly prominent beyond submedians, are very small and sharp. Last teeth are on a somewhat higher plane; their tips are rounded and their inner margins distinctly granular. Lateral borders are far less distinctly granular and run slightly outwards posteriorly and they are separated from sharp lateral teeth by a V-shaped incision. Lateral teeth are triangular and their margins are granular. Anterolateral border is cut into six teeth, which are all exceedingly broad, blunt and truncate, except posterior one, which is small and sharp. Anterior tooth is broad with a granular outer margin. It has a rather sharp pointed anterior edge and its posterior edge is rounded. Second tooth is much smaller. Third, fourth and fifth are of about the same size and shape. They are exceedingly blunt and short; their anterior tip is nearly obsolete and their margins are granular. Sixth tooth is short and sharp and is much smaller than the preceding ones. Posterolateral borders of carapace converge posteriorly and are smooth. Posterior border is slightly curved and it forms an angular junction with the posterolateral borders. Of the abdomen third up to the fifth terga are fused. Sixth tergum is shorter than its breadth at anterior border. Anterior border of merus of cheliped is rather sharply granular. Palm shows

seven granular costae and three spines. Chelipeds are subequal and both movable Fingers are about as long as the palm. Carapace is measuring 38.5 mm to 41.5 mm in length and 56.5 mm to 61.0 mm in breadth.

**Biology:** Not studied.

### *Charybdis erythrodactyla* (Lamarck, 1818)

**Common name:** Rainbow swimming crab
**Distribution:** Hawaii and Indo-Pacific.
**Habitat:** Shallow waters; benthic; hard bottom; and rocky and rubble.
**Description:** Carapace of this species is glabrous. Transverse granular lines are found on frontal, protogastric, and mesogastric regions. Epibranchial line interrupted only at the cervical groove. Front is with six triangular teeth. Antero-lateral borders are with 7 acute teeth. Among these teeth, second and fourth are very much smaller than remainder. Fifth is largest. Posterolateral junctions are rounded. Antennal flagellum is excluded from orbit. Cheliped merus is with three strong spines on anterior border and posterior border is granulate. Carpus is with a strong internal spine and its outer border is with 3 spinules. Palm has 4 spines on upper border. Merus of swimming leg is with a strong subdistal posterior spine and propodus is with about 10 denticles on posterior border. It attains a carapace width of 7.5 cm.

**Biology:** Not known.

## *Charybdis feriatus* (Linnaeus, 1758) *(= Charybdis cruciata)*

**Dorsal view**

**Frontal view**

**Common name:** Crucifix swimming crab; Asian paddle crab; Pakistan coral crab.

**Distribution:** tropical Indo-West Pacific from East Africa the Persian Gulf through to Indonesia and Japan, and throughout most of Australia.

**Habitat:** Sandy to muddy substratum at depths of 10–30 m; and young specimens on the bell of large scyphozoan jellyfish.

**Description:** Carapace is convex and smooth. Protogastric, mesogastric and metagastric and epibranchial ridges are faintly granular. There are six frontal teeth. Median teeth are slightly projecting beyond submedians. Laterals are triangular. Among six anterolateral teeth, first is truncate and strongly notched; second to fifth are increasing in size broadly and last is spiniform and directed laterally. Chelipeds are smooth and unequal. Anterior border of cheliped merus is with three strong spines. Carpus is with strong spine on inner angle and three spinules are at outer angle. Fingers are slender. Penultimate segment of male abdomen is broader than long with lateral borders which

Profile of Portunid Crabs (Family: Portunidae)

are evenly convex. Carapace is cream colored and is mottled with red patches. Ventral surface is pale yellow or whitish. Chelipeds and ambulatory legs are mottled red. This species usually has a characteristic yellowish cross on the gastric region of the carapace. Carapace width is 5–7 cm to 20 cm.

**Biology**

*Food and feeding:* This species is omnivorous and it mainly preys on crustaceans and fish.

*Reproduction:* During reproduction cycle, the period of premolt guarding, molting, pre-copulatory guarding, copulation and post-copulatory guarding was found to be 92.0 hours, 4.0 hours, 3.6 hours, 7.28 hours, and 12 hours, respectively. The total mating sequence lasted for 119 hours and the spawning of the female took 17.62 days after copulation (Soundarapandian et al., 2013). In another finding, dominance of male was noticed in almost all the size groups. The fecundity varied from 35,635 to 3,49,939 eggs with an average of 1,42,012 eggs. The productivity index was found to increase from a low of 0.35 in 80–90 mm CW to a maximum of 2.78 in 110–120 mm CW (Babu et al., 2006). This crab has been to reported to occur with the jellyfish species, *Stomolophus meleagris* and *Rhopilema esculenta* (Morton, 1989).

*Spawning:* In this species spawning takes place invariably during night hours. The newly spawned eggs are bright yellow/orange in color and subsequently with the progress of embryonic development, the color changes to dull brown and on the penultimate day of hatching, the color changes to deep grey. The duration of incubation varies between 9 and 11 days depending on the berry size and temperature of the rearing water. The size of the newly spawned eggs is 298 μm and by the time of hatching, the size increases to 369.3 μm.

*Larval Stages*

The larval development of this species includes six zoeal stages and a megalopa stage, which metamorphoses into the crab stage. The total duration of the zoeal phase varies between 19 and 26 days during different trials. Details of different zoeal stages are given below:

**Zoea-I:** Eyes are sessile. Abdomen is five segmented plus the telson. Telson is forked and each fork is bearing one inner and one dorsal spine. Inner margin of each fork bears three, long serrated setae. Duration of this stage is 4–5 days (The duration refers to the maximum number of zoeae that metamorphosed to the next stages).

*Zoea-II:* Development of stalked eyes is seen in this stage. Abdomen is five segmented as in the previous stage. Abdominal somites 3–5 bear more distinct lateral spines. Telson bears a pair of short, plumose setae on median margin of cleft part. Duration of this stage is 3–4 days.

*Zoea-III:* Abdomen develops 6 segments and lateral spines on 3–5 somites are longer. Telson is similar to that of previous stage. Duration of this stage is 3–4 days.

*Zoea-IV:* Abdomen as in the previous stage. Pleopod buds just started appearing at the ventral posterior end of somites 2–5. Telson adds one additional short seta on inner margin. Duration of this stage is 3–4 days.

*Zoea-V:* Pleopod buds are seen on the abdominal somites second to sixth. Telson has developed additional short setae on the inner margin. Duration of this stage is 3–4 days.

*Zoea-VI:* Abdominal somites and telson are similar to previous stage. Pleopod buds are well developed; biramous on somites 2–5 and uniramous on somite 6. Telson setation same as in the previous stage. Duration of this stage is 3–4 days.

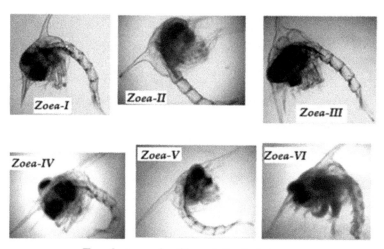

**Zoeal stages in *Charybdis feriatus***

*Holy crab:* The crucifix crab gets its name from the cross-shaped pattern on the carapace. The scientific name feriatus, literally means "holy day." The cross of this crab is the subject of Christian mythology. Some people believe that these crabs were blessed by the missionary, St Francis Xavier. The Jesuit story is that in February 1546, St Francis was caught in a storm in Eastern Indonesia. In an attempt to calm the storm, he took his crucifix and dipped it into the sea, but it slipped from his grasp and fell into the water. The next morning, as he reached the shores of Seram Island, a crab crawled up to him carrying his crucifix (cross) between its pincers. Xavier knelt down, retrieved the cross, and blessed the crab. Since this time the crabs have had the cross pattern on their carapace. Some Catholic people believe these crucifix crabs as holy and their shells are often sold as religious curios or possibly as good luck charms (http://museum.wa.gov.au/explore/blogs/museummarine/creature-feature-holy-crab-crucifix-crab-charybdis-feriata-linneaus-1758; Wild Fact sheets – http://www.wildsingapore.com/wildfacts/crustacea/crab/portunidae/feriatus.htm).

*Fisheries:* Major commercial fisheries exists for this species in China. This species is the commercially most important of the *Charybdis* species. It is more delicate than mud crabs (*Scylla* spp.), and is usually sold frozen in East Asia.

### *Charybdis goaensis* (Padate et al., 2010)

**Common name:** Not designated.
**Distribution:** Only from Goa, West coast of India.
**Habitat:** 6–7 m depth localities.
**Description:** Carapace of this species is evenly tomentose/pilose, with the exception of transverse ridges and swellings. Median frontal teeth are bluntly triangular and sub-medians and lateral frontal teeth are separated by wide U-shaped notch. Transverse ridges of carapace are distinctly granular. These ridges include two proto-gastric, one meso-gastric, a medially divided meta-gastric, and epibranchials along the level of the last antero-lateral teeth. There are no transverse ridges beyond the level of the last antero-lateral teeth. Chelipeds are subequal in both sexes, rough, and unevenly tomentose. Cheliped merus is with two well-developed spines and five to six granules on anterior margin. Propodus of natatory leg is smooth and its merus bears prominent spine on postero-distal margin. Sixth abdominal segment in male is broader than long. Carapace is light brown over tomentose regions and granular transverse ridges are with polished surfaces. Chelipeds are dark brown and other pereiopods are light brown.
**Biology:** Not known as it is a recently discovered species.

### *Charybdis granulata* (De Haan, 1833) *(= Charybdis moretonensis)*

**Common name:** Particles crab; stone crab.
**Distribution:** East Africa, Hong Kong, Japan, Singapore and Malay Peninsula.
**Habitat:** Bottoms of rock or sand at depths of 15–35 m.
**Description:** Carapace surface is convex and is covered unevenly with dense pile. Ratio of carapace breadth to length is 1.45 times. Frontal ridges

are short and granular. Protogastric and mesogastric ridges are with markedly granular outlines. Front is cut into six narrow lobes. Median lobes are slightly protruding and are separated by a V-shaped Notch. Submedians of equal size, bluntly pointed, directed anterolaterally and are set on a much higher plane than the other frontals. Lateral lobes are narrowest, sharp and separated from submedians by a deeper notch. There are six anterolateral teeth. Among these teeth, first to fifth teeth are stout and are increasing in size from front to rear. Outer borders of each tooth are convex and forms an acute tip with the anterior border. Sixth tooth is smallest and narrowest. Chelipeds are unequal, strongly granular and pubescent. Anterior border of cheliped merus bears three sharp spines and several spiniform tubercles, in between spines and near proximal end of border. Posterior border is with vertical granulated rows. Fingers are slender and are deeply grooved. Posterior border of the natatory leg is with the usual spine on the merus and with serrations on the propodus. Surface of abdomen and sternum are pubescent and finely granular on the anterior segments of the latter. Penultimate segment of the male abdomen is with lateral borders which are convex distally giving it a swollen appearance. Ultimate segment of abdomen is triangular and is slightly longer than broad. Second to fourth segment are keeled. Carapace is mosaic green. Large male measures 45.7 mm by 67.0 mm.

**Biology:** Not studied.

### *Charybdis hawaiensis* (Edmondson, 1954)

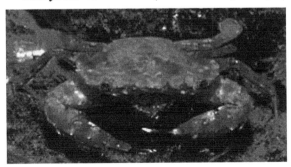

**Common name:** Hawaiian swimming crab
**Distribution:** Central Pacific, Polynesia, Hawaii, Society Islands, and Tuamotus.
**Habitat:** Benthic; shallow waters; 1–18 m depths.

**Description:** In this species, a spine on the posterior margin of the arm of cheliped; hand has five strong spines; inner surface of the hand is costate and granular; and merus of the natatory leg is 1.5 times as long as broad; 5 strong teeth on the upper surface of claw, and in the carination and granulation of the inner surface of manus. Instead of the existing- In male pleopod 1, long spines are seen on the outer surface. There is no spine on the posterior margin of cheliped merus. Length/width ratio in merus of last pereopod is 1.75. It attains a carapace width of 8 cm.

**Biology:** Not studied.

### *Charybdis hellerii* (Milne-Edwards, 1867)

**Common name:** Indo-Pacific swimming crab

**Distribution:** Mediterranean, East Africa, Red Sea, Persian Gulf, Pakistan, India, Andaman, China, Japan, Malaysia, Singapore, Australia and New Caledonia.

**Habitat:** Intertidal zones; under rocks and stones; among coral reefs and up to depths beyond 30 m.

**Description:** Carapace is densely pilose. All anterior carapace ridges are present and granular. There are six frontal teeth. Of them, medians are elliptical and are projecting beyond submedians Laterals are acutely triangular and are separated from submedians by deep V-shaped notch. There are six anterolateral teeth. Among them first and second are closer and subequal in size. Last one is elongated and spiniform, projecting beyond preceding tooth. Chelipeds are stout and unequal and their surface is finely pubescent. Anterior border of cheliped merus is with three spines and a spinule at distal end. Carpus is with strong spine on inner angle and three spinules at outer angle. Propodus of natatory leg is serrated on posterior border. Penultimate segment of male abdomen is with lateral borders which are parallel and are then converging distally. This species show color changes associated with growth. Larger specimens have a orange to cream colored carapace and purplish legs. Smaller specimens have reddish patches on either side of median line anteriorly and on entire posterior surface of carapace. Chelipeds and legs are mottled and banded. Fingers are red basally and dark brown distally with white tips. Maximum male and female carapace lengths have been reported to be 63.7 and 51.0 mm, respectively.

**Biology:** The female abdomen of this species has been reported to change in morphology and color with sexual maturity, e.g., becoming wider and orange-colored. However, immature males do not show morphological or color differences in their abdomen (Watanabe et al., 2015). Eggs of this species are bright yellow in color, with spherical diameters ranging from 0.224 to 0.322 mm A positive correlation has been observed between carapace width and egg number. Mean number of eggs in three size classes (carapace length: 31–40; 41–50; and 51–60 mm) of this species were found to be 40,203; 67,648 and 148,249 respectively. All larval, postlarval as well as juvenile through adult stages of this species have been successfully reared in the laboratory. *Chelonibia patula*, the turtle barnacle, was found associated with this species. Australian specimens of *Charbdis hellerii* showed *Sacculina* infection.

### *Charybdis hongkongensis* (Shen, 1934)

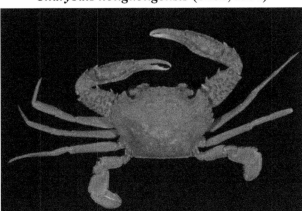

**Common name:** Hong Kong crab
**Distribution:** China, Hong Kong, Thailand, Malaysia, Sumatra and Banda Sea.
**Habitat:** Sandy or sandy muddy bottoms at 30–400 m depths.
**Description:** Carapace of this species is pilose. Anterior carapace ridges are granular and distinct. Metagastric, cardiac and mesobranchial regions are marked with granular patches. Granules are seen on orbital borders, frontals and anterolateral teeth. There are six frontal teeth and among them, medians are prominent, on lower plane and are separated by shallow notch. Laterals are acute and narrowest and are separated from rounded submedians by deeper notch. Chelipeds are unequal and are covered by granular squamiform markings. Anterior border of merus is with three spines and a spinule is found at distal end of posterior border. Cheliped carpus is with usual spines. Propodus of natatory leg is serrated on posterior border and merus of this leg is with spine on posterior border. Penultimate segment of male abdomen is with lateral borders which are convex and converging distally. Carapace is brown. Male measures 32.7 mm by 51.3 mm.
**Biology:** Not studied.
**Fisheries:** Major commercial fisheries exists for this species in China.

### *Charybdis hoplites* (Wood-Mason, 1877)

**Common name:** Not designated.
**Distribution:** India, Thailand, Indonesia and Australia.
**Habitat:** Bottoms of mud or sand; depth 30 to 60 m.

Profile of Portunid Crabs (Family: Portunidae) 139

**Description:** Length of carapace is not much more than half the extreme breadth (between the tips of the last spine of the antero-lateral borders). Carapace is covered with a dense short tomentum; convex and regions are well defined and fairly well areolated. Gastric region is divided into three sub-regions. There is a very pronounced and independent swelling on the inner part of either branchial region. Antero-lateral borders are cut into six teeth of which last is a Portunus like spine which is at least twice as long as those in front of it. Posterior border of the dorsum of carapace forms a strong dogs eared angle of junction with posterolateral borders. Front is cut into eight lobes arranged in four pairs, of which outer most pair of teeth on either side are rather sharp. Chelipeds are 2½ times the length of the carapace and all their surfaces are covered with granular transverse squamiform markings. Sixth tergum at the abdomen of the male is truncate triangular, without curve to the sides. Anterior male abdominal appendages are extremely broad at the base. Tip is spoon-like and bears hairs along both borders.

**Biology:** Not known.

### *Charybdis ihlei* (Leene & Buitendijk, 1949)
Image not available.

**Common name:** Not designated.

**Distribution:** Sarawak, Malaysia

**Habitat:** Mud flats

**Description:** Céphalothorax is crossed by the following ridges: a rather curved one between the posterior antero-lateral teeth, which is interrupted by the cervical groove; an unbroken and curved one on the nasogastric region; a short one on each protogastric region; and one on the frontal region, which is interrupted in the middle. Cardiac region moreover bears a faint ridge-like and granular elevation. Front is prominent beyond inner supra-orbital angles and is cut into six teeth of which median teeth are rather broad and blunt with granular margins. Antero-lateral border is cut into six teeth of which anterior tooth is very distinctly bifid and next four are triangular. Fifth tooth is the smallest. Anterior as well as posterior borders of these teeth are granular. Anterior border of sharp, spine-like posterior tooth is granula. Its posterior border is smooth. Posterolateral margin is rather short and strongly convergent. Posterior border forms a curve with the posterolateral borders. Orbital borders are rather strongly granular. Of the abdomen, third up to fifth terga are fused. Lateral margins of sixth tergum are gradually convergent. Boundary between fifth and sixth segments is faintly curved anteriorly. Chelipeds are subequal. Granular anterior border of arm has two rather short, but sharp spines. Both fingers are slender and movable one is longer than upper border of palm. Merus of natatory leg is rather short and broad with usual spine at its posterior border.

**Biology:** Not known.

### *Charybdis incisa* (Rathbun, 1923)
No image available
**Common name:** Not designated.
**Distribution:** Queensland, Australia.
**Habitat:** Marine
**Description:** This species is characterized by their frontal teeth which are rounded. Inner surface of hand of cheliped is smooth. Other features are: Carapace is hirsute; presence of a pair of frontal ridges; fifth anterolateral tooth is smaller than remainder (excluding second tooth); and cheliped arm's upper surface is sparsely granular and under surface is very finely granular. Wrist of cheliped upper surface is with spiniform carina terminating near spine at wrist articulation.

**Biology:** Not known.

### *Charybdis japonica* (Milne-Edwards, 1861)

Profile of Portunid Crabs (Family: Portunidae) 141

**Common name:** Japanese swimming crab; Asian paddle crab; shore swimming crab.

**Distribution:** Native to South East Asia; Red Sea, China, Taiwan, Japan, Korea, Thailand and Malaysia.

**Habitat:** Littorial zone; estuaries where there is firm sand or muddy fine sand; up to 10–15 m in depth.

**Description:** Carapace is pilose and is marked by finely granular protogastrics, mesogastric, epibranchial and metagastric ridges. Of six frontal teeth which are triangular and sharp, medians are directed outwards projecting beyond narrower submedians. Laterals are acute and narrowest. Of six lateral teeth which are directed forwards and sharp, first tooth is with slightly sinuous outer border; second to fifth are subequal and last is narrowest and spiniform. Merus of third maxillipeds is with outer distal angle which is more or less produced. Fingers are longer than manus. Propodus of natatory leg is smooth on posterior border and its merus is with spine on posterior border. Penultimate segment of male abdomen is broader than long with lateral borders evenly convex. These crabs range in color from pale green through olive green, to a deep chestnut brown with purplish markings on the carapace (shell). They may also have yellow-orange and brown-orange markings on the shell and legs. Claws are white-tipped. Adults have a carapace width of 12 cm.

**Biology:** Adult paddle crabs can produce hundreds of thousands of offspring. The larvae can float in the water for three to four weeks, during which time they can be moved large distances by tides and currents. Adults are also capable of swimming large distances. Paralytic shellfish poisoning toxin has been reported in this species. This species is also expected to be a possible vector species, because small quantities of the toxins have been detected in eight specimens of the crab (Oikawa et al., 2004).

**Fisheries:** Minor commercial fisheries exists for this species in China.

### *Charybdis jaubertensis* (Rathbun, 1924)

Image not available.

**Common name:** Not designated.

**Distribution:** Western Australia

**Habitat:** Not known.

**Description:** This species is diagnosed by the following: anterolateral teeth in which sixth tooth is sharpest; carapace: length: breadth ratio

0.72–0.75; chelipeds: arm—anterior border is with three or four spines, if four, second most proximal is always smaller than remainder; wrist—inner spine is less stout in juveniles; hand—upper surface is with inner carina composed of transverse granular rows.

**Biology:** Unknown.

### *Charybdis longicollis* (Leene, 1938)

**Common name:** Not designated.

**Distribution:** Indo-Pacific: Red Sea, Persian Gulf and Madagascar.

**Habitat:** Sandy-mud bottoms at 10–80 m and occasionally to a depth of 135 m; and juveniles found in Brackish waters.

**Description:** Front has 6 spines and there are 6 spines on anterolateral margin. Among these spines, first 4 anterior spines are serrated, square cut, and are separated by deep notches, Size of these crabs varies from 2.5 to 3 cm CL.

**Biology:** It is a benthophagic species with preference for infaunal and slow moving prey. Most frequent food items of this species are molluscs, crustaceans and fish.

Copper, chromium cadmium, iron and zinc concentrations were determined in the liver, gill and muscle samples of this species. The order of the metal concentrations found in muscle of crab was: $Cu > Cr > Fe > Zn > Cd$, The highest Cu, Cr, Cd, Fe and Zn concentrations were found in the liver, and this was followed by the gill and muscle. The levels of all metals in a given tissue were always higher (Firat et al., 2008).

### *Charybdis lucifera* (Fabricius, 1798)

**Common name:** Box crab
**Distribution:** India, Sri Lanka, Taiwan, Japan, Australia, Thailand, Malaysia and Indonesia.
**Habitat:** Stony and muddy substrates up to 2 m in depth.
**Description:** Carapace of this species is very much broader and its length is much less than two-thirds its breadth. A sharply dentiform lobule is seen at the outer end of the lower border of orbit. Chelipeds in male are not very much more than twice the length of carapace. Posterior border of the propodite of last pair of legs are serrated throughout. Sixth male abdominal segment has its sides parallel or even slightly divergent in at least two-thirds of its extent. In anterior male abdominal appendages, there is no bend near fringe of hairs on the distal part of outer margin also. Fingers of larger cheliped are shorter than palm. Color of body is yellowish-brown with large white spots on either branchial region. Chelipeds are scarlet pink; tips are light brown and extreme tips are whitish. Carapace length of male is 66.0 mm and breadth is 95.0 mm.
**Biology:** Not studied.

## *Charybdis miles* (De Haan, 1835)

**Common name:** Soldier swimming crab
**Distribution:** Western Indian Ocean: Persian Gulf, India, China, Taiwan, Hong Kong, Japan, Philippines, Burma, Malaysia, Singapore, Indonesia and Australia.
**Habitat:** Sandy or muddy bottoms at 20–200 m in depth.
**Description:** Carapace surface is convex and densely pilose. Anterior carapace ridges are present and granular. Frontals, cardiac and mesobranchial regions are with granular patches. There are six frontal teeth which are which are sharp and acute. Among these teeth, medians are projecting beyond laterally. Laterals are narrowest and separated from submedians by deep U shaped notch. Of six anterolateral teeth, first is notched and last is spiniform and is slightly projecting beyond preceding tooth. Chelipeds are elongated and its surface is finely pubescent. Anterior border of cheliped merus is with four spines and a distal spinule is seen at ventral border. Cheliped carpus is with strong spine on inner angle and three spinules at outer angle. Fingers are slender and deeply grooved. Propodus of natatory leg is with two to four denticles on its posterior border. Penultimate segment of male abdomen is with convex lateral borders. Its second to fourth segments are keeled. Color of carapace is red. Chelipeds are mottled red. Fingers are banded dark and light red. Large male measures 67.1 mm by 93.8 mm.
**Biology:** Not studied.
**Fisheries:** A commercial fisheries exists for this species

## *Charybdis natator* (Herbst, 1794)

**Common name:** Ridged swimming crab
**Distribution:** East Africa, Madagascar, Red Sea, India, China, Japan, Philippines, Thailand, Malaysia, Singapore, Indonesia and Australia.
**Habitat:** Bottom of rocks, pebbles or sand at depths of 1535 m in depth.
**Description:** Carapace of this species is uniformly pilose with sparse granules on anterolateral surface. Anterior carapace ridges are present except frontals. Of six frontal lobes, medians are on lower plane and are projecting beyond equally broad submedians. Lateral lobes are acute and are separated from sub medians by deeper V-shaped notch. Of six anterolateral teeth, first tooth is truncate; second to fourth are subequal with acute tips and last is spiniform and least prominent. Chelipeds are unequal, granular and pilose. Anterior border of merus is with three to four spines and carpus is with strong spine on inner angle and three spinules at outer angle. Fingers are stout and are deeply grooved. Propodus of natatory leg is serrated on posterior border. Second to fifth segment of male abdomen are keeled and abdomen's penultimate segment is with lateral borders which are parallel then converging distally. Dorsal surface is brownish and granules are bright red. Ventral surface is bluish, mottled with white and pale red. Largest specimen of a male measures 75.5 mm by 109.5 mm.

### Biology

***Reproduction:*** Female *Charybdis natator* showed two spawning peaks; a major peak in spring and a secondary peak in autumn. They were also capable of at least three ovulations per maturity instar. Examination of ovaries indicated little reproductive activity during winter. Fecundity estimates of ovigerous females ranged from 181,000 to 976,000 (eggs). The reproductive biology of this species was found to be similar to that of other subtropical portunids found in the region (Sumpton, 1990).

***Larval development:*** In the studies relating to the larval development of the swimming crab *Charybdis natator*, six zoeal stages and one megalopal stage have been identified. At an average salinity and temperature of 311% and 23°C, the megalopa stage attained in 38 days after hatching. Morphologically, the first zoea of this species is very similar to that of other species of *Charybdis* in having lateral spines on the carapace; three pairs of serrate setae on the posterior margin of the telson furca; one plus six setae on the endopod of the maxillule; and five setae on the endopod of the maxilla (Islam et al., 2000).

***Associated microbes:*** A total of 40 species bacteria and 50 species of fungi were isolated and identified from various tissues of this crab. There was a significant variation noticed in bacterial and fungal population among the tissues. The dominant bacterial flora were *Acinetobacter baumanii, Bacillus subtilus, Escherichia coli, Klebsiella pneumonia, Micrococcus luteus, Proteus mirabilis, Pseudomonas aeruginosa, Staphylococcus aureus, Salmonella typhi,* and *Vibro cholera* and the fungal flora constituted were *Aspergillus fumigatus, Aspergillus flavus, Aspergillus niger, Curvularia senegalensis, Cladosporium* sp. *Fusarium moniliforme, Penicilium citrinum, Rhizopus nigricans, Trichoderma viridae,* and *Verticillium* sp. The presence of large number of bacteria and fungi observed in the alimentary canal and Carapace of this species was largely due to the decomposition of food and environmental contamination. Hence the study suggested that this edible crab should be collected from uncontaminated waters for consumption (Kannathasan and Rajendran, 2010).

**Fisheries**: Commercial fisheries exists for this species.

## *Charybdis obtusifrons* (Leene, 1936)

**Common name:** Not designated.
**Distribution:** Indo-west Pacific; Madagascar, Red Sea, India, China, Japan, Malaysia, Australia and Melanesia.
**Habitat:** Shallow waters on coral reef.
**Description:** Carapace of this species is pilose. All carapace ridges are present and are granular. Of six frontal teeth, medians are truncated and submedians are with anterior edges sloping outwards. Laterals are with tip bluntly round and are separated from submedians by deeper V-shaped notch. Of seven anterolateral teeth, second and fourth are rudimentary and last is spiniform, projecting laterally. Chelipeds are pilose. Anterior border of cheliped merus is with three spines and its posterior border is granular. Carpus is with strong spine on inner angle and three spinules at outer angle. Merus of swimming leg is with a strong sub-distal posterior spine and propodus is with about 10 denticles on posterior border.with spine on posterior border. Penultimate segment of male abdomen is with lateral borders which are parallel then converging distally. Color of crab is not known. Females measure 21.0 mm by 32.5 mm.
**Biology:** Not known.

## *Charybdis omanensis* (Leene, 1938)

**Common name:** Not designated as it is a new species.
**Distribution:** Indian Ocean; from Red Sea to Gulf of Oman.
**Habitat:** Deep sea
**Description:** Carapace (in specimens with CL > 15 mm) is 1.5–1.68 times as broad as long and covered with short pile. Carapace regions are well defined and mesobranchial areas may be swollen. Frontals are as small groups of granules. Front (excluding inner supraorbital lobes) is about 1.5 times shorter than posterior border. Frontal teeth are with tips which are narrowing distally and are pointed. Front's medians and submedians are broadly subtriangular and laterals are separated from latter by a broad U-shaped incision. There are six truncated anterolateral teeth with granular edges. Of them first two are nearly bifid; second is smallest; third to fifth are subequal and broadest; and sixth is longest, but not much longer than others and is curved anteriorly. Chelipeds are granular and their anterior border of merus is with two or three blunt teeth and additional tubercles. Posterior and inferior borders of merus are ending in spinules. Propodus of natatory legs is with four to six very small spinules on posterior border. Abdomen of male and female is with fourth tergum which is keeled for more than half of its maximum width. Penultimate tergum of male abdomen is with outer edges converging posteriorly. Female genital opening is rounded and is shifted anteriorly to edge of sternite. It is bordered posteriorly and laterally by thickened cuticle.
**Biology:** Not studied.

## *Charybdis orientalis* (Dana, 1852)

**Common name:** Not designated.

**Distribution:** East Africa, Madagascar, Red Sea, India, Sri Lanka, China, Japan, Philippines, Malaysia, now Singapore, Indonesia and Australia.

**Habitat:** From intertidal rocky shore to sandy or muddy bottoms of 20 to 50 m in depth.

**Description:** Carapace is densely pilose except elevated ridges. All anterior carapace ridges are present and are granular. However, there no ridges behind epibranchials. Of six frontal teeth, medians are blunt, and are projecting beyond triangular submedians. Laterals are narrowest and sharply acute. Of six anterolateral teeth which are spine tipped, second is rudimentary and is attached to posterior border of first. Last tooth is produced slightly sideways. Chelipeds are unequal and their outer surface is pubescent. Anterior border of cheliped merus is with three spines and a spinule at distal end. Its carpus is with strong spine on inner angle and three spinules at outer angle. Propodus of natatory leg is serrated on posterior border and its merus is with spine on posterior border. Penultimate segment of male abdomen is with lateral borders which are parallel then converging distally. Carapace is brownish-grey. Fingers of chelipeds are dark red and their extreme tips are white. Legs are with alternate dark brown and light grey stripes. Largest female specimen measures 38.7 mm by 57.0 mm.

**Biology:** Unknown.

150                        Biology and Culture of Portunid Crabs of World Seas

## *Charybdis padadiana* (Ward, 1941)

Image not available.

**Common name:** Not designated.

**Distribution:** All oceans

**Habitat:** Not known.

**Description:** Carapace of this species is broader than long and its surfaces are uneven, and are crossed by four ridges. Front is broader and is composed of six teeth, excluding the orbital angles. Teeth are all short, broad and rounded. Anterolateral margins are armed with six teeth. Of these teeth, first two are blunt, and following three each is capped by a spine. Second and fifth teeth are smallest. Posterolateral margins are thick with a line of granules which do not reach posterior margin. Orbits are small and their upper margin is with two closed fissures and one straight and narrow below the external angle. Lower orbital border is not visible throughout its length from dorsal view. Walking legs are long, and their distal articles are fringed with long hair. Sternum appears smooth but is finely granulated and punctuate under a lens. Carapace width of female measures 17 mm.

**Biology:** Not studied.

## *Charybdis padangensis* (Leene & Buitendijk, 1952)

Image not available.

**Common name:** Not designated as it is a new species

**Distribution:** Not known.

**Habitat:** Not known.

**Description:** This species has a rather strongly convex and smooth cephalothorax and regions of carapace are ill defined, while only some traces of granular ridges are left. Ridge between the posterior anterolateral teeth is most distinct. Slightly prominent front is cut into six teeth. Median teeth are blunt and separated from each other by a very shallow incision; and they are less broad than the equally blunt submedian teeth. Anterolateral border is cut into six teeth; and anterior of these is strongly bilobed, and each lobe is blunt. Following four teeth are of about equal breadth, though their shape is not same. Second tooth is blunt and its outer margin is only slightly curved. In the third tooth, outer margin is more curved, while curve in fourth tooth is still stronger.

Fifth tooth is quite different in shape: its outer angle is sharply pointed; posterior anterolateral tooth is much larger than others, being spine-like and directed outwards. Margins of all these teeth, except posterior margin of last tooth, are very indistinctly granular. Posterolateral borders are strongly convergent. Posterior margin is straight and forms an eared junction with posterolateral borders. Ventral surface of he cephalothorax is granular and hairy. Third, to fifth tergum of male abdomen are fused and second and third are transversely keeled. Anterior border of the sixth tergum is nearly straight; posterior border is slightly anteriorly curved, and lateral borders are convergent. Length of sixth tergum is about 3/4 of its greatest breadth. Chelipeds are unequal and right is more slender than left; anterior border of merus is granular and hairy and it is armed with two teeth; and upper surface of merus is granular. Palm is smooth and bears three low and smooth ridges on its outer surface. Finger of larger cheliped is broken and that of smaller cheliped is slightly longer than palm.

**Biology:** Not known.

### *Charybdis philippinensis* **(Ward, 1941)**

**Common name:** Not designated as it is a new species

**Distribution:** Not known.
**Habitat:** Not known.
**Description:** Carapace is broader than long; bare and glossy and is granulated under lens. Four ridges are seen across the carapace. While first ridge is obsolete and the others are less developed. Front is equal to one-fourth the width of carapace and is cut into six teeth excluding inner orbital angles. Submedian teeth are broad, truncated and are sloping toward median teeth. Pair of lateral teeth are narrow obtuse and are sloping inward. Anterolateral margins are armed with six spines, including outer orbital angle; spines become larger posteriorly; and last is transversely directed. Posterolateral margins are strongly convergent and outlined by a fine ridge. Chelae are subequal. Cheliped merus is armed on anterior margin with two spines and granules. Cheliped carpus is armed with four spines, largest of which is on inner angle. Its upper surface is granulated. Males measure 35 mm.
**Biology:** Not known.

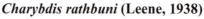

### *Charybdis rathbuni* (Leene, 1938)

**Common name:** No common name
**Distribution:** Eastern Central Pacific: French Polynesia, Sri Lanka
**Habitat:** Benthic; depth range 9–120 m.
**Description:** Granular ridge between posterior anterolateral teeth is interrupted by cervical groove, but no interruption is found in middle of the ridge. Median frontal teeth are broader, and therefore difference

in breadth between median and submedian teeth is less distinct. Lateral frontal teeth are broader too. Fifth anterolateral tooth is much less broad than fourth and of a rather different shape, being more spine-like. Inner part of the upper orbital border is granular, but this granulation is indistinct. Median of the three spinules on the wrist of chela is rather large. There are only five granular ridges on he palm. Second and third tergum of abdomen in the male are carinate; fourth and fifth terga are fused and sixth is broader than long. Lateral borders of the sixth tergum are rather strongly curved; posterior border is straight and anterior is curved.

**Biology:** Not known.

### *Charybdis riversandersoni* (Alcock, 1899)

**Common name:** Light palm ( Chinese name)
**Distribution:** Arabian Sea; India; East China Sea; Southern Taiwan
**Habitat:** No information yet.
**Description:** Branchial regions are swollen. Frontal teeth are blunt. Second anterolateral tooth is larger than first (and not vice versa). Anterior

border of cheliped merus has four spines. Merus of fifth leg is long. Size of crab is 50.7 × 34.9 mm.

**Biology:** Not known.

### *Charybdis rostrata* (Milne-Edwards, 1861)

**Common name:** Not designated.
**Distribution:** Not known.
**Habitat:** Coastal waters
**Description:** Carapace of this species is about four-fifth as long as broad. It is more or less convex with granular ridges. Its median frontal teeth are comparatively large and prominent. Teeth of the next pair are broad and slope outwards, while those of the outermost pair are narrow and almost straight. The last spine on the antero-lateral borders is proportionately longer in the female than in the male Chelipeds are nearly smooth and is less than twice as long as the carapace. Palm is with two spines on anterior border and none on he posterior border. Wrist is with a strong spine at inner angle and two-three spinules at he outer angles. Fingers are longer than palm in smaller cheliped and as long as palm of major cheliped. Sixth segment of male abdomen is broader than long and is gently curved and convergent at sides. Carapace is brownish purple and

chelipeds are creamy to brownish chocolate colored. Values of width and length of carapace is 25 mm and 20 mm, respectively.

**Biology:** Unknown.

### *Charybdis sagamiensis* (Parisi, 1916)

**Common name:** Not designated.

**Distribution:** Japan, Taiwan, the East China Sea, South China Sea, Vietnam and South Korea.

**Habitat:** Reef bottom at a depth of 20–150 m; sandy bottom; shells

**Description:** Antennal flagellum of this species is excluded from orbital hiatus. Posterior border of cephalothorax is curved, forming curved posterolateral junction. Of 6 antero-lateral teeth, first antero-lateral tooth is truncate and median and submedian frontal teeth are relatively stout. Transverse granular ridges on carapace are well developed and there are no distinct cardiac ridges. Fine diffused granular patches are seen on carapace. Branchial regions are at most faintly swollen. Anterior border of arm of cheliped is with 3 or more spines. There are 4 spines on hand of cheliped. Hand of cheliped is not swollen and its under surface is either smooth or with a hint of squamiform markings. Carapace width is 8.7 cm.

**Biology:** Not known.

Size: 8.7 cm.

## *Charybdis smithii* (MacLeay, 1838)

**Common name:** No common names
**Distribution:** Indian Ocean, Pacific and Atlantic Oceans; Southern Somali Basin, Omani Basin and Central Arabian Sea.
**Habitat:** Deep-sea
**Description:** Carapace of this species is evenly convex longitudinally and transversely and its surface is smooth. Very short pubescence is present on anterolateral and frontal regions, and along carapace grooves and depressions. Only epibranchial crista is visible and all other ridges absent. Front is cut into 6 acutely triangular teeth; lateral teeth are largest; margins of all teeth are lined with small rounded granules. Median teeth are separated from submedian teeth by distinct, shallow, broad U shaped cleft. Submedian teeth are sharp and are separated from lateral teeth by deep V-shaped cleft. Anterolateral margin is arcuate, with 5 teeth, which are all separated by distinct deep, U-shaped clefts; first to fourth teeth are truncate; second tooth is smallest; and fifth tooth is triangular. Posterolateral margins are smooth, gently concave and are converging towards posterior margin of carapace. Posterior carapace margin is forming distinct angular junction with end of posterolateral margin. Posterior carapace margin is entire and is gently convex posteriorly. Chelipeds are subequal in size and are between 2–3 times length of carapace. Outer anterior surface of cheliped merus is rugose; anterior margin is finely denticulated, with 2 stronger spines; posterior margin is finely granulated; and 2 anterior surfaces are gently rugose to smooth. Carpus is with granulated ridge ending in strong inner spine. Cutting margins of fingers are with numerous teeth and denticles. Movable fingers are shorter than manus. Surfaces of thoracic sternites are slightly

Profile of Portunid Crabs (Family: Portunidae)

rugose to smooth and glabrous. Male abdomen is triangular; third to fifth segments are completely fused without trace of sutures and lateral margin of third and fourth segments are separated by deep V-shaped cleft. Fourth segment is with low, median transverse ridge and sixth segment is trapezoidal. Telson is triangular, and its lateral margins are gently convex outwards. Carapace width of males is 58.3.1 mm and carapace length is 41.2 mm. The corresponding values for females are 58.3 mm and 41.2 mm, respectively.

**Biology**

*Food and feeding:* Natural diet of this species comprised a wide variety of invertebrates and myctophid fishes. Pelagic juveniles and subadults of this crab were found to feed actively than benthic adults. Diet of juveniles were mainly composed of pelagic shrimps (20% by volume) and crabs (17 %). Stomach contents of subadult crabs were dominated by teleost fishes and decapod crustaceans (64%) whereas in adults, crabs were the major food (42%). Pronounced cannibalistic tendency was observed in this species especially among juveniles.

*Reproductive biology:* There were 5 stages of sexual development viz. immature, early maturing, late maturing, ripe and spent were recorded. Size at first maturity at 50% level for females was 48.7 mm (CW). Individual fecundity was relatively low (1343–42209 eggs) compared to other brachyuran crabs but eggs were found larger. Crab population occupying pelagic realm was exclusively in non-breeding phase and breeding stock was found only at bottom. One-year life cycle of this species showed. Aggregation of the crabs on the continental shelf precedes their breeding from about October to January (NE Monsoon). The larvae hatching in shelf waters disperse offshore and, after metamorphosis, form dense pelagic swarms from about July to August (SW monsoon) (Balasubramanian and Suseelan, 1998).

*Swarms:* Surface "swarms" of these crabs are still considered as an unusual phenomenon in the open Indian Ocean, although their dense pelagic aggregations were already reported in waters off the Indian coast and in the northern Arabian Sea. Based on an extensive large-scale data series taken over 45 years, it is demonstrated that this species is common in the pelagic provinces of the western Indian Ocean driven by the wind monsoon regime. They are found are dispersed by the monsoon currents throughout the equatorial Indian Ocean. They aggregate at night in the upper 150-m layer. *C. smithii* is an important prey for more than 30 species of abundant epipelagic top predators. In turn, it feeds on mesopelagic species. This swimming crab is a major species of the intermediate trophic levels and represents a crucial seasonal trophic link in the open ocean ecosystem of the western Indian Ocean (Evgeny et al., 2009).

***Charybdis truncata* (Fabricius, 1798)**

**Common name:** Not designated.

**Distribution:** World distribution: India to Japan including eastern and northern Australia.

**Habitat:** Muddy bottoms of 10–100 m in depth.

**Description:** Carapace of this species is densely pilose. All anterior carapace ridges are granular and distinct. Metagastric, cardiac and mesobranchial regions are marked with granular patches. Granules are seen on orbital borders, frontals and anterolateral teeth. Of six frontal teeth, medians are prominent and are separated by shallow notch. Laterals are acute and narrowest and are separated from rounded submedians by deeper notch. Of six anterolateral teeth with borders serrated, first two are closely set and third to last are decreasing in size. Chelipeds are unequal and are covered by granular squamiform markings. Anterior border of cheliped merus is with three spines and a spinule at distal end of posterior border. Larger cheliped fingers are as long as manus and smaller cheliped fingers are slightly longer than manus. Propodus of natatory leg is serrated on posterior border and its merus is with spine on posterior border. Penultimate

Profile of Portunid Crabs (Family: Portunidae) 159

segment of male abdomen is with lateral borders which are convex and converging distally. Carapace is dirty green and legs are with transverse bars or patches of reddish brown. Ventral surface is white. Largest male specimen measures 33.5 mm by 49.1 mm.

**Biology:** The midgut of this species accounts for half of the postgastric intestinal tract. The paired anterior midgut caeca start just behind the pyloric stomach, on either side of the midgut. The unpaired posterior midgut caecum arises dorsally at the rear end of the midgut, where this joins the hindgut. The midgut and its caeca largely help in the digestive absorption of food. The hindgut is of ectodermal origin and is lined with chitin of a collagenous nature. The connective tissue of the anterior part of the hindgut is packed with tegmental glands secretion of which contains both sulphated and weakly acidic mucosubstances. The latter facilitate the passage of fecal matter and help to bind food particles. The digestive gland – hepatopancreas – opens into the anterior part of the midgut, below the anterior midgut caeca. Histologically, its tubules are known to contain three different types of cells – "F", "R" and "B" cells (Devi et al., 1989).

**Fisheries:** Minor commercial fisheries exists for this species in China

### *Charybdis vadorum* (Alcock, 1899)

Image not available.

**Common name:** Not designated.

**Distribution:** Red Sea, Persian Gulf, India, China, Hong Kong, Taiwan, Philippines, Thailand, now Malaysia and Indonesia.

**Habitat:** Sandy bottoms at depths of 10–80 m.

**Description:** Carapace of this species is pilose. Granular patches are seen on protogastrics and mesobranchial region; Y-shaped on metagastric and cardiac regions; and mesogastric ridge is sinuous. Of six frontal teeth, medians are on lower plane; submedians are with inner edge sloping inwards overlapping medians; and laterals are narrowest and separated from submedians by deep notch. Of six anterolateral teeth with serrated borders, first is acutely pointed; second to fourth are gradually increasing in size; and last is elongate and spiniforrn, projecting laterally beyond preceding tooth. Chelipeds are covered with squamiforrn markings. Anterior border of cheliped merus is with two spines and a spinule at distal end of posterior border. Carpus is with strong spine on inner angle and two spinules at outer angle. Fingers are shorter than manus. Propodus of natatory leg is serrated on posterior border and its merus is with spine on posterior border. Penultimate segment of male abdomen is with lateral borders which are strongly convex and abdomen's second to fourth segments are keeled. Color of carapace is not known. Largest male specimen measures 18.7 mm by 30.7 mm.

**Biology:** Not studied.

**Fisheries:** Minor commercial fisheries exists for this species in China.

### *Charybdis vannamei* (Ward, 1941)

Image not available.

**Description:** Carapace of this species is broader than long, smooth and glossy. It is finely granulated when it is seen under a lens. There are four granulated transverse ridges on anterior half of the carapace. Cardiac region is separated from gastric and branchial regions by an H-shaped depression. Hepatic regions are slightly sunken below level of the rest of carapace and there are a few scattered punctuations. Anterolateral margins are divided into six subequal teeth. First and last are smallest and third and fourth are largest. Posterolateral margins are convex with an indistinct line of granules outlining margin. Posterior margin is developed into a sharp granular ridge. Front is divided into six teeth, excluding orbital angles; median pair of teeth are rounded and are separated by a narrow V-shaped incision; submedian pair are more triangulate with obtuse tips; and outer pair are the narrowest and directed slightly outward. Chelae are subequal and its merus is armed with three spines on the anterior margin. Other surfaces of merus are quite smooth. Carpus is armed with four spines, largest of which is upon inner angle. Abdomen of the male is narrow. Penultimate segment is as broad as long and anterior angles are abruptly rounded. Carapace width of male crab measures 54 mm.

Nothing is known about its other ecological and biological features.

### *Charybdis variegata* (Fabricius, 1798)

**Common name:** Not designated.
**Distribution:** Western Atlantic coast; Madagascar, Red Sea, Persian Gulf, India, China, Hong Kong, Taiwan, Japan, Philippines, Thailand, Malaysia and Australia.
**Habitat:** Bottoms of mud or sand at 30–50 m in depth.
**Description:** Carapace is densely pilose. All anterior carapace ridges are present and granular. Two pairs of mesobrancbial ridges are seen and cardiac ridge is interrupted. Of six frontal teeth, medians are triangular projecting beyond laterally directed submedians. Laterals are acute and narrowest and are separated from submedians by deeper notch. Of six anterolateral teeth, first is notched; second is smallest; and last is spiniform and directed laterally, projecting beyond preceding tooth. Chelipeds are finely pubescent, with squamiform markings. Anterior border of cheliped merus is with three spines and distal is widely separated from proximal two. Carpus is with strong spine on inner angle and two spinules at outer angle. Fingers are shorter than or equal to manus. Propodus of natatory leg is with spinules on posterior border and its merus is with spine on posterior border. Penultimate segment of male abdomen is with lateral borders which are strongly convex. Color of carapace is not known. Largest female specimen measures 30.5 mm by 17.6 mm.
**Biology:** Chandran (http://link.springer.com/article/10.1007/BF0305-3904#page-1) reported on the breeding periodicities of this species. This crab is a biannual breeder. Males have two peaks in their gonadal cycle. Increase in gonad index coincides or just follows a drop in hepatic index indicating probable utilization of reserved nutrients. Accumulation of nutrients by a rise in hepatic index precedes second peak in gonad index.
**Fisheries:** Minor commercial fisheries exists for this species in China

## *Charybdis yaldwyni* (Rees & Stephenson, 1967)

Image not available.

**Common name:** Not designated.

**Distribution:** Australia

**Habitat:** Not known.

**Description:** Front has six lobes with medians and submedians which are rounded. Lateral lobes of front are narrow, and inner orbital lobes are broadly triangular. In small specimens median and submedian frontal lobes almost square-cut. Of six anterolateral teeth, second is smaller than first; third is large and other teeth are small. Carapace.is relatively narrow, hirsute and mostly minutely pitted, with granular ridges. Posterior-posterolateral junctions are rounded. Chelipeds are either subequal or one is larger than other. Arm of upper surface is proximally smooth and distally it is with transverse granular rows. Under surface is squamiform, and with spine or tubercle near wrist articulation. Finger is shorter than palm and is deeply grooved, with inwardly curved tips. Fifth leg merus is moderately broad, with large spine on posterior border. Posterior border of propodus is bearing spinules. Antepenultimate segment of male abdomen is with concave borders and penultimate segment is with convex borders. Ultimate segment of male abdomen is longer than broad.

**Biology:** Not studied.

## *Thalamita acanthophallus* (Chen & Yang, 2008)

Image not available.

**Common name:** Not designated as it is a new species

**Distribution:** South China Sea, Chenhang Island, and Xisha.

**Habitat:** Sand and rocky bottom along coast; and coral reef; shallow water Islands.

**Description:** Carapace is about 1.5 times as broad as long. All anterior carapace ridges are present. Cardiac region is with sinuous ridge. Front is cut into 6 lobes of which median ones are rounded and are separated by a V-shaped notch, lying on a lower plane and distinctly narrower than first lateral lobe; and second lateral lobe is narrowest and bluntly angular. Supraorbital lobes short, arched. Anterolateral margin is with 4 sharp teeth (including outer orbital tooth); decreasing from front to rear in breadth; and last one is smallest and directed anterolaterally. Posterolateral margin is longer than anterolateral margin and is slightly converging backwards. Posterior margin is straight and is joining posterolateral margin as curved margin. Third maxilliped is about 05 times breadth of ischium. Its merus is shorter than ischium and is subquadrate,. Chelipeds are covered with granules on outer surface. Cheliped carpus is with 3 spines on outer surface and a stout, short spine at inner distal angle. Palm is armed with 5 spines on dorsal surface, of which distal spine is smaller; and with 2 longitudinal ridges on outer surface. Fingers are shorter than palm and cutting edges are with unequal blunt teeth. Propodus of natatory leg is bearing 5 spines and dactylus is ovate with spiniform tip. Male abdomen is with third to fifth segments which are completely fused. Sixth segment is trapezoidal and is about 1.5 times as broad as long. Telson is triangular and is about as long as broad, with blunt tip. Size of crab is 7.5 mm by 11.5 mm.

**Biology:** Not studied.

### *Thalamita admete* (Herbst, 1803) *(=Thalamita savignyi)*

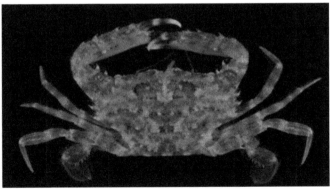

Photo courtesy of Tadafumi Maenosono

**Common name:** Not designated.

**Distribution:** East Africa, Red Sea, Indian Ocean, Malaysia, China, Japan, Australia, Tahiti and Hawaii.

**Habitat:** Rocky shores and reef flats within the intertidal areas.

**Description:** Carapace of this species is finely pilose and broader than long. All anterior carapace ridges are present. Cardiac and mesobranchial ridges are less granular. Two frontal lobes are separated by a distinct notch and anterior border is with square cut profile. Of five anterolateral teeth, fourth tooth is rudimentary. Chelipeds are unequal and their merus is with three to four spines on anterior border. Posterior border is granulated. Carpus is with four usual spines. Upper surface of manus is granulated and pilose. Fingers are short and stumpy. Propodus of natatory leg is with five to eight spines on posterior border. Penultimate segment of male abdomen is with parallel lateral borders. Carapace is with reddish brown patches. Brown and cream bandings are present on legs and fingers of chelipeds. Largest male specimen measures 18.2 mm by 31.4 mm.

**Biology:** Not studied.

### *Thalamita auauensis* (Rathbun, 1906)

**Common name:** Not designated.

**Distribution:** Philippines, Mariana, China, Japan, Hawaii Island, and Samoa.

**Habitat:** Corals; depth of 24–335 m.

**Description:** Carapace of this species is reddish-orange with cream-colored spots of varying sizes. Legs exhibit a banding pattern, which alternates between reddish and cream colors. Tips of chelipeds are black and white. While ovigerous female measures 12.9 × 7.3 mm, male and female measure 18.4 × 11.2 mm and 14.1 × 9.9 mm, respectively.

**Biology:** Not known.

## *Thalamita bevisi* (Stebbing, 1921) *(= Thalamita medipacifica, Thalamita dakini)*

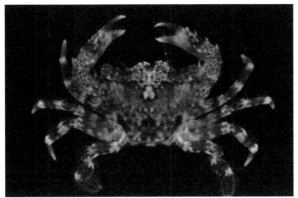

Photo courtesy of Tadafumi Maenosono

**Common name:** Not designated.

**Distribution:** Abrolhos Island, Western Australia, Sri Lanka, Oman, South Africa, Society Island, Gilbert Island, Marianas, Samoa, French Polynesia, Hawaii, and Easter Island.

**Habitat:** Shallow waters; intertidal zone.

**Description:** Carapace of this species is dark brown to greenish with blotches of lighter brown and cream. Legs are also brown with spots of cream color. Tips of lateral carapace teeth are orange. While male measures 15.2 × 10.8 mm., female measures 15.4 × 10.5 mm.

**Biology:** Not studied.

## *Thalamita bouvieri* (Nobili, 1906)

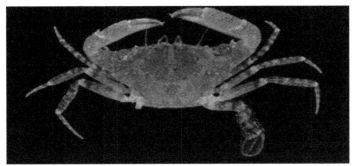

Photo courtesy of Tadafumi Maenosono

**Common name:** Not designated.

**Distribution:** Mozambique; Mayotte; Madagascar; Amirante Islands; Cargados Carajos Islands; Japan; Indonesia; Australia; Lord Howe Island; Kiribati; and Tuamotu Archipelago.

**Habitat:** Depth range, 35–85 m.

**Description:** Carapace is only with epibranchial ridge. This last one is interrupted medially. Front is 4-lobed of which median frontal lobes are much broader than laterals. Of 4 antero-lateral teeth, third is rudimentary to reduced. Basal antennal joint is with short smooth curved crest. Upper surface of cheliped palm is with one median spine and 3 blunt tubercles near distal end of posterior border. Similar tubercles are also seen on anterior border. Inner face is smooth. Posterior border of propodus of swimming leg is with 4–10 sharp spines. Male first pleopod is very stout. It is only slightly curved, and its tip is ending obliquely and abruptly. Inner surface is with 6–7 terminal spiniform bristles and outer surface is with 12–15 terminal bristles.

**Biology:** Not studied.

### *Thalamita cerasma* (Wee & Ng, 1995)

**Common name:** Powder blue-clawed swimming crab
**Distribution:** Singapore, Japan
**Habitat:** 5 m depth.

**Description:** Carapace is convex and smooth. All carapace ridges are finely granular. Front is straight and cut into six lobes of which medians are closely set and lying on a lower plane, and are separated by a short incision. Submedians are broadest and are separated from the medians by a shallow notch. Lateral lobes are narrowest with rounded anterior borders and are separated from submedians by a V-shaped notch. Of five anterolateral teeth which are increasing in size from first to third, fourth and fifth are subequal and smaller than first three teeth. Fifth tooth is slightly stouter than fourth. All spines are ending in a sharp black tip. Chelipeds are unequal and swollen in larger cheliped. Fingers of larger cheliped are stout and blunt, and of the smaller cheliped, are straight and slender. All fingers are deeply grooved except for lower surface of immovable fingers. Propodus and dactylus of ambulatory legs are deeply grooved on the anterior and posterior borders. Merus of natatory leg is with spine at posterodistal angle in front of the usual spine at end of posterior border. Posterior border of propodus is finely serrated. Sternum and abdominal surface is smooth. Penultimate segment of the male abdomen is broader than long and its lateral borders are parallel for proximal half and then they are slanting gently inwards towards distal end. Ultimate segment is obtusely triangular, as broad as long. Carapace dorsal surface is basically bright olive; and margin and dorsal ridges are vermilion. Manus of chela is orange and both movable and immovable fingers are red in proximal 2/3 and black in distal 1/3. Dactylus of ambulatory legs is whitish and other segments are dark green. Natatory leg is also basically dark green. Propodus and dactylus are surrounded by brown hairs. Size of crab is 37 mm (CL) and 58 (CW).

**Biology:** Not studied.

### *Thalamita chaptali* (Milne Edwards, 1861)

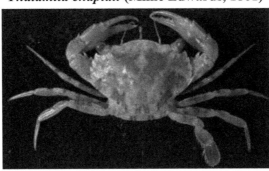

**Common name:** Not designated.

**Distribution:** Red sea, India, Sri Lanka, China, Thailand, Malaysia and Australia.

**Habitat:** Crevices of coral reef or sandy stony bottoms of 5–40 m in depth.

**Description:** Carapace is sparsely pilose and all anterior ridges are present including pair of mesobranchials. Metagastrics are widely separated and cardiac ridge is absent. Two frontal lobes are separated by faint notch. Of five anterolateral teeth, first three are broad and square cut; fourth is smallest; and fifth is sharpest and most protruding. Chelipeds are unequal and their merus is with three spines on anterior border. Carpus is with strong spine on inner angle and usual three spines on outer surface. Fingers are stout and sharp. Propodus of natatory leg is smooth along posterior border. Penultimate segment of male abdomen is with lateral borders which are more or less parallel. Ultimate segment is triangular with borders which are slightly concave. Color of specimens is overall dirty white with very small, scattered darker flecks. Male specimen measures 8.0 mm by 11.7 mm.

**Biology**: Not studied.

### *Thalamita coeruleipes* (Jacquinot and Lucas, 1853)

**Common name:** Not designated.

**Distribution:** Indian Ocean: Mauritius; China Sea; Philippines; and Central Pacific: French Polynesia.

**Habitat:** Littoral; intertidal waters

**Description:** Carapace is with 3 pairs of gastric ridges. Median cardiac and mesobranchial ridges are absent. Front is with 6 rounded lobes. Of the 5 antero-lateral teeth, fourth and fifth are shortest and subequal.

Upper surface of cheliped palm is with 4 remarkably sharp spines. Posterior border of propodus of swimming leg is with 4 large and 3 small spines. Male first pleopod is tapering quite strongly to tip; inner surface is with 2 indistinct rows of moderately sized bristles; and outer surface is with about 15 bristles and 3 large, blunt, rounded tubercles immediately behind tip.

**Biology:** Not studied.

### *Thalamita cooperi* (Borradaile, 1902)

**Common name:** Not designated.

**Distribution:** From East Africa and Laccadives to the China Seas, Philippines, and Australia; and Moluccas.

**Habitat:** Depth range, 4–20 m; muddy sand with many sponges, and outer reef.

**Description:** This species has the following characteristics: 4 anterolateral teeth, a pilose sternal surface, a characteristic basal antennal joint, straight frontal edge, a length/width ratio of 0.64, lateral to median frontal lobes' ratio of 0.30 and length/width ratio in merus of P5 of 2.6. Carapace has a length of 5 mm and width of 7.8 mm.

**Biology:** This species is the host for the crustacean parasite, *Sacculina anomala.*

## *Thalamita corrugata* (Stephenson & Rees, 1961)

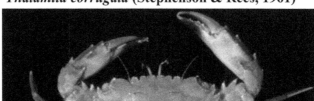

**Common name:** Not designated.
**Distribution:** Kenya, Japan, China, South China Sea, Indonesia, Australia, Gilbert Islands), and Tuamotu Archipelago.
**Habitat:** Intertidal zone to 11 m.
**Description:** Front of carapace is 4-lobed of which median frontal lobes are much broader than laterals. Of the 4 antero-lateral teeth, third antero-lateral tooth is of same size as second. Carapace is with small transverse ridges or corrugations in addition to ridges normally present. Female measures $8.2 \times 5.5$ mm.
**Biology:** Not known.

## *Thalamita crenata* (Rüppell, 1830)

**Common name:** Crenate swimming crab.

**Distribution:** South Africa, Madagascar, Red Sea, Persian Gulf, India, China, Malaysia, Singapore, Korea, Japan, Australia, Tuamotu, Tonga and Hawaii.

**Habitat:** Rocky shores and reef flats of broken down corals within the intertidal zones; mangrove creeks.

**Description:** Carapace surface is smooth and sparsely pilose. Carapace ridges are faintly distinct, and none behind epibranchial ridges. Of six frontal lobes which are broadly rounded, medians are lying on a slightly lower plane and laterals are narrowest. Five anterolateral teeth which are subequal are decreasing slightly in size from front to rear. Chelipeds are is unequal and their merus is bearing three to four spines on anterior border. Carpus is armed with strong spine at inner angle and three spinules at outer angle. Propodus of natatory leg is with serrulations along distal half of posterior border. Penultimate segment of male abdomen is with lateral borders which are slightly convergent distally. Color of this crab consists of uniform greenish grey carapace and pinkish claws with dark brown tips and white extreme tips. Largest male specimen measures 52 mm by 80 mm.

**Biology**

*Food and feeding:* This species is a generalistic predator and its diet is mainly composed of bivalves and slow-moving crustaceans. Both the stomach fullness and the relative presence of animal prey in the contents were found to be significantly higher in crabs collected at sunset than in those caught at dawn. Stomach fullness seems to depend also on the tidal rhythm and it is higher during spring tide periods. Females stomachs are slightly fuller than those of males, while there is no difference in diet between juveniles and older specimens. This species forages more actively during daytime, thus differing from the majority of swimming crabs. Both the great abundance of this species and its diet, based on a wide range of slow-moving or sessile species, testify to the importance of the role played by this predator in the mangrove ecosystem (Cannicci et al., 1996).

*Breeding cycle:* Studies on the breeding cycle of this species was studied by Sigana (2002) and it was found that there was no significant difference in male to female ratio between months. But the overall sex ratio was significantly different from 1:1. Size heterogeneity between sexes was also significantly different. The existence of continuous breeding was also reported in this species.

**Fisheries:** Commercial fisheries exists for this species.

### *Thalamita crosnieri* (Vannini, 1983)

Image not available.

**Common name:** Not designated.

**Distribution:** East African coast, Madagascar, Andaman Islands, India, Sri Lanka and Mauritius.

**Habitat:** Sandy or stony pools; on or among the roots of sea grass; under stones and in crevices; and exposed and semi-exposed zones.

**Description:** It is a variable species. Fourth anterolateral tooth is more or less reduced or even completely absent. Depth of notch separating median frontal lobes from lateral ones is variable and is occasionally absent. Carapace has a faded pink color and tip of fingers dark. Size of crab is 16.8 × 11.4 mm.

**Biology:** Not studied.

### *Thalamita danae* (Stimpson, 1858) *(=Thalamita stimpsoni)*

**Common name:** Blue swimming crab; blue mottled swimmer crab.

**Distribution:** Mozambique, Red Sea, India, China, Hong Kong, Japan, Philippines, Malaysia, Singapore, Indonesia, Australia, New Caledonia, Marshall Islands and Fiji and Samoa.

**Habitat:** Sandy to rocky shores in the intertidal zone; coral rubble and reefs.

**Description:** Carapace is densely pilose except for the raised transverse ridges Front is cut into six lobes, of which medians are with truncate anterior borders. Submedians are with inner border directed obliquely inwards and overlapping medians. Laterals are as broad as submedians. Anterior border is bluntly round and separated from submedians by a V-shaped notch. Of five anterolateral teeth which are all stout, first three are subequal and similar; fourth and fifth are smaller than those preceding and fourth tooth is smaller than fifth. Chelipeds are only slightly unequal. Anterior border of cheliped merus bears three spines on distal half and several large granules on proximal half. Posterior border is with granules forming a wrinkled surface. Carpus bears a strong spine at the inner angle and three spinules at its outer angle. Merus of natatory leg bears a triangular tooth at postero-distal angle, in front of usual spine on posterior border. Propodus is serrated along its posterior boarder. Penultimate segment of male abdomen is about as long as broad, with lateral borders which are parallel for 3/4 of the way and then converging distally. Carapace is dark purplish red or brick red above and much lighter on ventral surface. Largest male specimen le measures 41.8 mm by 66.6 mm.

**Biology:** This species has been reported to be active at night. Under laboratory conditions, first zoea of this species became first crab instar after six molts, taking at least 26 days (Krishnan and Kannupandi, 1988).

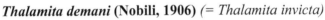
***Thalamita demani*** **(Nobili, 1906)** *(= Thalamita invicta)*

Photo courtesy of Tadafumi Maenosono

**Common name:** Misuji Beni haunt crab

**Distribution:** Red Sea; Kenya; Madagascar; Mauritius; Sumatra; Taiwan; China; Australia; and New Caledonia.

**Habitat:** Intertidal; tide pools, exposed and semi-exposed areas.

**Description:** Carapace of this species is sparsely covered with long hairs. Front is with 4 lobes of which medians are much broader than the laterals. There are 4 antero-lateral teeth and, a rudimentary denticle is usually present between third and fourth but this may also be absent. Basal antennal joint is with an acute ridge whose edge is formed of about 15 fused rounded granules. Upper surface of cheliped palm is with 2 spines on anterior and 2 spines on posterior border. Posterior border of propodus of swimming leg is with 5–6 spines. Male first pleopod is short; slightly curved; tip is blunt; club-shaped, and is slightly swollen. Inner surface is with about 20 stout bristles and outer surface is with about 25 terminal bristles. Color of body is pale beige. Dark transversal bands are seen on legs 2–5. Size of crab is 9.0 mm × 6.2 mm.

**Biology:** Not known.

### *Thalamita difficilis* (Crosnier, 2002)

Image not available

**Common name:** Not designated.

**Distribution:** South Pacific Ocean: French Polynesia.

**Habitat:** Deep waters at depths of more than 100 m; hard bottom: rocky; rubble.

**Description:** Carapace is wide (width /length equal to 1.67). Front is bilobed, although prominent, it is slightly sinuous (concave towards the middle). Anterolateral edges of shell are cut into four teeth and a large granule is observed between third and fourth teeth. Rear edge of carapace, has a length of just over three times of carapace width. Pereiopods fifths have a merus which is relatively small (2.75 times longer than wide) and whose posterior edge is armed with a subdistal strong, very acute tooth. Abdomen has a sixth segment which is 1.2 times wider than long and its lateral edges are subparallel in their basal third and become very slightly sinuous. First male pléopode which is truncated at its end has six large and long bristles.

### *Thalamlta gatavakensis* (Nobili, 1906)
Image not available
**Common name:** Not designated.
**Distribution:** Madagascar, Seychelles, Malaysian, Philippines, Indonesia, Western Australia, Saipan and Tuamotu.
**Habitat:** Coral reefs at 10–20 m in depth
**Description:** Carapace surface is pilose. All anterior ridges are present and obvious including a pair of mesobranchials. Cardiac ridge is as long as frontal lobes. Two frontal lobes are separated by a distinct notch. Of five anterolateral teeth, first is largest and fourth is small to rudimentary. Chelipeds are short and stout. Cheliped merus and carpus are normal. Fingers are short and stumpy. Propodus of natatory leg is with six to nine spines on the posterior border. Penultimate segment of male abdomen is with lateral borders which are parallel. Color of carapace is not known. Male measures 9.1 mm by 15.0 mm.
**Biology:** Not known.

### *Thalamita gloriensis* (Crosnier, 1962)

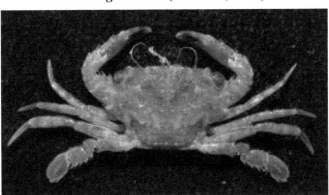

**Common name:** Not designated.
**Distribution:** Tanzania; Madagascar – Glorieuses; Taiwan and Marianas.
**Habitat:** Intertidal
**Description:** Front is 2-lobed. Tips of chelipeds are pointed and not spoon-shaped. Basal antennal joint is granular or tubercular. 5 or only 4 antero-lateral teeth are seen. Mesobranchial and lateral mesobranchial

ridges are present. Male first pleopod is with tip which is neither flared nor recurved, with 4 to 6 stout protruding bristles on one side of tip only.

**Biology:** Not known.

### *Thalamita granosimana* **(Borradaile, 1903)**

Image not available.

**Common name:** Not designated.

**Distribution:** From East Africa, Madagascar, and Maldives to Philippines; and Moluccas.

**Habitat:** Muddy substrate with many stones; sand (to 75 m depth) sandy, silty and weedy coast.

**Description:** Female crab measures 15 mm. No other information is available.

**Biology:** Not known.

### *Thalamita holthuisi* **(Stephenson, 1975)**

Image not available.

**Common name:** Khodorkovsky paddle crab

**Distribution:** Taiwan and Indonesia

**Habitat:** Depth range, 0–5 m; rocks or coral reefs.

**Description:** Carapace, apart from ridges, is covered with fine pile of short hairs. Front of carapace is six-lobed; median notch is deep and wide; median lobes are moderately rounded and protruding; submedian lobes are slightly truncate with moderately inclined inner margins and lateral lobes are rounded and less broad than others. Inner orbital lobes are broad and slightly curved. Of 5 antero-lateral teeth, 1–3 are large and stout with 1 largest and 4 and 5 are smaller, thinner with sharper tips and of equal size. Right cheliped is larger than left, hairy and granular on all but under surface. Anterior border of arm is with 3 usual spines. Wrist is with inner and 3 outer spines. Fifth leg is with ca. 12 spines on posterior border of propodus. Ultimate segment of male abdomen is approximately equilateral triangle with slightly concave margins. Penultimate segment is broader than long and is slightly swollen distally. Male first pleopod is elongate, thin and is gradually tapering and distinctly curved behind tip.

**Biology:** Not known.

## *Thalamita integra* (Dana, 1852)

**Common name:** Not designated.
**Distribution:** East Africa, Madagascar, Red Sea, India, China, Japan, Malaysia, Australia, Tahiti and Hawaii.
**Habitat:** From shallow waters at low tidal mark up to 40 m in depth; sandy bottoms and coral Reef.
**Description:** Carapace is convex and smooth. There are two broad frontal lobes and inner orbital lobes are straight and as broad as frontals. Of five anterolateral teeth, first tooth is largest and fourth is smallest. Chelipeds are unequal and their merus bears two to three spines on anterior border. Carpus is with stout spine on inner angle and rounded tubercles on outer surface. Finger of larger chela is short and stout. Propodus of natatory leg bears six to nine spines on posterior border. Penultimate segment of male abdomen is broader than long and lateral border is convex. Color of the crab is not known. Male specimen measures 17.8 mm by 28.0 mm.

**Biology:** Among a large number of specimens of the above species collected from Palk Bay, near Mandapam, some showed abnormalities, most important among them was the total absence of a walking leg from the right side. Another abnormality noted in some specimens was the absence of some of the antero-lateral teeth (Sankarankutty, http://eprints.cmfri.org.in/1608/1/Article_19.pdf).

## *Thalamita kagoshimaensis* (Sakai, 1939)

**Common name:** Not designated.
**Distribution:** Japan; Taiwan and South China Sea.
**Habitat:** Depth range, 10–53 m.
**Description:** Carapace is covered with a fine pile. A pair of frontal ridges and a pair of protogastrics, are seen. Mesogastric is interrupted in mid line and epibranchial ridges are interrupted by cervical grooves and medially. Cardiac ridge is short and medially interrupted. There are 2 short mesobranchial ridges. Front is with 4 lobes of which medians are narrower than laterals. Of 5 antero-lateral teeth, fourth is smallest. Upper surface of cheliped palm is with 2 spines on anterior and 2 on posterior borders, respectively. Lower border is with squamiform markings. Size of crab measures 19 mm CW.
**Biology:** Not known.

## *Thalamita longifrons* (Milne-Edwards, 1869)
*(=Thalamita yoronensis)*

Photo courtesy of Tadafumi Maenosono

**Common name:** Yoron Beni haunt crab
**Distribution:** From Japan to tropical Indo-Pacific; Sudanese Red Sea.
**Habitat:** Coral reefs
**Description:** Carapace is little more than two-thirds as long as broad. Dorsal surface is covered with thin hairs and mottled with eight brown color patterns in fresh specimen. There are 6 frontal teeth, of which median 2 are close together, rounded at the tip and well-produced forwards. Submedian ones are well-separated and short but also rounded at tip. Lateral ones are again projected forward and their tips rounded. The five anterolateral teeth are uniformly acuminate—first and second projecting forward, third and fourth obliquely forward, of which the fourth is not distinctively reduced in size; and fifth is prominent, projecting strongly sideways. Posterior border is very narrow. Chelipeds are slightly asymmetrical; merus is unarmed; carpus is with its inner angle armed with a long sharp tooth and its outer extremity with three spinules. Propodus is slightly swollen in middle and its upper border has five teeth, of which one is proximal, two are on the inner margin and rest are on outer margin. Ambulatory legs are banded with dark green color and posterior border of propodus of last pair is spinulated. Penultimate segment of male abdomen is subquadrate in outline and lateral borders are parallel and obtusely angular at anterolateral angle. Length of carapace is 12.7 mm and width is 19.0 mm.
**Biology:** Not studied.

### *Thalamita macropus* (Montgomery, 1931)

180 Biology and Culture of Portunid Crabs of World Seas

**Common name:** Not designated.
**Distribution:** Japan and Australia.
**Habitat:** Depth range, 10–30 m.
**Description:** Carapace is closely beset with fine hairs and frontal ridges are short. There is merely a cluster of granules on carapace. Front is with six lobes. Of five antero-lateral teeth, fourth is smallest. Basal antennal joint is with a curved ridge bearing rounded granules. Upper surface of cheliped palm is with two spines on anterior and posterior borders, respectively. Posterior border of propodus of swimming leg is with 5–10 spines. Male first pleopod is regularly curved and sickle-shaped; its inner surface is with 2 small spinules set back from tip; and outer surface is with 2 separate and parallel rows of bristles with about 18 in each row. Carapace width is 27–74 mm.
**Biology:** Not known.

### *Thalamita malaccensis* (Gordon, 1938)

Image not available.
**Common name:** Not designated.
**Distribution:** Malay Peninsula and Java Sea.
**Habitat:** At depth of up to 60 m; amongst coral and clay.
**Description:** Carapace surface is pilose. Of four frontal lobes which are broad, submedians are with shallow concavity in anterior border forming a narrow lobe on outer angle. Of five anterolateral teeth, fourth is smallest. Chelipeds are subequal and their surface is covered with squamiforrn markings; merus is armed with three spines. Cheliped carpus is with usual spines. Female abdomen covers most of cephalothoracic sterna and sutures distinct, with long median crest on each of segments two to four. Color of carapace is not known. Female specimen measures 15.5 mm by 22.3 mm.
**Biology:** Not known.

### *Thalamita minuscula* (Nobili, 1906)

Image not available.
**Common name:** Not designated.
**Distribution:** Eastern Central Pacific: French Polynesia.
**Habitat:** Shallow-waters (0–100 m); soft bottom (mud or sand).

**Description:** It is a very small species with a maximum width of only 4 mm. Largest specimen measures 4.5 × 3.2 mm and smallest juveniles measure 3.3 × 2.6 mm. Live colors are unknown.
**Biology:** Not known.

### *Thalamita mitsiensis* **(Crosnier, 1962)**

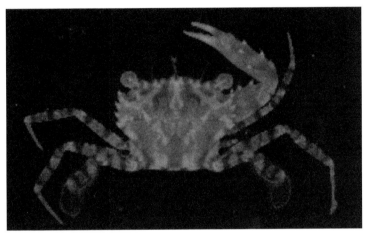

**Common name:** Not designated.
**Distribution:** Mitsio Isles, Madagascar, Malaysia, Philippines and Japan.
**Habitat:** Sandy bottoms at depths of 35–80 m.
**Description:** Carapace of this species is glabrose. Of six frontal lobes, medians are protruding; submedians are broadest and laterals are narrowest. Of four anterolateral teeth, first is stoutest and second to fourth are subequal. Merus of cheliped bears three spines on anterior border. Carpus bears one large spine on internal angle and one spine and two granules on the external surface. Propodus of natatory leg is with denticles on posterior border. Penultimate segment of abdomen is with lateral borders which are convex and regularly convergent. Color of carapace is not known. Male measures 7.3 mm by 10.0 mm.
**Biology:** Not known.

## *Thalamita murinae* (Zarenkov, 1971)

**Common name:** Not designated.
**Distribution:** Sudanese Red Sea
**Habitat:** Coral-inhabiting swimming crab
**Description:** Carapace is about 1.5 times as broad as long, hairy. Frontal, mesogastric, metagastric, epibranchial, cardiac and mesobranchial carapace ridges are present and finely granular. Front (excluding inner supraorbital lobes) about as wide as, or little wider than, posterior margin, somewhat projecting, divided in six lobes, medians and submedians truncated, the former not much produced beyond others. Five anterolateral teeth with sharp tips, fourth smallest but not vestigial, fifth longest but not much longer than others. Cheliped carpus with an additional spine on upper face; manus with 5 spines on upper face, outer face with upper two costae granular, the uppermost ending in a small spine, inner face with a median granular costa, lower face with squamiform marking. Propodus of natatory legs broader than dactylus, with posterior margin bearing three spinules, dactylus lanceolate. Lateral margins of male abdomen slightly diverging over proximal two thirds of length, then converging. Abdomen of female is with transverse keel on fourth tergite extending for most of segment width. Female genital opening is located nearly medially, ovoid, bordered laterally by thickened cuticle. Alcohol preserved specimen shows paired brick red spots on meso- and meta-gastric, epibranchial and mesobranchial areas. Fresh specimens has a brownish background color of the carapace, a pair of large bright red spots resembling ink patches which extend to mesogastric, metagastric and epibranchial areas, and a pair of smaller spots on mesobranchial areas. Female measures 13.0×20.0 mm, and males 14.0 × 20.5 mm and 4.1 × 6.2 mm.

**Biology:** No information

### *Thalamita muusi* (Serène & Soh, 1976)
Image not available.
**Common name:** Not known.
**Distribution:** Not known.
**Habitat:** Not known.
**Description:** Front of carapace is divided into four lobes (inner orbital lobes excluded). This species is mainly characterized by five anterolateral teeth (external, orbital angle included). Three posterior teeth are smaller than two anterior; and teeth 3 and 4 are smaller than fifth. Submedian frontal lobes are little salient. Chelipeds are with a strong, long, acute tooth at the inner angle of the carpus and only one strong spine on superior border of palm. Pereiopod 5 is with a spine on posterior border of merus and no denticles are seen on posterior border of propodus. Male abdomen is very wide and its segment 6 is wider than long and subquadrate with the lateral sides which are not gradually convergent distally but forming a round angle.
**Biology:** Not known.

### *Thalamita oculea* (Alcock, 1899)

**Common name:** Not designated.
**Distribution:** Madagascar, Seychelles, Saya de Malha Bank, India, Maldives, Sri Lanka, Andaman Sea and Japan.
**Habitat:** Depth range, 15–50 m.
**Description:** Carapace is closely beset with hairs and its frontal ridges are short. Protogastrics is evident; mesogastrics is straight and uninterrupted; epibranchial ridge interrupted by cervical groove only; and 2 pairs

of mesobranchial, and 2 cardiac ridges are seen. Front is with 2 lobes. Of 5 antero-lateral teeth, fourth and fifth are subequal and are smaller than others. Upper surface of cheliped palm is with 1 spine on anterior and 2 on posterior borders, respectively. Posterior border of propodus of swimming leg is with about 10 very small spinules. Male first pleopod is stout and is regularly curved. Its tip is flared; inner surface is without conspicuous bristles; and outer surface is with 18 bristles.

**Biology:** Not known.

### *Thalamita parvidens* (Rathbun, 1907)

**Common name:** Not designated.

**Distribution:** Madagascar; Sumatra; Japan; Philippines; Caroline Islands; Marshall Islands and Australia.

**Habitat:** Depth range, 10–45 m.

**Description:** Carapace is with short sparsely arranged hairs underlaid by small granules. Frontal and 3 pairs of gastric ridges are present. Epibranchial ridge is interrupted medially; cardiac ridge is represented by 2 very small broadly separated indistinct crests; and mesobranchial ridges are markedly prominent; front with 2 broad lobes separated by a distinct notch. Of 5 antero-lateral teeth, fourth is shortest and fifth is sharpest and most protruding. Upper surface of cheliped palm is with 2 anterior and 2 posterior spines. Inner face is usually smooth and polished, sometimes with traces of squamiform markings. Posterior border of propodus of swimming leg is with 4–6 spines. Male first pleopod is relatively stout, and is curving smoothly and evenly to slightly flared, strongly recurved tip. Inner surface is with 3–9 stout recurved terminal bristles and outer surface is with 3–5 stout bristles. Size of carapace is 32.25 mm CW.

**Biology:** Not studied.

### *Thalamita pelsarti* (Montgomery, 1931)

Photo courtesy of Tadafumi Maenosono

**Common name:** Not designated.

**Distribution:** China, Japan, Malaysia, Singapore, Indonesia and Australia.

**Habitat:** From muddy to rocky shores within the intertidal zone; under coral heads of reef flats.

**Description:** Carapace is broader than long and ratio of breadth to length is 1.6 times. Dorsal and ventral surface are densely pilose except on raised carapace ridges. All carapace ridges are distinct and granular. Frontals are straight and are cut into six truncated lobes. Of these lobes, medians are set on a slightly lower plane and are separated by a narrow notch. Submedians are broadest, with inner border sloping inwards anterolaterally and overlapping medians. Laterals are narrowest with anterior borders rounded, and are separated from submedians by an open and deeper notch. Of five anterolateral teeth, first three are large increasing in size from front to rear; fourth tooth is rudimentary and fifth tooth is smaller than third. Chelipeds are slightly unequal, granular and densely pilose. Anterior border of cheliped merus is bearing three sharp spines and several spiniform tubercles in between. Posterior border is granular and finely pilose. Upper surface of carpus bears widely spaced rounded granules. Fingers are slender and deeply grooved. Merus of natatory leg is with two grooves bearing fine felt of hairs on the dorsal surface. Posterior border of propodus is serrated. Penultimate segment of male abdomen is slightly broader than long and its lateral borders are parallel for proximal half then converge gently to distal end. Ultimate segment is acutely triangular. Dorsal surface of carapace is dark green with reddish tinge at the joints of appendages. Large males measure 40.2 mm by 63.9 mm.

**Biology:** In this species larval development studies were made in laboratory by Islam et al. (2005). Five zoeal stages and one megalopa stage were identified. At an average temperature and salinity of 24°C and 34 ppt the megalopa stage was attained 17 days after hatching.

### *Thalamita philippinensis* (Stephenson & Rees, 1967)
Image not available.
**Common name:** Not designated.
**Distribution:** Indian Ocean: Seychelles, Indonesia; West Pacific: Philippines, New Caledonia; Central Pacific: French Polynesia.
**Habitat:** Sandy bottom; pearl oysters and corals; and depth range, 15–48 m.
**Description:** Not known.
**Biology:** Not known.

### *Thalamita picta* (Stimpson, 1858)

**Common name:** Not designated.
**Distribution:** East Africa, Red Sea, India, China, Taiwan, Japan, Philippines, Malaysia, Australia, Marianas, Marshall, New Caledonia, Tuamotus, Samoa, Hawaii and Clipperton Island.
**Habitat:** Intertidal zones of rocky shores to fringing coral reefs.

**Description:** Carapace surface is pilose and its ridges, including cardiac and a pair of mesobranchial ridge are distinct. Of six frontal lobes, medians are most projecting, rounded and distinctly separate. Laterals are narrowest and separated from broad submedians by deep notch. Of five anterolateral teeth, fourth is smallest and fifth is sharpest and slightly protruding. Chelipeds are slightly unequal and their merus is bearing three to four spines on anterior border. Carpus is armed with strong spine at inner angle and three spinules at outer angle. Propodus of natatory leg is with five to seven spines along posterior border. Penultimate segment of male abdomen is with lateral borders which are parallel for two thirds of its length and then converge distally. Color of body is mottled greenish yellow, but exhibiting color variations. There are two differing color morphs. Largest male is measuring 10.9 mm by 15.9 mm.

**Biology:** Not known.

### *Thalamita pilumnoides* (Borradaile, 1902)

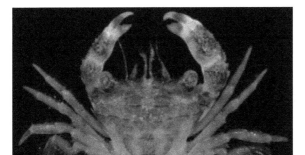

Photo courtesy of Tadafumi Maenosono

**Common name:** Not designated.

**Distribution:** Maldives, Lakshadweep, Madagascar, Marianas Islands, Tuamotu, Society Islands, and East African coast.

**Habitat:** Reef margins

**Description:** Two frontal lobes are present. Claw propodus is massive and is fully covered by spines and granules. Color of carapace is pale yellow and is irregularly marbled. Size of crab is 6.9 × 5.0 mm.

**Biology:** Not known.

## *Thalamita prymna* (Herbst, 1803)

Photo courtesy of Tadafumi Maenosono

**Common name:** Blue-spined swimming crab

**Distribution:** East Africa, Madagascar, India, Andaman Islands, Japan, Hong Kong, Malaysia, Singapore, and Australia.

**Habitat:** Rocky shores at low tidal levels.

**Description:** Carapace is broader than long and ratio of breadth to length is 1.6 times. Dorsal surface is smooth and shiny and pubescence is restricted only to bases of anterolateral spines and in front of lateral parts of the carapace ridges. Frontals are straight, and are cut into six short and truncated lobes which are in close contact with one another. Medians are set on a slightly lower plane and are divided by a narrow incision. Submedians are broadest and are often fused to medians. Laterals are slightly narrower than medians; bluntly rounded and separated from submedians by a very shallow notch. Of five anterolateral teeth, first three are large and are increasing in size from front to rear; fourth tooth is Rudimentary and fifth tooth is smaller than third. Chelipeds are unequal, stout and somewhat tumid. Anterior border of cheliped merus is bearing three sharp spines and a spinule each on the distal corner of anterior and ventral borders. Faint granules are seen on upper surface of posterior border and lower surface of merus is smooth. Upper surface of carpus bears widely spaced rounded granules. Fingers are stout in larger chela, and are long and sharper in smaller cheliped. Under surface of the immovable finger is without groove. Merus of ambulatory and natatory legs is smooth and is without hairs on the posterior surface. Merus of natatory leg is with usual spine at posterior border' and another at the distal end. Propodus is with posterior border serrated. Penultimate segment of male abdomen is longer than broad and its lateral borders converge gently to distal end.

Ultimate segment is acutely triangular. Third, fourth and fifth segments are smoothly fused. Ambulatory and natatory legs are bluish green, including spines and fingers of chelipeds. Dorsal surface of carapace is yellowish green with orange brown patches on ridges. Ventral surface is pale orange. Largest male specimen is measuring 46.3 by 71.7 mm.

**Biology:** Content of mercury in hepatopancreas, muscle and exoskeleton of blue crab *Thalamita prymna* from Musa estuary and their relationship with size of the organism and season were determined by Hosseini et al. (2013). Mean concentration of mercury in hepatopancreas, muscle and exoskeleton were found to be 1.23, 0.61 and 0.41µg/g (dry weight) respectively . There was a significant difference ($p < 0.05$) in mercury levels depending on the sex with greater mercury contents in female crabs. Mean concentrations of mercury in tissues of crab during summer and winter varied and it was greater in summer than winter 0.75 µg/g (dry weight), resepctively.

### *Thalamita pseudopelsarti* (Crosnier, 2002)

**Common name:** Not designated.
**Distribution:** West and Central Pacific: Guam and French Polynesia.
**Habitat:** Littoral; reef margin at night.
**Description:** Carapace is of moderate width (width/length between 1.45, pubescent, and adorned with transverse lines of granules. Front is divided into six lobes which are quite regularly rounded. Median and have submedian lobes are of very similar widths and lateral are slightly wider. Anterolateral edges of shell are cut into five teeth. All these teeth are acute with exception of fourth which is vestigial and more or less reduced to state of

tuber. Rear edge of carapace is slightly convex and has a length close to one third of that carapace. Chelipeds have a merus which has three sharp teeth on its forward edge. Size of these teeth increases from base to tip. Carapace is purple and is dotted with many white dots. Chela are also purple with a fixed finger colored black on most of its length except at its end. Other pereiopods are yellow-orange and more or less washed purple near joints and at dactyls.
**Biology**: Not known.

### *Thalamita pseudopoissoni* (Stephenson & Rees, 1967)
Image not available.
**Common name:** Not designated.
**Distribution:** Japan; Philippines; Palau and Papua and New Guinea.
**Habitat:** Sand mud bottom; depth range, 3–10 m.
**Description:** Front of carapace is 2-lobed. Tips of chelipeds are pointed and are not spoon-shaped. Basal antennal joint is granular or tubercular. There are 5 antero-lateral teeth. Male first pleopod is with strongly flared tip which is shaped like a foot in profile view. No spinules are seen on posterior border of propodus of fifth leg. There is no transverse sculpture on surface of sternum and abdomen.
**Biology**: Not known.

### *Thalamita quadrilobata* (Miers, 1884)

**Common name:** Not designated.

**Distribution:** Madagascar, Persian Gulf, Gulf of Mannar, Arakan Coast Andaman Islands, Japan, Philippines, Malaysia, Java Sea, Australia, Tongatabu and Honolulu.
**Habitat:** Sandy bottoms at 20–80 m in depth.
**Description:** Carapace of this species is covered with short hairs. There 3 pairs of gastric ridges; epibranchial ridges are interrupted by cervical grooves and also mesially; 2 broad cardiac ridges are interrupted mesially; and mesobranchial ridges are short but conspicuous. Front is with 2 lobes which are anteriorly concave. Front may thus be interpreted as bi- or quadrilobate. Of 5 antero-lateral teeth, fourth is smallest. Basal antennal joint is with a raised crest bearing 3–4 spines. Upper surface of cheliped palm is with 2 spines on anterior and 2 on posterior borders. Posterior border of propodus of swimming leg is with 8 irregularly spaced teeth. Male first pleopod is short, stout, and is slightly curved. Inner surface is with 4 very long and stout terminal bristles, followed by 2 hairs. Outer surface is with about 14 terminal bristles. Color of body is unknown. This crab measures 5.1 by 6.2 mm.
**Biology:** Unknown.

***Thalamita seurati*** **(Nobili, 1906)** *(=Thalamita wakensis)*

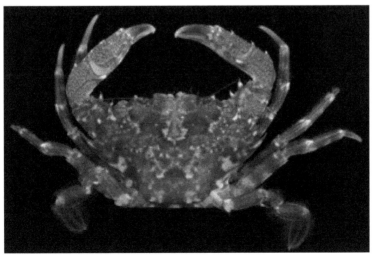

Photo courtesy of Tadafumi Maenosono
**Dorsal view (entire)**

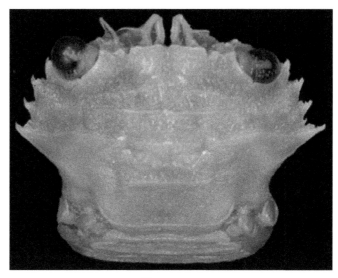

**Carapace**

**Common name:** Not designated.
**Distribution:** Japan, Taiwan, Wake Island and Hawaiian Islands.
**Habitat:** Shallow waters; mud or sand.
**Description:** Carapace is covered with a fine pile. One pair of frontal ridges and one pair of protogastrics are seen. Msogastric is uninterrupted or very inconspicuously interrupted in mid line; epibranchial ridges is interrupted by cervical grooves only; a broad median cardiac ridge is seen; and a pair of mesobranchial ridges join to last antero-lateral tooth. Front is with 6 lobes of very unequal shape. Of 5 antero-lateral teeth, fourth is smallest, sometimes missing. Basal antennal joint is proximally with 3 prominent tubercles and distally with a raised crest made of smaller tubercles. Upper surface of cheliped palm is with 2 spines on anterior and 2 on posterior borders. Inner face is inflated but sparsely granulate. Posterior border of propodus of swimming leg is with 7 spinules. Male first pleopod is stout, with flared tip. Live color of carapace is unknown. Female and ovigerous female measure 8.2 × 5.6 mm and 14.0 × 9.8 mm, respectively.

**Biology:** This species collected from coral reefs of the Great Barrier Reef, Australia, yielded extracts of water soluble toxins lethal to mice by i.p. injection (Llewellyn and Endean, 1988).

### *Thalamita sexlobata* (Miers, 1886)

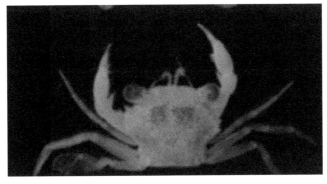

**Common name:** Not designated.
**Distribution:** Red Sea, Madagascar, Persian Gulf, Gulf of Oman India, Andaman Islands, Burma, Japan, Sulu Archipelago, Indonesia, Makassar, Australia, Tonga and Hawaii.
**Habitat:** Sandy bottom up to 50 m depth.
**Description:** Front of carapace is 4-lobed and sinuous. Median frontal lobes are fairly prominent narrower than laterals. Of 5 antero-lateral teeth, fourth antero-lateral tooth is much smaller than fifth. Posterior border of propodus of fifth leg is without spines or denticles.
**Biology:** Unknown.

### *Thalamita sima* (Milne Edwards, 1834)

**Common name:** Four-lobed swimming crab

**Distribution:** East Africa, Red Sea, Sri Lanka, China, Taiwan, Japan, Thailand, Malaysia, Singapore, Indonesia, Australia, New Zealand, New Caledonia and Hawaii.

**Habitat:** Deep waters, up to 30 m in depth; intertidal zones of mud flats and rocky shores.

**Description:** Carapace surface is pilose. Carapace ridges are present in frontal, gastric and branchial regions. Mesobranchial ridge are less distinct and cardiac ridge is interrupted in middle. Two frontal lobes are seen and are separated by shallow notch. Anterior border is slightly sinuous. Last of the five anterolateral teeth is distinctly longer than the rest. Cheliped is stout and unequal and is covered with conspicuous squamifonn tubercles. Cheliped merus is with three spines on anterior border. Carpus is armed with strong spine at inner angle and three spinules at outer angle. Propodus of natatory leg is with posterior border smooth. Penultimate segment of male abdomen is with lateral borders which are parallel for three quarters of length, only converging distally. Ultimate segment of abdomen is acutely triangular. Dorsal surface of carapace is obscured by hairs, blackish and whitish mottled. Ventral surface is bluish. Mouth parts are reddish; chelipeds have black blotch at middle; fingers are with bluish band on movable finger and finger tips are white. Large male measures 36.2 by 57.8 mm and gravid female measures only 8.9 by 12.6 mm.

**Biology:** Studies on reproductive biology have been made by Norman (1996). Captive males of this species were found copulated readily with intermolt ovigerous and non-ovigerous females. Copulation was brief, for example, for a period of only 1–2 minutes. Brooding occurred over an extended period.

### *Thalamita spinicarpa* (Wee & Ng, 1995)

Image not available.

**Common name:** Not designated.

**Distribution:** Singapore.

**Habitat:** Intertidal zone on sandy to rocky shores at tide levels of 0.1– 0.5 m.

**Description:** Carapace surface is densely pilose except on raised ridges. Ratio of carapace breadth to length is 1.65 times. Frontal ridges are prominent and arched. Front is cut into six truncated lobes. Medians are lying on a lower plane separated by a narrow notch. Submedians are slightly broader, with inner edges sloping gradually inwards. Lateral lobes are separated from submedians by an open notch. Anterior border is truncated in larger specimens. Of five anterolateral teeth, first tooth is stoutest; second and third are of equal size; both are slightly smaller than first; fourth tooth is rudimentary; and fifth tooth is slightly smaller than third. Chelipeds are unequal, granular and finely pilose. Anterior border of cheliped merus bears three to four sharp spines and several small rounded tubercles. Posterior border is with dispersed granules. Movable finger of larger cheliped is stout and blunt, and of smaller cheliped is more straight and slender. Penultimate segment of male abdomen is a slightly broader than long and lateral borders are parallel for proximal two thirds and then slanting gently inwards towards distal end. Ultimate segment of abdomen is acutely triangular, longer than broad. Second to fourth segment are keeled. Dorsal surface of carapace and chelipeds is light green with mottled brown patches. Brown bandings are seen on dactylus of ambulatory legs, anterior border of frontal lobes and lateral borders of female abdomen. Large male measures 31.4 by 52.2 mm.

**Biology:** Not known as it is a new species

### *Thalamita spinifera* (Borradaile, 1902)

**Common name:** Not designated.

**Distribution:** Madagascar, Laccadives, Maldives, Thailand, Malaysia, Indonesia, Philippines, Japan and Hawaii.
**Habitat:** Coral and sand beyond depths of 100 m.
**Description:** Carapace surface is pilose. Mesogastric and cervical groove are interrupted in middle. Cardiac and a pair of mesobrianchial ridges present. Of six frontal lobes, medians are prominent, and set on lower plane; submedians are broader than former; and laterals are acute and widely separated from submedian by deep notch. Of six anterolateral teeth, first is largest and is bearing a subsidiary basal tooth. Basal antenna I segment is bearing seven to eight granules. Cheliped is granular on upper surface and lower to inner surface is smooth. Cheliped merus is with three to four spines on anterior boarder. Propodus of natatory leg is serrated at posterior border. Penultimate segment of abdomen of lateral borders are parallel for 2/3 of its length and converge gradually at distal end. Color of body is not known. Male crab measures 10.0 by 27.0 mm.
**Biology:** Not known.

### *Thalamita spinimana* (Dana, 1852)

Photo courtesy of Tadafumi Maenosono

**Common name:** Spiny claw swimming crab
**Distribution:** Hainan (China), Thailand, Malaysia, Singapore, Indonesia, Philippines, Pulao, Mariana, Australia, New Caledonia and Fiji.
**Habitat:** Rocky shores and coral reef flats at 0.0–0.1 m tide levels.

**Description:** Carapace is pilose and broader than long. Carapace ridges of anterior to last anterolateral tooth are granular and distinct. Cardiac and mesobranchial ridges are smooth. Front is cut into six lobes of which medians and submedians are truncate; and laterals are narrowest, rounded at the apex and directed slightly obliquely. Five anterolateral teeth are decreasing in stoutness and increasing in sharpness from front to rear. Basal antennal segment is armed with three to four spinules. Cheliped is granular and its merus is with three to four spines on the anterior border and a smaller one at the distal end of the same border. Fingers are sharp and deeply grooved. Propodus of natatory leg is serrated on posterior border. Penultimate segment of male abdomen is with lateral borders which are parallel and slightly convergent. Two widely differing body colorations are seen in this species. Large male is measuring 56.8 by 90.3 mm.

**Biology:** Unknown.

### *Thalamita spinimera* (Stephenson and Rees, 1967)

**Common name:** Thorn foot paddle crab.
**Distribution:** Sudanes Red Sea
**Habitat:** Coral-inhabiting species

It is an endangered species and other aspects are not known about this recently created species.

## *Thalamita stephensoni* (Crosnier, 1962)

**Common name:** Shi paddle crab
**Distribution:** Madagascar; Taiwan; China; Indonesia; New Georgia; Solomon Islands; Samoa and Melanesia.
**Habitat:** Intertidal; Reef habitat in 0–5 m.
**Description:** Carapace is not very broad. Front of this species is 2-lobed. 5 antero-lateral teeth are seen. Tips of chelipeds are spoon-shaped. Spines on hands of chelipeds are of normal length. Carapace is lime colored.
**Biology:** Not studied.

## *Thalamita xishaensis* (Chen & Yang, 2008)

**Common name:** Not designated.
**Distribution:** South China Sea; Yongxing Island and Xisha Islands.
**Habitat:** Sand and rocky bottom along coast, in coral reef, and in shallow water.
**Description:** Carapace is 1.7 times as broad as long and is densely covered with granules and setae. Front is cut into 4 lobes of which median lobes are straight, inclined laterally outward and separated by a deep V-shaped notch; and lateral lobes are triangular and much narrower than median lobes. Supraorbital lobes are narrow and arched. Anterolateral margin is with 5 sharp teeth (including outer orbital tooth), of which second tooth is largest, fourth is smallest and last tooth is longest and clearly produced beyond other teeth. Posterolateral margin is longer than anterolateral one, concave, and is

converging posteriorly. Posterior margin is slightly convex, forming a blunt angle with posterolateral margin. Chelipeds are unequal and are covered with coarse granules on dorsal and outer lateral surfaces. Cheliped merus is with inner margin bearing 3 acute spines medially; proximal portion is serrated; and distal portion is smooth. Fingers are stout; immovable finger is shorter than palm; and movable finger is equal to latter. Natatory leg is with propodus bearing 5 spines. Dactylus is slender with 20 spines on anterior border and 4 spines on posterior border and tip is spiniform. Male abdomen is with third to fifth segments completely fused. Sixth segment is trapezoidal and is clearly broader than long. Telson triangular and its tip is blunt. This crab measures 8.1 by 13.5 mm.

**Biology:** Not known.

### *Thalamitoides quadridens* (Milne Edwards, 1869)

**Common name:** Not designated.

**Distribution:** Red Sea, Madagascar, Japan, South China, Guam, Indonesia, Australia, Fiji, Samoa, Johnston Island, Hawaiian Islands, and Tuamotu Archipelago.

**Habitat:** Coral reefs and shallow waters.

**Description:** Carapace is much broader than long (CB/CL about 1.85). Epibranchial crests are interrupted medially, but in median part small crests are still extant. Antero-lateral borders are with 4 teeth which are decreasing in size fore to aft. Basal antennal joint is with a low granulate crest occupying its whole length. Cheliped merus is granulate on posterior border and anterior border is armed with 3 spines. Carpus is with 5 small and acute spines on its external border. Upper face of palm is with 4 spines on anterior and 3 on posterior borders. Male first pleopod is short, stout and is gently curved with 2 rounded membranous lobes on inner surface.
**Biology:** Not studied.

### *Thalamitoides tridens* (Milne Edwards, 1869)
*(= Thalamitoides spinigera)*

Photo courtesy of Tadafumi Maenosono

**Common name:** Not designated.
**Distribution:** Red Sea, Madagascar, Mauritius, Japan, China, Philippines, Guam, Indonesia, Australia, Fiji, Samoa, Johnston Island and Hawaiian Islands.
**Habitat:** Coral reefs and shallow waters.
**Description:** Carapace is much broader than long (CB/CL is 2.1). Epibranchial crests are broadly interrupted medially. Antero-lateral borders are with 3 teeth which are decreasing in size fore to aft. Basal antennal joint is with a short granulate crest occupying about 2/5 of its length. Cheliped merus is smooth on posterior border and anterior border is armed

with 4 spines. Carpus is with 3 small and acute spines on external border. Upper face of palm is with 3 spines on anterior and 4 on posterior borders. Male first pleopod is short, stout, and gently curved, with swollen and truncate tip.

**Biology:** Not known.

### *Gonioinfradens paucidentata* (Milne Edward, 1869)

**Common name:** Red swimming crab

**Distribution:** Red Sea, Kenya, Seychelles, Aldabra Islands, Coëtivy Islands, Réunion, Mauritius, Persian Gulf, Gulf of Oman, Japan, New Caledonia, Marquesas, and Tuamotu Archipelago.

**Habitat:** Marine, littoral to 100 m.

**Description:** Carapace of this species is bare and smooth. It is rather faint with transverse granular lines on frontal, protogastric and mesogastric regions. Epibranchial line is interrupted at the cervical groove and across midline. Front is with 6 teeth. Medianteeth and submedian teeth are truncated and laterals are triangular with rounded tips. Antero-lateral borders are with 6 teeth of which second and fourth are very small and sixth is acute and is smaller than the preceding one. Posterolateral junctions are rounded. Antennal flagellum is excluded from orbit. Chelipeds are slightly heterochelous and their merus is with 3 strong spines on anterior border. Posterior border is smooth. Carpus has a strong internal spine and its outer border is with 3 spinules. Palm is smooth, with 3 spines on upper border.

Merus of swimming leg is with a subdistal posterior spine. Propodus is with a row of spinules on posterior border.
**Biology:** Not studied.

## 3.4 SUBFAMILY PODOPHTHALMINAE

### *3.4.1 CHARACTERISTICS OF THE SUBFAMILY PODOPHTHALMINAE*

Portunidae with carapace subtrapezoidal; widest anteriorly; Front is extremely narrow and T-shaped. Orbits and eyestalks are enormously long. Supraorbital fissures are reduced. Antero-lateral border is divided into 2–4 teeth. Chelipeds are much longer than ambulatory legs, bearing a set of spines on merus, carpus and palm. Last pair of legs are with paddle-shaped propodus and dactylus.

### *Euphylax dovii* (Stimpson, 1860)

**Common name:** Not designated.
**Distribution:** West coast of Mexico.
**Habitat:** Epibenthic swimming
**Description:** Carapace of this species is transversely subovate. Front is narrow and broadly T-shaped. Short, almost straight anterolateral margins are much shorter than the posterolateral margins, which are nearly straight and converge to deep coxigeal embayments. There are five small teeth on an anterolateral margin. First anterolateral spine is minute and

attached to the base of the outer orbital spine; it is followed by three stout, obliquely directed, isosceles-triangular spines increasing in length posteriorly. Eyes are extremely long and palms are carinated. Carapace is coppery iridescence in color. Chela and ambulatory legs are light red. Eyes are deep brown.

**Biology:** Not studied.

### *Euphylax robustus* (Milne-Edwards, 1874)

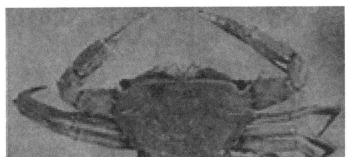

**Common name:** Robust swim crab
**Distribution:** Pacific: Mexico.
**Habitat:** Sandy muddy flats between 10 and 70 m.
**Description:** These crabs have a streamlined profile that allows for rapid swimming and long pincers armed with sharp spines to snag food. Three strong spines on anterolateral margin of carapace, with a small spine at the immediate base of the ocular spine. Carapace is rectangular and 1.6 times wider than long; and is with granulated lines and prominences. Lower orbital plate is coarsely granulated. Front is narrow and broadly T-shaped. Eye peduncle is extremely long. Inner surface of movable finger is smooth. Both margins of merus are lacking hair. Orbital spine is directed forward, and curving outward at tip. Carapace is light brown with iridescent coppery tints. Chela are deep pink, with red blotches at articulations, at base of spines and at anterolateral margin of merus. Lower dactyli are white. These crabs reach a maximum size of 12 cm CW (males) and 10 cm CW (females).

**Biology:** This species mainly feeds on clams, fish, snails, worms and other crustaceans. These are active crabs that are among the few crabs that are swift and agile swimmers. They swim sideways utilizing their paddle-like fifth pair (dactyl) of legs that rotate like propeller blades when they swim.

## *Podophthalmus minabensis* (Sakai, 1961)

**Common name**: Petal-eyed swimming crab; stock-eyed swimmer crab.
**Distribution:** Japan
**Habitat:** 15–30 m depths.
**Description:** Carapace is about 1.65 times as broad as long and its surface is uneven. A rather deep groove extends from the median region to the base of the second antero-lateral tooth. Front is T-shaped and is very narrow between eyestalks. Its anterior extremity is broad and thick. Eyestalks are bearing wing-like expansions. Second antero-lateral tooth is much smaller than first. First segment of eyestalk is markedly flattened with a pterygoid expansion on its anterior distal end. Chelipeds are asymmetrical and right is slightly heavier than left. Merus is with 2 spines on anterior and posterior borders. Carpus is with exterior spinule. Palm has 7–8 sharp denticles on upper border.
**Biology:** Not studied.

## *Podophthalmus nacreus* (Alcock, 1899)

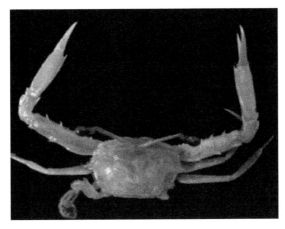

**Common name**: Not designated.
**Distribution:** Andaman Islands and the Gulf of Martaban (Western Australia).
**Habitat:** Soft sediments; 127 m deep
**Description:** Carapace of this species is broadly hexagonal, width 1.7–2 times length; regions are well defined; fronto-orbital grooves are transverse, but lateral portion abruptly angled posteriorly to receive cornea; and orbital angle is notched. Anterolateral teeth are acute, produced laterally; and second anterolateral tooth is very small. Eye-stalk is narrow, cylindrical; and distal segment is swollen. Cheliped merus anterior margin is with row of 5–7 short proximal spines followed by two larger curved, distal spines; posterior margin is with two curved spines in distal half; and dorsal surface is with short distal crista. Propodus is with 4 longitudinal carinae; and internal surface is smooth. Male gonopod 1 is strongly arcuate and evenly tapering, with scattered sharp spinules apically. Wet specimens still retain their nacreous sheen. Male measures 24.7 × 13.5 mm. Ovigerous females measure 12.4 × 7.0–26.4 × 13.3 mm.
**Biology:** Not studied.

## *Podophthalmus vigil* (Weber, 1795)

**Common name:** Long eyed swimming crab; sentinel crab.

**Distribution:** Hawaii and the Indo-Pacific; Red Sea; Gulf of Aden; South Africa; Madagascar; Mauritius; Gulf of Oman; Pakistan; India; Nicobar Islands; Japan; Taiwan; China; Thailand; Malaysia; Singapore; Philippines; Indonesia and Australia.

**Habitat:** 10–70 m depths; mud and silty sand; burrows in bays, harbors, fishponds, and estuaries.

**Description:** Carapace is about 2.5 times as broad as long. Surface of carapace is uneven and, regions are not very well defined. A rather deep groove extends from the median region to the base of the second antero-lateral tooth. Front is entire and deflexed. Second antero-lateral tooth is much smaller than first. Eyestalk is without wing-like expansions. Chelipeds are asymmetrical and right is slightly heavier than left. Merus of cheliped is with 3 spines on anterior and 2 on posterior border; and carpus is with exterior tooth. Male measures 127.9 × 52.1 mm and female 108.2 × 46.5 mm.

### Biology

***Sexual maturity:*** In this species, sexual maturity has been achieved at 11.51 cm for male and 9.53 cm for female. Mean fecundity reported was 312,613 eggs. Size distribution of male and female was found to be slightly different with males having a size range from 8.0 to 18.9 cm CW, while females ranged between 7.0 and 11.4 cm CW. Mean CW and BW of males of this species were significantly greater than that of females. The sexual ratio for male to female was 1: 0.65 (Ikhwanuddin et al., 2015).

***Larval development:*** The complete larval development of this species consist of five zoeal and one megalopal stages. Total duration for zoeae I to III is 3–4 days, while zoeae IV and V take 2–3 days and V zoea requires 3 days to metamorphose into megalopa stage. The complete larval development takes a span of 12–15 days.

## 3.5   SUBFAMILY CARCININAE

### *3.5.1   CHARACTERISTICS OF THE SUBFAMILY CARCININAE*

Last leg of crabs of this subfamily is more leg-like than paddle-like. These crabs are therefore poor swimmers. Carapace is subhexagonal, as wide as long or moderately wider than long. It is not obviously convex in both directions. Eyestalks are short. 5 or fewer anterolateral teeth are present. Basal antennal joint is narrow, cylindrical and longer than wide. Lobe on maxilliped 1 ('portunid lobe') is inconspicuous. Chelipeds which are long and stout are usually without spines except for inner carpal spine.

***Portumnus latipes** (**Pennant, 1777**) (= Portunus latipes)*

**Common name:** Pennant's swimming crab

**Distribution:** Moray Firth, Aberdeen, Firth of Forth, Yorkshire, Wash, Thames, East Channel, Wight, Portland, Channel Isles, Plymouth, Bristol Channel, Liverpool Bay, Dublin, Belfast, Clyde & Argyll, Donegal Bay, Galway Bay and German North Sea.

**Habitat:** Intertidal and wash zone of sandy, exposed beaches.

**Description:** Carapace is heart-shaped and its dorsal surface is smooth. Frontal region is slightly projecting, and median and submedian lobes are subacute. Orbits are wide. Antero-lateral margins of carapace are with five teeth. Fourth tooth is often very small. Chelipeds are slightly unequal and somewhat compressed. Dorsal margin of carpus-dactylus is cristate and setose. Second to fifth pereiopods are compressed, cristate and setose. Dactylus is lanceolate and dactylus of fifth pereiopod is broadly lanceolate. Carapace is reddish brown and is variegated with dark brown. Length of carapace is up to 27 mm.

**Biology:** This species occurs during warmer periods and, thus, is a good indicator for warming of North Sea coastal waters. This species seems to be an omnivorous mainly feeding on crustaceans, polychaetes and macroalgae. It ascends to the midlittoral zone in order to feed. Stomach fullness percentages indicated a higher feeding activity of this species in winter and spring than in summer and autumn. This is probably due to the necessity of accumulating energy resources for the spawning period, which begins in winter. In berried females, three egg development (I: white, II: orange, and III: grey coloration) stages were observed. It has been reported that the fecundity decreased with egg development stage from orange to grey phase, indicating egg loss over time. Mean fecundities per gram of female for stages I, II and III were 7711.1, 10004.0 and 4837.6 eggs, respectively (Cores and Erzini, 2015).

**Female *Portumnus latipes* with orange (Stage II) eggs**

***Xaiva biguttata* (Risso, 1816)** *(=Portunus biguttatus, Portumnus nasutus)*

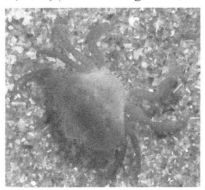

**Common name:** Not designated.
**Distribution:** Mediterranean, Atlantic, and English Channel.
**Habitat:** From low water spring tides in coarse shell and fine gravel to meters.
**Description:** Carapace is posteriorly elevated and its dorsal surface is smooth. Frontal region is produced well in advance of first pair of antero-lateral teeth. Median and sub-median lobes are sometimes developed but obtuse. Each dorso-orbital margin is with a small lobe. Antero-lateral margins of carapace are with five teeth. Chelipeds are equal in size and are compressed. Inner margin of carpus and propodus have setose longitudinal carinae. Second to fifth pereiopods are moderately stout and compressed. Anterior margins of propodus and carpus of second to fourth pereiopods

are cristate. Dactylus of fifth lanceolate is with a median carina. Carapace is uniform dull greenish-yellow to white and dactylus of second to fourth pereiopods is pale brown to violet. Carapace length is up to 20 mm

**Biology:** This species has been reported to be responsible for ship's hull fouling in the coasts of Galicia ( NW Spain). Further it is hard to find this species as fishing by-catch because of its littoral habitat and small size.

### *Echinolatus bullatum* **(Balss, 1923)** *(=Nectocarcinus bullatus)*
Image not available.

**Common name:** Not designated.

**Distribution:** Only known from the Juan Fernandez Archipelago, off Chile, in depths between 10 and 40 m.

**Habitat:** Not known.

**Description:** Carapace is subhexagonal and is slightly broader than long. Front is produced well in advance of small inner orbital lobes. Supraorbital margin is with two fissures. Anterolateral margins of carapace are slightly convex. Exorbital lobe is well developed and bluntly pointed, with convex outer margin. Three following teeth are acute and are slightly decreasing in size from first to third. Chelipeds are equal and merus is with granular superior surface. Carpus is with conspicuous, very strong, long, acute tooth at anterointernal angle. Chela is with inner superior surface bearing a projecting granular ridge ending in denticle. Fingers are almost as long as palm. Male abdomen is 7-segmented. Sixth segment is quadrangular. Telson is triangular and is, apically rounded. Coloration is unknown.

**Biology:** Unknown.

### *Echinolatus caledonicum* **(Moosa, 1996)** *(=Nectocarcinus caledoni)*
Image not available.

**Common name:** Not designated.

**Distribution:** Known only from New Caledonia and the Chesterfield Islands.

**Habitat:** Between 200 and 580 m depths

**Description:** Carapace is wider than long and is finely but sparsely granular. Front is quadrilobate. Median lobes are rounded and are much smaller than laterals. Lateral lobes are triangular and pointed. Anterolateral margin is convex, with four teeth. First tooth is large and blunt. Second to third are as sharp spines and are about equal in size. Fourth is larger than preceding two, with one or two small, accessory spines at or near base. Third, maxillipeds are narrow and its merus is much longer than broad. Carpus is placed mediointernally. Chelipeds are subequal. Anterior border of its merus is granular, without spine on either margin, but is covered with setae. Carpus is broad and its dorsal surface sparsely granulated. Male abdomen is 7-segmented and telson is heart-shaped.

**Biology:** Not known.

### *Echinolatus proximus* (Davie & Crosnier, 2006)

**Common name:** Not designated.
**Distribution:** French Polynesia
**Habitat:** Unknown.
**Description:** Carapace is subhexagonal and is distinctly broader than long in adults. Front is produced well in advance of small inner orbital lobes. Supraorbital margin is with two fissures. Anterolateral margins of carapace are moderately convex. Exorbital lobe is well developed and is bluntly pointed, with convex outer margin. Three following teeth are spinous. Chelipeds are subequal (right chela is slightly stouter than left. Merus is little ornamented except for a small spine. Carpus is with conspicuous, very strong, long, acute tooth at anterointernal angle. Dactylus is lanceolate. Male abdomen is 7-segmented and sixth segment is quadrangular. Telson is triangular and as broad as long. Its lateral margins are strongly convergent and are apically rounded (Davie and Crosnier, 2006).

**Biology:** Unknown.

### *Echinolatus poorei* (Davie & Crosnier, 2006)
Image not available.
**Common name:** Not designated.
**Distribution:** Only known from off southern Australia between 130° 36.59′E and 149° 34.02′E longitude.
**Habitat:** Depths from 134 to 520 m.
**Description:** Carapace is subhexagonal and is distinctly broader than long. Front is produced well in advance of small inner orbital lobes. Anterolateral margins of carapace are convex and margins are bearing tiny denticles over entire length. Exorbital lobe is well developed, triangular, and bluntly pointed. Three following teeth are spiniform and their size is increasing from first to third. First tooth is pointed anteriorly; second tooth is directed anterolaterally and third tooth is pointed laterally. Posterolateral margins are concave, with dentiform granules. Posterior margin is slightly sinuous. Dorsal surface of carapace is granular over entire surface. Chelipeds are equal and is covered with coarse denticles and even low spines. Merus is granular on external and internal surfaces. Carpus is with conspicuous, very strong, long and acute tooth at anterointernal angle. Chelae are covered in many dentiform granules. Fingers are almost as long as palm. They are with triangular cutting teeth which are often tricuspid. Dactylus is lanceolate, from 2.2–2.3 times longer than broad (excluding peduncle). Its surface is with low, rounded median longitudinal swelling. Male abdomen is 7-segmented. Its sixth segment is quadrangular. Telson is triangular, broader than long and is apically rounded.
**Biology:** Not known.

### *Carcinus maenas* (Linnaeus, 1758)

Profile of Portunid Crabs (Family: Portunidae) 213

**Common name:** Shore crab; green crab.

**Distribution:** Native to the north-east Atlantic Ocean and Baltic Sea; colonized in Australia, South Africa, South America and both Atlantic and Pacific coasts of North America.

**Habitat:** All types of protected and semi-protected marine and estuarine habitats, including habitats with mud, sand, or rock substrates, submerged aquatic vegetation, and emergent marsh; prefer soft bottoms; depth range, from high shore pools to about 60 m.

**Description:** Carapace is finely granular, about as long as broad with 5 acuminate antero-lateral teeth. Antero-lateral teeth are tipped with yellow. Frontal area between orbits is with three rounded teeth. Carapace color is variable, usually mottled, dark brown to dark green and granules for the most part are yellow. Female ground color is orange in place of green with whitish granules. Claws are similar with small black spots which are arranged longitudinally on palm. Fingers and adjacent portion of palmare light blue with black stripes in grooves. Walking legs are green and are speckled with black. Second and third pair of walking legs are longest and are about 1 and 2/3 times the length of the carapace. Fourth pair of legs is shorter than the first. Last pair of legs is somewhat flattened and are with setae. Adult green crabs range in size from 6 to 10 cm in carapace width.

**Biology:** Some large crabs of this species have red limbs and undersides rather than the usual green. This is related to the breeding period and prolonged intermolt, and is caused by photodegradation of the green exoskeletal pigment. Red morphs of *Carcinus maenas* are found to have a thicker carapace for greater protection during intraspecific conflict for mates. Green crabs of this species are mainly found sheltering under algae where their color blends-in with the background. Red *Carcinus maenas* appear brown against a brown background in deep water and are mostly found in the shallow sublittoral where red light does not penetrate. This species increases its body size by 20–33% per molt and takes about 10 molts to reach 20 mm carapace width (CW) in its first year, if conditions are favorable. Though it may molt more than once per year after the first year, molting rate slows once maturity is reached and is probably about once per year post maturity. This species is considered an omnivore and consumes plants, algae, molluscs, arthropods (including their own species), annelids and carrion. The diet of large *Carcinus maenas* mainly consists of molluscs and the common mussel *Mytilus edulis* is the most important of these. Smaller crabs (<30 mm CW) have more plant matter and arthropods in their diet. On rocky shores, adults consume more gastropods especially the dog whelk *Nucella lapillus* and winkles *Littorina* sp. Females can produce up to 185,000 eggs, and larvae develop offshore in several stages before their final molt to juvenile crabs in the intertidal zone. Young crabs live among seaweeds and seagrasses, such as *Posidonia oceanica*, until they reach adulthood.

**Fisheries:** It is fished commercially.

### *Carcinus aestuarii* (Nardo, 1847)

**Common name:** Mediterranean shore crab; littoral crab; green crab.

**Distribution:** Mediterranean to Southeast Atlantic; introduced in Japan.

**Habitat:** Littoral species, common in the fields with Zostera. Depth range, 0–200 m; and inhabits brackish lagoons also.

**Description:** It is a large-sized crab. Carapace is smooth and more or less hexagonal. Posterolateral margin is more or less straight, without armature. There are 5 prominent sharps anterolateral teeth. Front is produced into 3 low teeth. Short antennules are with 25 articles. Antennae are with a very thin flagella. Dactylus of the fifth pleopods is normal with a long, cylindrical and slightly flattened craw. Top of the first pleopods is straight or very slightly bent. Color of body is variable. Upper side is generally deep green in adults while the under side is tinged with yellow or red. Length and width of this species are 63 mm and 80 mm, respectively.

**Biology:** It is a carnivorous species, eating small fishes, shrimps and corpses. It is known for its high prolificacy with a long period of reproduction, from May until December. In the period of reproduction, it makes migrations to the shore. The planktonic larvae (zoea and megalopa) are the most resistant of all decapods of the Black Sea. It is euryterm and euryhaline species. This species has a commercial fishery.

**Fisheries:** Commercial fisheries exists for this species.

## 3.6 SUBFAMILY POLYBIINAE

### 3.6.1 CHARACTERISTICS OF SUBFAMILY POLYBIINAE

Carapace of the crabs of this subfamily are only moderately wide and not obviously convex in both directions. Eyestalks are of normal length. There are 4–5 anterolateral teeth. Antenna is arising from orbital hiatus and is lying almost along longitudinal axis of the carapace. Antennal flagellum is in orbital hiatus. Chelipeds and walking legs are stout and long. First pair of legs is as long as chelipeds and last legs are with rounded or moderately lanceolate dactylus.

***Bathynectes longipes* (Risso, 1816)** *(= Portunus longipes)*

**Common name:** Not designated.
**Distribution:** East Channel, Channel Isles, Plymouth, Scilly Isles and Bristol Channel.

**Habitat:** Depth range from 20 to 226 m.

**Description:** Frontal margin is with a very shallow pair of sub-median lobes and median lobe is absent. Antero-lateral teeth are strongly curved forward; alternating slightly in size and first and second are lobate. Chelipeds are unequal in size and dactylus of smaller cheliped is with four carinae which are, usually absent on larger cheliped. Second to fourth pereiopods somewhat compressed. Carapace is brilliant reddish brown and sometimes with dark blotches. Carapace length is up to 34 mm.

**Biology:** Not studied.

### *Bathynectes longispina* (Stimpson, 1879)

**Common name:** Bathyal swimming crab

**Distribution:** Gulf of Mexico

**Habitat:** Deep-water swimming crab; seamounts and knolls; and at depths as great as 600 m.

**Description:** This species has reduced swimming dactyls and long walking legs. As an adaptation to deep water, it has large eyes. Front is with four teeth. Basal antennal joint is free and not united with the lateral subfrontal process. While the carapace length and breadth of collected male measured 41.5 mm and 72.9 mm respectively, the values of female were found to be 37.2 mm and 68.2 mm respectively.

**Biology:** Not studied.

### *Bathynectes maravigna* (Prestandrea, 1839)

**Common name:** Not designated.
**Distribution:** Mediterranean Sea; Alborán Sea; southern Adriatic and northwestern Thyrrenian Se; Ionian Sea; Alborán Sea and Italian peninsula.
**Habitat:** Depth range 245–786 m.
**Description:** This species has a polygonal shell. In the face, it has 4 teeth and anterolateral edge has 5 teeth which are much larger and connected to the opposite edge by a transverse hull. First pair of walking legs is with tweezers. It is chelated with 5 keels, 3 of which are toothed and prickly. Body has reddish color with white spots. Size of males and females ranges from 9–51 mm CL and 12–51 mm CL, respectively. Ovigerous females have a size range of 40–41 mm CL.
**Biology:** Eighty three percent of both males and females of this species are right-handed. Sexual dimorphism is present in cheliped length with males having longer chelae than females. This species appears to be much commoner in those areas where Atlantic influence is the highest.

### *Coenophthalmus tridentatus* (Milne Edwards, 1879)

**Common name:** Not designated.
**Distribution:** Endemic species of Southwestern Atlantic, from southern Brazil to northern Patagonia.
**Habitat:** Shallow, coastal water (10–40 m deep) species.
**Description:** Length of carapace is 10–40 mm. It is an occasional bycatch in shrimp fisheries.
**Biology:** Nothing is known about its Biology.

### *Liocarcinus corrugatus* (Pennant, 1777)

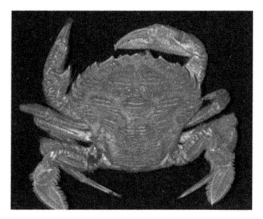

**Common name:** Wrinkled swimming crab
**Distribution:** British Isles to the Mediterranean.
**Habitat:** Depth range up to 100 m; coarse sand and gravel.
**Description:** Carapace is suboval and is much broader than long. Dorsal surface is moderately convex with numerous long, transverse, setose carinae. A pair of very broad concave submedian lobes and an advanced, broad, subacute to rounded median lobe are present. Outer ventral margin of orbit is with moderately narrow incision. Antero-lateral margins of carapace is with prominent equally-developed 5 teeth. Chelipeds are equal in size and with carinae. Pereiopods are stout and setose. Carpus-dactylus of second to fourth pereiopods are with longitudinal carinae. Dactylus of second to fourth pereiopods are styliform and that of fifth is broadly lanceolate with pronounced median carina. It is reddish-brown in color, sometimes with red or yellow patches. Carapace reaches up to 43 mm in length and 41 mm in breadth.

**Biology:** Not studied.

### *Liocarcinus depurator* (Linnaeus, 1758)

**Common name:** Blue-leg swimming crab; harbor crab; sandy swimming crab.

**Distribution:** North Sea, Atlantic Ocean, Mediterranean Sea, and Black Sea.

**Habitat:** Lower shore and sublittoral on fine, muddy sand and gravel; and depth range, −5 m to −300 m.

**Description:** Carapace is broader than long and is relatively flat, with numerous transverse and hairy, crenulations. Antero-lateral margins of the carapace have 5 pointed teeth. Wide orbits and three similar-sized rounded lobes are seen between eyes. Front of carapace is with a median lobe which is slightly more prominent than two similar flanking lobes. Chelipeds are equal and stout. Pereopod 5 is with strongly paddled dactylus. This species is immediately recognized by its violet-tinted paddle of the fifth leg. Carapace which is pale reddish-brown with transverse rows of hairs measures 51 mm wide and 40 mm long.

**Biology**

*Food and feeding:* It feeds on polychaetes, crustaceans, molluscs, ophiuroids and fishes. It serves as a host for polychaete worm *Iphitime cuenoti* and parasitic nemertean *Carcinonemertes carcinophila* that live in its branchial chambers. It is gonochoristic (dioecious) and it matures at the age of 1 year. Females with eggs occur all year although a maximum proportion of ovigerous females has been observed indicating the existence of an annual reproductive cycle. Number of eggs (fecundity) carried by ovigerous females ranges from about 30,000 to 230,000 (http://www.marlin.ac.uk/species/detail/1175).

***Lectin content:*** A lectin that recognized sialic acids and agglutinated mouse erythrocytes has been purified from hemolymph of this species. Lectins are widely used as biochemicals tools in many types of research (Fragkiadakis and Stratakis, 1997). Lectins have been reported to manifest a diversity of activities including anti-insect activities, anti-parasitic, anti-tumor, immunomodulatory, antimicrobial and HIV-1 reverse transcriptase inhibitory, which may find applications in many therapeutic areas. A small number of lectins demonstrate anti-parasitic activities (Hamid et al., 2013).

*Liocarcinus holsatus* **(Fabricius, 1798)**

**Common name:** Flying crab

**Distribution:** All British and Irish coasts.

**Habitat:** Rock pools, in the shallow sublittoral and offshore; on sand, gravel, mixed and hard substrata; depth range to a 73 m.

**Description:** This species is very similar to Liocarcinus depurator but lacking any ridges. Body is longer than it is wide and is more or less smooth, without bristles. Carapace has three blunt teeth between eyes and middle one is sometimes shortest. Chelipeds are equal and slender, with a sharp tooth on the outer edge of the carpus. Orbits are smaller, and posterior edge of carapace is narrow. Body is brownish-grey and is tinged with green and grows up to 4 cm long.

**Biology:** The flying crab has striking flattened legs making it a good swimmer. Its build is light and it moves readily through the water, practically flying. It also uses these paddles to burrow itself into the sandy bottom. Flying crabs catch food while swimming. They pinch rapidly and painfully. This crab doesn't always actively hunt its prey. A lazy crab just waits till its prey swims past. Diet of this species comprises crustaceans, especially juvenile Crangon, molluscs such as Spisula elliptica, and fish. Flying crabs are often infected with parasitic barnacle species, *Sacculina carcini*.

## *Liocarcinus maculatus* (Risso, 1827)

**Common name:** Not designated.
**Distribution:** Restricted to the Mediterranean Sea.
**Habitat:** Sublittoral habitats in depths between 5 and 73 m; prefers coastal muddy sand in depths between 10 and 25 m.
**Description:** Carapace is slightly wider than long (ratio CB./CL. 1.08–1.26). Front is advanced beyond orbit, with three teeth or lobes, of which median lobe is slightly longer and more acute than laterals. Front orbital angle is poorly marked. Carapace surface is granulate, and regions are well marked. Anterior margin of cardiac region is rising steeply. On anterolateral margin of carapace, fifth, fourth, and usually also third tooth are sharply pointed and first two with apex are more or less rounded. Fourth tooth is smaller than fifth. Antennal flagellum is about one-third carapace length. Carpus of cheliped is with strong spine on antero-internal border, and is with pointed tubercle. In third and fourth pereiopods, carpus is shorter than propodus. Carapace length of males is 4.2–12.5 mm and of females is 4.4–11.2 mm.
**Biology:** Not studied.

## *Liocarcinus marmoreus* (Leach, 1814)

**Common name:** Marbled swimming crab

**Distribution:** Northern Atlantic Ocean and North Sea; westernmost section of the Mediterranean Sea.

**Habitat:** Sand gravel in the sublittoral and lower littoral zones, down to a depth of 84 m.

**Description:** Carapace of this species is suboval and is broader than long. Dorsal surface is relatively flat and smooth. Median and sub-median lobes are equally developed and subacute to rounded. Sub-medians are often slightly in advance of median. Outer ventral margin of orbit is with a narrow incision. Antero-lateral margins of carapace is with first and second teeth which are usually slightly obtuse. Third to fifth teeth are spinose and fourth is at the most as long as the third. Chelipeds are equal in size and stout. Carpal process of cheliped is acute and broad. Second to fourth pereiopods are moderately thin. Dactylus of fifth pereiopod is broadly lanceolate. Color of carapace is varying from variegated light brown to dark brown with white reticulations. Pereiopods are paler and dactylus of fifth is tinged with blue. Carapace length is up to 35 mm.

Profile of Portunid Crabs (Family: Portunidae)

**Biology:** This species is sometimes parasitized by the barnacle Sacculina.

### *Liocarcinus mcleayi* (Barnard, 1947)

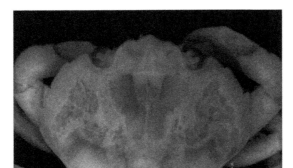

**Common name:** Not designated.
**Distribution:** South Europe
**Habitat:** Coarse sand and fine gravel under bottom currents at depths of 13 to 24 m.
**Description:** Carapace is as long as broad and smooth. Frontal margin of carapace is triangular and is projected forward. Front is with 1 undulation on both sides. There is no small tooth between supraorbital fissures. Chelipeds are armed with ridges on carpus and lateral surface of palm. Palm is with 1 ridge on outer surface. Dactylus of fifth leg is broader and more flattened than that of fourth leg and is paddle-like. Color of body is variable. Carapace is from light beige with a few orange points or greenish with lighter margins, to pale pink (Raso and Manjon-Cabeza, 1996).
**Biology:** Not studied.

## *Liocarcinus navigator* (Herbst, 1794) *(= Liocarcinus arcuatus)*

**Common name:** Arch-fronted swimming crab

**Distribution:** British Isles; north-east Atlantic Ocean, including the North Sea as far north as Trøndelag in Norway, and as far south as Mauritania; and Mediterranean Sea.

**Habitat: Habitat:** It is found on mixed sandy substrata from the shallow sublittoral to offshore i.e. from 2 to 108 m.

**Description:** Carapace of this species is suboval and is much broader than long. Dorsal surface is moderately convex with minute transverse (sometimes setose) carinae. Margin of frontal region are entire and setose. Outer ventral margin of orbit is with a narrow incision. Antero-lateral margin of carapace is with 5 teeth alternating slightly in size. Fourth tooth is smallest and sometimes almost obsolete. Chelipeds are equal in size and is relatively smooth. Propodus of cheliped is with longitudinal (usually obtuse) carinae. Pereiopods are moderately stout and carpus-dactylus of second to fourth is with longitudinal carinae. Dactylus of fifth pereiopod is broadly lanceolate with very indistinct median carina. Carapace is dark brown in color and the limbs are somewhat paler. Carapace length is up to 29 mm and carapace width is up to 33 mm.

**Biology:** Not studied.

## *Liocarcinus pusillus* (Leach, 1815)

**Common name:** Dwarf swimming crab

**Distribution:** Atlantic coast of Europe from northern Norway to Portugal, possibly on NW African coast.

**Habitat:** Gravel or stony substrate; from the intertidal zone and down to 100 m depth.

**Description:** Carapace is slightly wider than long (ratio CB./CL. 1.1–1.3). Front is advanced beyond orbits and is produced into three lobes. Median lobe is slightly more prominent than the laterals. Front-orbital corner is distinguishable. Carapace surface is more or less granulate. Anterior margin of cardiac region is rising gently. On anterolateral margin of carapace, fourth tooth is smaller than fifth and, second tooth is very small. Fifth tooth is only with pointed tip. Length of antennal flagellum is less than or equal to one-fourth carapace length. Carpus of chelipeds is with anteroexternal border which is rounded or sinuate, with two slight prominences. Third and fourth pereiopods are with carpus which is shorter than propodus. Maximum carapace length and width is 23 mm and 25 mm, respectively.

**Biology:** Not studied.

### *Liocarcinus vernalis* (Risso, 1816)

**Common name:** Vernal crab, grey swimming crab, and oval swimming crab.

**Distribution:** Mediterranean species; West Africa to the southern North Sea.

**Habitat:** Marine; shallow-water crab; and depth range (m): 0.5–13.

**Description:** These crabs are of small size and with front projecting beyond orbits.

**Biology:** Not studied.

### *Liocarcinus zariquieyi* (Gordon, 1968)

**Common name:** Not designated.

**Distribution:** Mediterranean Sea; Canary Islands; and England.

**Habitat:** Sublittoral habitats in depths between 10 and 60 m; prefers substrates of coarse sand and gravel.

**Description:** Carapace is slightly wider than long (ratio CB./CL. 1.02–1.25). Front is advanced beyond orbits, with three teeth or lobes. Median lobe is slightly longer than Laterals which are markedly more obtuse. No prominence is visible between internal margin of orbit and external margin of front. Carapace surface is smooth, with regions well marked. Anterior margin of cardiac region is rising gently. Anterolateral margin of carapace is with teeth decreasing in size from third to fifth. All teeth are with rounded apices. Antennal flagellum is short and is less than one-fourth carapace length. Dorsal ridge of chela is not prolonged into distal tubercle. Carpus of cheliped is with spine only at antero-internal border. Third, and fourth pereiopods are with carpus which is longer than or subequal to propodus. Carapace length of males is ranged from 7.4 to 12.7 mm and that of female is 10.5 mm.

**Biology:** Not studied.

### Necora puber (Linnaeus, 1767) *(=Macropipus puber.)*

**Common name:** Devil crab, velvet swimmer crab, and witch crab.

**Distribution:** North-west Europe; common around all coasts of Britain.

**Habitat:** Rocky shores at low water; depths of 80 m; fairly sheltered shores.

**Description:** This species has a flattened carapace, which is wider than it is long. Upper surface is blue but has a reddish-brown velvety covering, which disguises blue coloration and earns its common name. Color of these eyes and the general aggressive nature of this species may explain the alternative names of Devil crab and witch crab. Pincers are equal in size and are also velvety and the eyes are bright red. Between the eyes there are around 10 narrow teeth on the edge of the carapace. Width of carapace is up to 90 mm.

**Biology:** Adults feed on brown seaweeds, molluscs and crustaceans, whereas juveniles feed mainly on crustaceans such as small crabs and barnacles. This species is a fast-moving and very aggressive species and can deliver a painful nip. Females carrying eggs may be found at all times of the year in Britain. Clutch size ranged from 5000 to 278,000 eggs. Heterochely and allometric changes related to sexual maturity were found to be the main factors determining the variability of the relative growth in both sexes. In males, the length of the first pleopod and the length of the first chela showed a change in allometry. Morphometric maturity size was similar for both sexes (53.3-mm carapace width in males and 52.3-mm in females). The maturity size was 54.8-mm carapace width in males and 49.8-mm in females. The size at onset of maturity based in morphometric and reproductive criteria coincided with a decrease in growth rate at molt in females (González-Gurriarán and Freire, 1994).

**Fisheries:** In some parts of Europe, this species is fished commercially.

### *Nectocarcinus bennetti* (Takeda and Miyake, 1969)
Image not available.
**Common name:** Not designated.
**Distribution:** New Zealand
**Habitat:** Depth range, 20–474 m; and coarser sediment.
**Description:** Carapace is hardly wider than long and naked, with rounded granules forming distinct ridges and structures on anterior half but more generally scattered on posterior half. Regions are distinct. Front is quadrilobate and is slightly upturned. Medial lobes are smaller, narrower, more acute, and closer to each other than to submedials. Anterolateral margin is with four subequally spaced teeth. 1st tooth is subacute and others are acute. 3rd and 4th are stronger and standing out more distinctly from margin than others. Eyestalk is short and is narrowed slightly near terminal subspherical cornea. Chelipeds are subequal in length and size. Right chela is armed with strong, rounded lobe-like teeth on fingers. Left

chela is with more numerous, smaller, more acute teeth on fingers. Walking legs are flattened and unspined. Second, third and fourth pereiopods are elongated, subequal in length and are slightly over-reaching chelipeds in length and similar in form. Male abdomen is with ultimate segment which is triangular, a little broader than long and is with bluntly rounded apex. Carapace and dorsal surface of chelipeds are mainly purplish-red with areas of pink iridescence. Some regularly pattened paler areas are seen especially on the posterior half of the carapace. Walking legs and ventral surface of body and appendages are pale off-white to dirty cream in color. Fingers of chela are not distinctly pigmented. Carapace length is 5.8–68.0 mm. Ovigerous females measure 36.1 mm CL.

**Biology:** Not studied.

## *Nectocarcinus antarcticus* (Jacquinot, 1853)

**Common name:** Hairy red swimming crab
**Distribution:** New Zealand
**Habitat:** Depth range, 0–550 m; coarser sediment.
**Description:** Carapace is hardly wider than long and is completely clothed in short, fine, dark tomentum, but with rounded granules forming distinct ridges and structures projecting through tomentum. Regions are distinct. Front is quadrilobate, in plane of carapace. Medial lobes are smaller, narrower, blunt and closer to each other than to submedials. Anterolateral margin is with four subequally spaced teeth. Among these teeth, first is blunt and others acute. First and second are not standing

out from anterolateral margin but third and fourth are stronger and standing out distinctly from margin. Margin between teeth is edged with small granules. Chelipeds are subequal in length and size. Both right and left chelae are with similar teeth on fingers. Male abdomen is with ultimate segment which is triangular and is a little broader than long, with apex, which is some what truncated and rounded. Penultimate segment of abdomen is with weakly concave lateral margins and length is subequal to distal width. Segment 1 is largely obscured by proximal expansion of segment 2. Segments 2 to 5 are with a distinct transverse ridge across middle of segment. Surface of abdomen, sternum and ventral surface of body and appendages are in general covered with short, fine, dark-colored tomentum. First pleopod of male is stout and straight. Carapace and dorsal surface of chelipeds and walking legs are mottled with dark red over a background of pinkish red. There is no trace of iridescence, but with some small white marks on various ridges and spines. Ventral surface of body and legs are pale with some regularly placed bands of red across sternum and coxa of each leg size range is 8.0–62.0 mm. Ovigerous females measure 8.8 mm carapace length.

**Biology:** Not studied.

*Nectocarcinus bullatus* **(Balss, 1924)** *(= Echinolatus bullatum)*
Image not available.

**Common name:** Not designated.

**Distribution:** Only known from the Juan Fernandez Archipelago, off Chile.

**Habitat**: Depths between 10 and 40 m; sand.

**Description:** Carapace is subhexagonal and is slightly broader than long (ratio of CB/CL, 1.08–1.14). Front is produced well in advance of small inner orbital lobes. Maximum width of carapace is about 3.2 times frontal width. Front is composed of 4 lobes and among them, lateral pair well developed and apically rounded. Median pair is much smaller and is separated medially by V-shaped notch and from lateral lobes by broad sulcus. Anterolateral margins of carapace is slightly convex. Exorbital lobe is well developed, bluntly pointed and is with convex outer margin. Three following teeth are acute and are slightly decreasing in size from first to third. First two are pointed anteriorly and last tooth is more anterolaterally.

Posterolateral borders of carapace are concave. Dorsal face of carapace is with some isolated granules on protogastric regions, anterior part of mesogastric region, and in vicinity of anterolateral teeth. Chelipeds are equal and their merus is with granular superior surface. Carpus is with conspicuous, very strong, long, acute tooth at anterointernal angle. Chela is with inner superior surface bearing a projecting granular ridge ending in denticle. Its superior face is with median longitudinal row of three large granules. Fingers are almost as long as palm and are with triangular cutting teeth which are often tricuspid; extremities pointed, and recurved. Superior margin of mobile finger is with several denticles. Pereiopods 2–4 are smooth and subequal. Fifth pereiopods are also smooth and are hardly shorter than P2–4 and propodi and dactyli are dorsoventrally flattened for swimming. Dactylus is lanceolate and its anterior margin is slightly convex. Male abdomen is 7-segmented. Sixth segment is quadrangular. Telson is triangular and is broader than long. It is apically rounded with lateral margins which are straight and strongly converging. Body coloration is unknown.

**Biology:** Not studied.

### *Nectocarcinus integrifrons* (Latreille, 1825)

**Common name:** Rough rock crab
**Distribution:** Southern Australia
**Habitat:** Seaweed, reef and sand areas; intertidal to 20 m.
**Description:** Carapace is convex, uninterrupted or sometimes with a very shallow and minute central notch. Front is with a single row of numerous small rounded tubercles (usually on soft bottoms on sheltered crabs). There are 4 anterolateral teeth. Chela palm outer surface is with tubercles in transverse rows. Front two thirds are purplish-brown and back

third is pale yellow-brown. Black claws are with molar-like teeth. Last segment of last leg is dagger shaped. Size of crab is 80 mm.

**Biology:** These crabs are active during the day. When a male finds a female to mate with, he carries her around between his claws until she molts and is ready to mate. It is a carnivore and it is not a commercial species.

### *Nectocarcinus spinifrons* (Stephenson, 1961)

Image not available.

**Common name:** Not designated.
**Distribution:** Shark Bay, Western Australia; South Australia
**Habitat:** Depth range 75–80 m.
**Description:** Carapace is uniformly covered with setae. Front which is bilobed is without row of tubercles Protogastric region is with granules. There are 4 anterolateral teeth. Chelied merus has 4–5 spines (excluding spinules) on lower outer border. Inner border is with single prominent spine near tip. Cheliped carpus inner border is bearing a larger spine with secondary spinules on dorsal surface. Chela palm outer surface is with irregular rows of spiniform granules. Inner margin of outer surface of chela palm has a large spine subdistally. Size of crab is 21 mm CW.
**Biology:** Not studied.

### *Nectocarcinus tuberculosus* (Milne Edwards, 1860)

**Common name:** Rough rock crab; velvet crab

**Distribution:** Victoria, Tasmania, and South Australia.

**Habitat:** 50 m depths; rock reef areas; and sometimes enters rock lobster pots.

**Description:** This crab earns its name due to furry coating (red hairs) of carapace. Carapace is densely pilose, particularly anteriorly. Frontal margin is almost straight, with a distinct narrow V shaped notch centrally, bordered by double row of prominent blunt tubercles. Epibranchial regions are with groups of tubercles merging with groups surrounding anterolateral teeth. There are 4 anterolateral teeth. Chela palm outer surface has tubercles in transverse rows. It is a carnivore. Body coloration is red. Size of crab is 9 mm CW.

**Biology:** It is a commercially important species.

### *Polybius henslowii* (Leach, 1820)

**Common name:** Henslow's swimming crab

**Distribution:** E. Atlantic, from British Isles to Morocco and W. Mediterranean (very common off the Spanish and Portuguese coasts).

**Habitat:** Pelagic species

**Description:** Adult carapace of this species is nearly circular, flat and a little wider than long. Posterolateral margin is little contracted. Dorsal surface of carapace is smooth, with indistinct regions. There are five small antero-lateral teeth which are somewhat projecting forward. Front of carapace is with three shallow teeth (between the inner orbital teeth). Chelipeds are equal and are little longer than width of carapace. Pereiopods 2–4 are compressed and are little shorter than P1. Fifth pereiopod (P5) is shorter than other pereiopods. Last two segments (propodus and dactylus) of P5 are broad and flat and dactylus is oval. All pereiopods are with moderately haired margins. Coloration of body is reddish brown and underside is pale. Carapace length is 40 mm and width is 48 mm.

**Biology:** The conspicuous pelagic swarming of predominantly males is most probably related to the reproduction cycle. These crabs have been reported to be the most important marine prey for yellow-legged gulls (*Larus cachinnans*) especially in the coasts of Galicia (north-western Spain) (Munilla, 1997).

*Raymanninus schmitti* **(Rathbun, 1931)** *(=Benthochascon schmitti)*

**Common name:** Sharpoar swimming crab
**Distribution:** Western Atlantic Ocean, Gulf of Mexico.
**Habitat:** Depth range, 210–511 m.
**Description:** Front of carapace of this species is axially notched. Epibranchial region is arcuate or keeled. Anterolateral is shorter than posterolateral. There are only two anterolateral teeth (unlike 4 of other polybiines) (excluding outer-orbital spine). Fifth pereiopods lack paddle-like elements. Chelae are with weak keels. Propodus of fifth pereiopods is slightly broadened and dactyls are lanceolate. In males, somites are fused but with very clear sutures.

**Biology:** Not studied.

## 3.7 SUBFAMILY MACROPIPINAE

### 3.7.1 CHARACTERISTICS OF SUBFAMILY MACROPIPINAE

Carapace is not very broad. Antero-lateral margins are with five teeth. Antennae are arising from orbital hiatus and are lying almost in longitudinal axis of carapace. Flagellum is standing inside the orbit. Eyestalks are of normal length. Chelipeds and legs are both long, and one pair of legs is as long as chelipeds. Last legs are typical swimming paddles.

***Benthochascon elongatum* (Sakai, 1969)** *(=Brusinia elongata)*

**Common name:** Not designated as it is a new species
**Distribution:** Japan
**Habitat:** Deep-water swimming crab
**Description:** Carapace is oblong, 1.2 times as long as broad. Dorsal surface is smooth and glabrous. Front is composed of three obtuse teeth, of which the median one is smaller than the lateral ones. Neither preorbital tooth or upper orbital fissures are present. Anterolateral borders are cut into four teeth, the first or external orbital tooth is very long, the second, third and fourth are smaller and subequal in size. Posterolateral borders are much longer than the anterolateral ones and slightly concave. Basal joint of the antenna is short but longer than broad and freely movable. Flagel-

lum is moderately long. Chelipeds are slightly robust. Merus is short and prismatic. Carpus is subquadrate in outline and is armed with a tooth at the inner angle. Propodus is high. Immovable finger is very short and is very broad proximally. Movable finger is long and slender and is strongly curved inward in the distal half. Length and width of carapace measure 6.8 mm and 5.5 mm, respectively.

**Biology:** Nothing is known about its biology as it is a new species.

### *Benthochascon hemingi* (Alcock & Anderson, 1899)

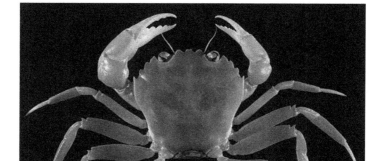

**Common name:** Not designated.

**Distribution:** Andaman Sea, Nicobar Islands, Japan, South China Sea Indonesia, Australia, New Caledonia, Vanuatu.

**Habitat:** Deep-water swimming crab inhabiting depths of 150–780 m.

**Description:** Carapace is about as broad as long and smooth. Front is with 3 lobes of which the middle one is bifid at tip. Antero-lateral borders are provided with 4 teeth of which last one is spiniform and is well separated from the others. Basal joint of antenna is short; longer than broad and is freely movable. Flagellum is moderately long and is situated in a narrow orbital hiatus. Eyestalks are of normal length. Chelipeds are somewhat heterochelous. Chelipeds and legs are long. One pair of legs is as long as chelipeds. Last legs are typical swimming paddles. Dactyl is oval and ambulatory legs are fairly stout.

**Biology:** Not studied.

### *Macropipus tuberculatus* (Roux, 1830)

**Common name:** Knobby swim crab

**Distribution:** Shetland, Viking, Plymouth, Dublin, Fastnet, Cork, and Sole.

**Habitat:** Depth range, just beyond 80 m.

**Description:** Carapace is suboval and is much broader than long. Dorsal surface is coarsely tuberculate and tubercles are often carinate. Median and sub-median lobes are acute. Dorsal orbital margin is with a V-shaped incision. Antero-lateral margins of carapace are setose, with five well-spaced curved teeth. Fifth tooth is longest. Chelipeds are equal in size and relatively thin. Their propodus is with longitudinal tuberculate carinae. Carpal process is long. Pereiopods are long and second to fourth are thin. Dactylus of fifth pereiopod is broadly lanceolate with a conspicuous longitudinal median carina. Dorsal surface of carapace is yellowish red and chelipeds are lighter with touches of red. Juveniles can live hollowed into the substrate. Size of males ranges from 10 to 38 mm and that of females from 15 to 36 mm.

**Biology:** This species predates opportunistically upon epifaunal crustaceans and small fish. Infaunal organisms such as echinoderms, polychaetes and bivalves form a smaller part of the diet. Pelagic shrimps are preyed upon during day-time hours. A clear annual reproductive (molt) cycle has been reported in the females of this species. Females carried the eggs for about one month. Males did not show any evidence of seasonal molting or maturity (Abellóa, 1989). Fecundity is up to 65,000 eggs.

### *Ovalipes australiensis* (Stephenson & Rees, 1968)

**Common name:** Sand crab; surf crab or two-spot crab
**Distribution:** Southern Australia; and extends from Western Australia to Queensland, including Tasmania.
**Habitat:** Soft substrates; sand areas and beaches, and to depth of 100 m.
**Description:** Carapace is smooth and is about as wide as long. Color of carapace is reddish-orange-grey with a pair of large deep red or purple spots posteriorly. Rear leg ends with flat rounded paddle. Front is divided into 2 short teeth. Five antero-lateral teeth are bordered by coarse granules. Chela palm's upper surface is with 3 distinct carinae. Basal antennal article is without flattened dorsal area and its upper surface with dense long hairs. Carapace is up to 20 cm wide.

**Biology:** It is a carnivore. These crabs are particularly agile and are able to bury rapidly into the sand by digging backwards when disturbed, leaving only their stalked eyes poking up into the water above. Though this species is not harmful but a nip from large claws could be painful. It has a commercial fisheries.

### *Ovalipes catharus* (White, 1843)

**Common name:** New Zealand paddle crab.

**Distribution:** New Zealand, the Chatham Islands, and in south-eastern parts of Australia.

**Habitat:** Shallow intertidal waters (0.1–0.5 m depths); off sandy beaches, and in harbors and estuaries.

**Description:** This species is a relatively large and fast growing species of *Ovalipes*. Front is divided into 2 short teeth. Upper orbital border is toothed. Inner supraorbital notch is deep. Lower surface of chela palm is with stridulating ridge. Mesobranchial area of carapace is without stridulating apparatus. Basal antennal article is with flattened dorsal area and upper surface is sparsely hirsulate. These crabs reach a maximum size of 150 mm carapace width after 13 postlarval molts.

**Biology:** Paddle crabs are mainly active in early evening or at night, when they move into the shallow intertidal zone to feed. They are versatile and opportunistic predators. They feed mainly on either molluscs or crustaceans, but also on polychaetes, several fish species, cumaceans, and occasionally on algae. Cannibalism is common, particularly on small crabs and during the winter molting season. Mating generally occurs during winter and spring (May to November) in sheltered inshore waters. Female paddle crabs can only mate when they are soft-shelled. Male crabs protect and carry pre-molt females to ensure copulation. Female crabs are thought to migrate to deeper water to spawn over the warmer months (September to March). After spawning the eggs are incubated until they hatch. This species has an extended larval life characterized by eight zoea stages and a (crab-like) megalopa. The larvae are thought to live offshore in deeper water, migrating inshore in the megalopa stage to settle from January to May. Brood fecundity estimates vary considerably geographically from between 82,000 and 2,122,807. The number of broods per year also varies geographically from 1.2 to 3.3. This species has commercial and recreational fisheries.

### *Ovalipes elongatus* (Stephenson & Rees, 1968)

Image not available.

**Common name:** Not designated.

**Distribution:** Known only from Lord Howe Island and Kermadec Island.

**Habitat:** Not known.

**Description:** Front is four toothed (including inner supraorbital lobes). Median teeth are close together, moderately sharp, broadly separated from laterals, and are produced slightly further forwards than laterals. Upper border of orbit is toothed; inner supraorbital notch is deep and outer supraorbital notch is present and suborbital border is hairy. There are 5 anterolateral teeth which are distinctly large and robust. They are broad based and narrowly Separated. These teeth are bordered by moderately coarse granules. First tooth is blunter than second and sharp with broad base; fourth is the largest and fifth is the smallest and sharpest. Carapace is relatively elongate. Areas of carapace are generally not distinct. Granulation of carapace is fairly coarse anteriorly and fine posteriorly. Chelipeds are relatively short, robust and subequal. Dactyl of first walking leg is with flattened grooved surfaces and carinate edges. Dactyl of third leg is with flattened grooved surfaces and carinate. Carpus of fifth leg is with sharp, hirsute anterior border. Male abdomen and first pleopod: There are no adult males in the present collection.

**Biology:** Unknown.

### *Ovalipes floridanus* (Hay & Shore, 1918)

**Common name:** Florida lady crab

**Distribution:** Western Atlantic ocean; Gulf of Mexico (from southwest Florida, to off Port Aransas, Texas)

**Habitat:** Shallow gulf and lower bays.

**Description:** Carapace is oval and compressed. Last pair of legs are flattened into paddle-shaped appendages. 5 pair of spines (including the outer orbital spines) are seen along the lateral edges of the carapace and are directed forward. 3 sharp teeth are found between the ocular orbits (eye sockets). Small spines are present on the inner margins of merus (2nd segment from the claws) and larger spines on inner and outer dorsal carpus (segment next to claw). Color of body is orangish brown to gray and joints are with darker orangish color. Claws are long and their color is brownish with some orange and blue. Fingers are brown on top and are mostly white interiorly. Size of crab is 8.5 cm carapace width.

**Biology:** These crabs are mostly gonochoric. Precopulatory courtship ritual is common (through olfactory and tactile cues). Indirect sperm transfer is common.

## *Ovalipes georgei* (Stephenson & Rees, 1968)

Image not available.
**Common name:** Not designated.
**Distribution:** Australia
**Habitat:** Not known.
**Description:** Front of this species is divided into 2 short teeth. There are 5 anterolateral teeth, bordered by coarse granules. Basal antennal article is with flattened dorsal area. Chela palm upper surface with 3 distinct carinae, innermost ending in a spine. It possesses a unique stridulating apparatus. It involves the modified leg 4 rubbing against the granular ridge on the mesobranchial area of the carapace.
**Biology:** Not studied.

## *Ovalipes iridescens* (Miers, 1886)

**Common name:** Not designated.
**Distribution:** South Africa; Madagascar; Japan; East China Sea; Taiwan; South China Sea; Philippines; Indonesia; Australia; Loyalty Islands; Chesterfield Islands and Kinmei Seamount.
**Habitat:** 120–520 m depths.
**Description:** Carapace is moderately broad (CB/CL is 1.28) and granulation is moderately developed in most of the carapace surface. Two oval, translucent areas with smooth surface and remarkably thin cuticle are seen on posterior half of carapace. There are 3 frontal teeth. Antero-lateral teeth are sharp, narrow based and broadly separated. Chelipeds are subequal and relatively long and thin. Distal part of anterior border of merus is with 2–5 large spines. Inner spine of carpus is moderately robust. Palm is

slightly swollen and its lower border is with granules tending towards a squamiform pattern. Carina on anterior border of carpus of swimming leg is projecting slightly at distal end. Female measures 36.3 × 28.9 mm and male measures 27.1 × 21.7 mm.

**Biology:** Not studied.

### *Ovalipes molleri* (Ward, 1933)

**Common name:** Not designated.
**Distribution:** Australia
**Habitat:** 150–270 m depths.
**Description:** Front is with 2 median teeth which are close together and fused. There are 5 antero-lateral teeth and area between teeth is bordered by very fine granules. Basal antennal article is protruding well forwards, with flattened, hirsute, dorsal area. Upper surface of chela palm is with 3 granular carinae. Size of crab is 62 mm CW.

**Biology:** Not studied.

### *Ovalipes ocellatus* (Herbst, 1799)

**Common name:** Ocellate lady crab; calico crab.
**Distribution:** Eastern North America; from Canada to Georgia
**Habitat:** Sand beaches; buries itself in the sand.
**Description:** Lady crabs have very sharp, powerful pinchers which are whitish in color with purple-spotted tips and jagged teeth. Three sharp points are present between the eye sockets of the lady crab, as well as five sharp points along the carapace that turn toward the eye sockets. Carapace is yellow-grey or light purplish with leopard like clusters of purple dots. Carapace of this species is slightly wider than long, at 8.9 cm wide, and 7.5 cm long.
**Biology:** The lady crab is a brightly colored and aggressive. In the water and under direct sunlight, this crab's coloring appears iridescent. The species is called the lady crab because of the beautiful color patterns on the shell. These crabs are excellent swimmers. This species lives mainly on molluscs, such as the Atlantic surf clam, *Spisula solidissima.* It has five larval stages, lasting a total of 18 days at 25°C (77°F) and a salinity of 30‰, and 26 days at 20°C (68°F) and 30‰. It is a nocturnal predator, which often buries itself in the sand. It has been described as "vicious" and "the crab most likely to pinch a wader's toes".

### *Ovalipes punctatus* (De Haan, 1833)

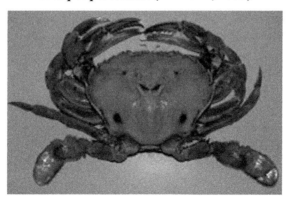

**Common name:** Three-spot swimming crab
**Distribution:** Japan; Korea; East China Sea; Taiwan and China.
**Habitat**: Swash zone of a sandy beach; intertidal zone, especially at night; and 15–115 m depths.

Profile of Portunid Crabs (Family: Portunidae) 245

**Description:** Carapace is broad (CB/CL is 1.33) and its surface granulation is fairly coarse anteriorly and fine posteriorly. Antero-lateral teeth are large and robust, broad based and narrowly separated. Chelipeds are subequal, relatively short and robust. Anterior border of merus is without spines or tubercles. Inner spine of carpus is long and robust. Palm is not swollen and its lower border is with 13–20 coarse, deep striae. Anterior border of carpus of swimming leg is sharp and hirsute. Size of crab measures 9.5 cm CW.

**Biology:** It mainly feeds on amphipod *Haustorioides japonicus*, mysid *Archaeomysis kokuboi*, isopod *Excirolana chiltoni* and *Crangon* sp. This species has an ontogenetic migration into deeper water. Female crabs spawn from mid-September until mid-November in offshore water 40–60 m deep. The incubation period of the eggs is about 20 days. After a planktonic phase, the larvae settle on sandy bottoms in March and April. Settled crabs of both sexes grow to carapace widths (CW) of 50–65 mm in the first year. In the second year, females and males grow to 70–90 and 80–85 mm CW, respectively. The life span of this crab is 2 to 2.5 years. The minimum size at maturity is estimated to be 45–50 mm CW but the spawning population consists mainly of 2-year-old crabs of 75–80 mm CW.

### *Ovalipes stephensoni* (**Williams, 1976**)

Image not available.

**Common name:** Coarsehand lady crab

**Distribution:** Virgina to near Biscayne Bay, Florida.

**Habitat:** Surface to 183 m.

**Description:** Carapace is with relatively coarse granulation behind frontal margin and inside anterolateral borders. Chelipeds with dorsal surface of carpus is finely and densely granular. Dorsal surface of palm between ridges is densely granular, mostly fine, but often with scattered enlarged granule. Dorsal surface of dactyl between ridges is very finely and densely granular. Outer surface of palm is finely to obsolescently granular. Fixed finger is becoming smooth or microscopically pitted. Inner surface of palm is similar to outer surface or smooth and microscopically granular ventrally, but becoming granular in very large males. Dactyl is smooth or microscopically pitted. Fixed finger is partly granular. Merus of third maxilliped is with distal projection which is relatively long and narrow. Its anterolateral border straight or very slightly curved.

**Biology:** A preliminary survey of 253 *Ovalipes stephensoni* collected off the southeast coast of the United States revealed a 94.5% incidence of infestation by the parasitic ciliate *Synophrya hypertrophica*. Analysis of 67 crabs indicated an infestation density ranging from 1 to 270 lesions per branchial chamber (Haefner and Spacher, 1985).

### *Ovalipes trimaculatus* (De Haan, 1833)

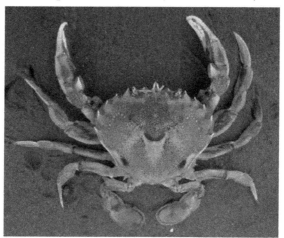

**Common name:** Paddlecrab ; three-spot swimming crab; rowing crab.

**Distribution:** Worldwide distribution; Atlantic Ocean – southeastern United States, Central America, Southeast of Brazil and Uruguay to Argentina Patagonia; South Africa in the Indian Ocean – South Africa; Australia and New Zealand; southeastern Pacific Ocean – Peru and Chile.

**Habitat:** Soft bottoms or buried in coastal marine waters, at depths between 3 and 65 m.

**Description:** Carapace has a smooth texture. Chelipeds (large pincers) are robust, elongated and equal in size regardless of sex or age. Color pattern composed of brick tones. Size of carapace is up to 9 cm diameter. Size of the carapace of females between 15 and 70 mm, while the size of the males is between 35 and 80 mm.

**Biology:** This species is commercially exploited because of their edible flesh, but as bycatch. Acoustic behavior of this species has been studied. It produced wide frequency band multi-pulse signals with significant differences between males and females. Males showed a lower 1st peak in frequency, with a higher amplitude and a higher bandwidth. Sound emission rate was significantly higher in grouped animals than in single individuals. In the trials involving pre-copulatory females, the total number of sounds was significantly higher compared to the trials with non-copulatory (control) females, and the sounds were not correlated with the agonistic events between males. These results suggest that in this species, sound emissions play a role in intraspecific communication related to sexual attraction (Buscaino et al., 2015).

### *Parathranites hexagonus* (Rathbun, 1906)

**Common name:** Not designated.
**Distribution:** Hawaii Islands only.
**Habitat:** 80 to 460 m depths.
**Description:** Carapace is broader than long, Front is of four blunt Teeth. Medians are a little narrower than the laterals, which are separated from the small inner supraorbital angles by a shallow notch. Of the six anterolateral teeth, first and second are blunt, and are separated by a slight notch. Three following teeth are broad and are acutely pointed. Last one is a long spine. Upturned spines of posterior border of carapace are large. Upper surface of carapace is somewhat irregular with tuberculate elevations, and small tubercles are found scattered in entire area. Surface of chelipeds is granular. Arm is with a sharp spine. Anterior border of carpus bears a long, stout spine. Outer border is granular with three or four

tubercles, one or two of which are sharp. Upper border of palm is with two granular crests, inner one bearing a distal tooth, and outer one with two teeth. Fingers are long, slende and are slightly curved inward. Propodus of fifth leg is without spinules on posterior border. Overall breadth of carapace is 21.5 mm.

**Biology:** Unknown.

### *Parathranites latibrachium* (Rathbun, 1906)

**Common name:** Not designated.
**Distribution:** Nihoa Island
**Habitat:** 40 to 60 m depths
**Description:** Front consists of four blunt, triangular teeth. Medians are narrower than the laterals, which slope down to a very inconspicuous inner supraorbital angle. Of the six anterolateral teeth, first is blunt, next four are sharp, decreasing in size, and sixth is a long straight spine. Upturned spines of posterior border of carapace are small. Upper surface of carapace is with tuberculate elevations. Microscopic tubercles are mainly confined to cardiac and branchial regions. Arm of cheliped region has three spines on anterior border and one minute spine on posterior distal border. Carpus is with a very long spine on inner border and a smaller one on outer border. Palm has two crests on upper border. Inner crest is terminating distally in a prominent spine. A sharp tooth is seen at junction of aim with carpus. Outer surface of palm is granular and pilose, bearing three crests. Propodus of fifth leg is without spinules. Carapace has an overall width of 10.8 mm.

**Biology:** Not studied.

### *Parathranites orientalis* (Miers, 1886)

**Male**

**Ovigerous female**

**Common name:** Not designated.

**Distribution:** South Africa, Seychelles, Madagascar, Mauritius, Chagos Archipelago, India, Andaman Islands, Japan, Yellow Sea, East China Sea, Taiwan, South China Sea, Philippines, Indonesia, Admiralty Islands, Australia, New Caledonia, Loyalty Islands, Chesterfield Islands, Matthew and Hunter Islands, Vanuatu, Wallis and Futuna Islands.

**Habitat:** 80–510 m depths.

**Description:** Carapace is hexagonal and is broader than long. Surface is granular with distinct tubercles on each protogastric, one in the middle of the mesogastric, two on cardiac, and one on each branchial region. Front is with four subequal pointed teeth. There are five antero-lateral teeth. Chelipeds are homochelous. Merus is with 1 spine on anterior and 1–2 on posterior border Carpus has spine on inner corner, and 2–3 small ones on outer face. Palm is with 3 spines. Posterior borders of joints of swimming leg are smooth. Size of crab is 66 mm CW.

**Biology**: Not studied.

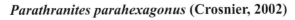

*Parathranites parahexagonus* (Crosnier, 2002)

**Common name:** Not designated.
**Distribution:** Eastern Central Pacific: French Polynesia.
**Habitat:** Benthic; depth range 101–240 m.
**Description:** Width ratio (last teeth anterolateral included)/carapace length of this species is slightly greater than 2. Dorsal carapace is covered with granules which are enough large, dense, relatively uniform size and average. Cardiac region is devoid shell.
**Biology:** Not studied.

*Parathranites tuberosus* **(Crosnier, 2002)**

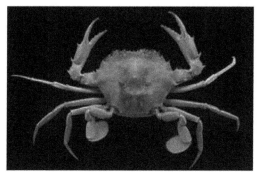

**Common name:** Not designated.
**Distribution:** South Pacific Ocean
**Habitat:** Not known.
**Description:** Carapace of this species is wider than long; Front is divided into four triangular teeth of the same size, with apex which is rounded and equidistant from each other. Anterolateral edges are each cut into five teeth. Of them, first (outer orbital lobe) ends in a short sharp point and has a rather strongly sinuous outer edge. Tooth 2 is acute. Teeth 2 and 4 are of the same size. Tooth 3 is a slightly larger size and fifth tooth which is also acute, is directed laterally and bent forward in its distal part. Dorsal surface of carapace has several large tubers. Upper edge of the orbit has two cracks which are well marked. Lower edge ends the inner side, by a sharp tooth. Maxillipeds have merus whose front is rounded. Fingers are rather tapered and are slightly longer the palm. Pereiopods 2–4 are smooth and slender. Male abdomen has strong cross-hull on the segments 2 and 3. Segments 3–5 are welded. Sixth segment has its side edges very slightly sinuous and convergent. Telson is 1.3 times longer than wide and 1.35 times longer that the

sixth segment First male pléopode is long, hairless, smooth (with the exception of a few tiny spinules) and hail. Size of carapace measures 65 mm CW.
**Biology:** Not studied.

### *Parathranites granosus* (Crosnier, 2002)

**Common name:** Not designated as it is a new species.
**Distribution:** Asia – South-East Asia-Indonesian, Philippine Indonesia, and Philippines.
**Habitat:** Not known.
**Description:** Width ratio (last teeth anterolateral included)/carapace length of this species is between 1.30 and 1.40. Dorsal side of carapace has granules which are tight, well visible and fairly homogenous in size. Region cardiac carapace region is devoid of tuber posteriorly. First pleopods are regularly bent over their male length.
**Biology:** Not known.

### *Parathranites tuberogranosus* (Crosnier, 2002)

**Common name:** Not designated as it is a new species
**Distribution:** Philippines
**Habitat:** At depths of 96–118 m.
**Description:** Width ratio (last teeth anterolateral included)/carapace length of this species is almost always between 1.39 and 1.45. Dorsum of carapace is covered with good embossed granules which are not very dense and with quite different sizes. Cardiac region of carapace is provided with a posterior tubercle. First pleopods of males are regularly curved over their entire length.
**Biology:** Not studied.

### *Parathranites ponens* (Crosnier, 2002)

**Common name:** Not designated as it is a new species.
**Distribution:** Chagos; Mauritius Island.
**Habitat:** At depths of 97–377 m.
**Description:** Width ratio (last teeth anterolateral included)/carapace length of this species is between 1.39 and 1.45. Dorsal side of carapace is covered with dense granules. Cardiac region of the shell is provided with a tuber posteriorly.
**Biology:** Not studied.

## *Parathranites intermedius* (Crosnier, 2002)

**Common name:** Not designated as it is a new species.
**Distribution:** New Caledonia
**Habitat:** 150–285 depths.
**Description:** Width ratio (last teeth anterolateral included)/carapace length of this species is almost always between 1.50 and 1.55. Dorsum of carapace is covered with dense granules of relatively uniform size. Cardiac region of carapace is provided with a posterior tubercle. Dent orbital is entire. First pleopods males are regularly curved over their entire length.
**Biology:** Not known.

### 3.8 SUBFAMILY: CARUPINAE

#### *3.8.1 CHARACTERISTICS OF SUBFAMILY CARUPINAE*

The crabs of this subfamily are with eyestalks of normal length. Carapace is very wide, transversely oval, subelliptical or indistinctly subhexagonal. Supraorbital fissures may be absent and infraorbital margin is variously modified. Front is much narrower than posterior border. Antero-lateral border is convex and toothed. Basal antennal segment is narrow, long and is lying obliquely. Lobe on first maxilliped ("portunid lobe") is usually present. Chelipeds are longer than ambulatory legs. Spines on cheliped segments are reduced in number or absent. Dactyli of last pair of legs are styliform or lanceolate.

## *Carupa ohashii* (Takeda, 1993)

**Common name:** Not designated.
**Distribution:** Indian ocean-Australia, Christmas Island.
**Habitat:** Not known.
**Description:** Carapace is narrowly oval or rather oblong hexagonal in its contour. Front orbital border exceeds posterior border by breadth of an orbit. Dorsal surface of carapace is smooth and moderately convex in both directions. Sparse short hairs are seen along front orbital and anterolateral borders. Frontal border is cut into two by a median, narrow U-shaped sinus. Orbit is deep and subdorsal in position. Anterolateral border is distinctly serrated, with seven sharp teeth including external orbital tooth, which are isolated from each other by six deep notches. Chelipeds are heavy, smooth, and different in size. Merus is protruded beyond anterolateral border of carapace for its most part and is armed with three sharp spines with dark-colored tips on its anterior border. Abdomen is smooth and composed of four pieces. First, segment is disguised under carapace for its most part. Second to fifth segments are completely fused to form one piece, with faint transverse depressions as vestigial indication of segmentation. A short lateral incision is seen between second and third segments. Penultimate and terminal segments are subequal in length and are tapering as a whole. First pleopod is slender; straight for its most part and curved inward just near terminal end.
**Biology:** Unknown.

## *Carupa tenuipes* (Dana, 1852) *(=Carupa laeviuscula)*

**Common name:** Violet-eyed swimming crab.
**Distribution:** Indo-Pacific, from the Red Sea, East Africa and Madagascar to Japan, Australia, French Polynesia, and Hawaii.
**Habitat:** Under stones of a rocky substrate with low vegetation coverage at 0.5 m; coral reef and coral rubble; and intertidal to 80 m.
**Description:** Carapace is transversely elliptical 1.4–1.5 times as broad as long, smooth and convex in both directions. Front is with median and submedian indentations of similar depth. Antero-lateral borders of carapace are with 7 teeth. Of them, first to fifth are rounded and lobular; sixth and seventh are acute; fifth and seventh are smallest and sixth is largest. Chelipeds are slightly heterochelous and smooth. Merus is with 3 strong teeth on anterior border. Its posterior border is smooth. Carpus is with a strong tooth at its internal border. Carapace length measures 12 mm and width, 17 mm.
**Biology:** Not known.

### Catoptrus marigondonensis (Takeda, 2010)

**Common name:** Not designated as it is a new species.
**Distribution:** Philippines
**Habitat:** Submarine caves.
**Description:** Carapace of this species is ovate. Its dorsal surface is smooth and is evenly convex in both directions, shining without granules and setae. Frontal margin is about one third of carapace breadth and is separated into two lobes by a median small notch. Margin of each lobe is very weakly convex forward along its main part, and is shallowly concave along lateral one fourth of frontal margin. Anterolateral margin of carapace is weakly arched, thin, and cut into four lobes by three small notches. Both chelipeds are long, stout, smooth and slightly unequal in shape. Inner margin of merus is armed with three short, strongly procurved spines. Carpus is smooth and shining, without granules and ridges. Palm is smooth and shining. Abdomen is wide, covering whole surface of eighth thoracic sternum. Live color of the paratype is wholly brick red, without special tone. Color of the holotype is similar to that of the paratype, but paler and more or less pinkish.
**Biology:** Unknown as it is a new species.

## *Catoptrus nitidus* (Milne Edwards, 1870)

**Common name:** No common name
**Distribution:** Seychelles, Tanzania, and from Madagascar to Hawaii, including the Arabian Gulf, Japan, China, Australia, New Guinea, Melanesia, Tuamotu, Marshall Islands and Marianas Islands, Timor and Somalia.
**Habitat: Exposed c**oral reef; sub littoral; low Intertidal; dead corals; and depth range, intertidal to 145 m.

**Description:** Carapace is 1.6 times as broad as long and its surface is smooth. Front is bilobate. Median incision is clear but not deep. Antero-lateral borders are with 6 subequal teeth. Antero-external angle of third maxilliped is not strongly produced and is rounded. Chelipeds are slightly heterochelous, long and heavy; they are glabrous without large teeth and granules. Pereiopods 2–5 are slender and fifth pereiopod is not modified for swimming. Last segment of male abdomen is longer than broad. Color of carapace is pale yellow with two brown spots in the anterior central area. Chelipeds are darker than the carapace and the tips of the fingers are brown. Limbs, from the second to the fifth, are violet. Size of the specimen measures 17.0 × 11.8 mm.
**Biology:** Not known.

***Laleonectes nipponensis*** **(Sakai, 1938)** *(=Portunus nipponensis)*

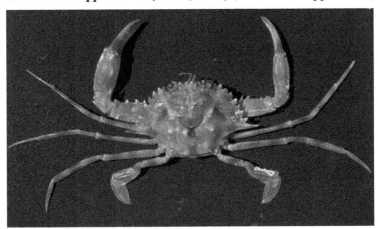

**Common name:** Japanese swimming crab
**Distribution:** Tropical; Hawai and Indo-Pacific.
**Habitat:** Hard bottom (rock and rubbles); and depth range 15–250 m.
**Description:** Not available. Male and female measure 10.5 × 21.2 mm and 31 × 47 mm, respectively.
**Biology:** Not known.
**Fisheries:** Not known.

## *Laleonectes vocans* (Milne Edwards, 1878) *(= Portunus vocans)*

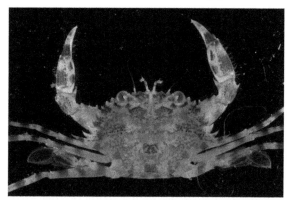

**Common name:** Not known.

**Distribution:** Occidental Atlantic: Gulf of Mexico, The West Indies and Brazil (Espírito Santo and Rio de Janeiro); Central Atlantic: Ascension Island; Oriental Atlantic: Madeira, Cape Verde and Annobon Islands.

**Habitat:** Depth range, 40–310 m.

**Description:** Carapace is broad and its upper surface is tuberculate; four tubercles are seen in a curved transverse line across the gastric area. Front is with 4 lobes; and median ones are more tooth-like and more advanced than the lateral ones. A sharp tooth at the posterolateral angle of the carapace is curved upward and forward. Merus of cheliped is with inner border which is serrated and spinous; a small sharp spine is seen at the distal end of the outer border; carpus is with a strong spine at the inner angle; outer and upper borders are strongly serrated; palm is with a tooth in front of the articulation with the carpus and one on the upper border near the distal end; and upper border of palm is serrated. Fingers are nearly as long as palm. Walking legs are long and slender; merus of last leg is with a tooth at the lower distal end. Telson of male is triangular and its segments 3 to 5 are fused. When alive the Hawaiian form is conspicuous by color bands' of red which encircle the chelipeds and walking legs. Except in the last leg the dactylus and carpus are each encircled by one red band and the propodus and· merus by two such bands. The merus of the last leg has but one color band. Size of the crab is 22 mm long and 43 mm broad.

**Biology:** Not known.

**Fisheries:** Not known.

### *Libystes edwardsi* (Alcock, 1900)
Image not available.
**Common name:** Not designated.
**Distribution:** Persian Gulf; Strait of Hormuz; India; Andaman Islands; Japan; China; Vietnam; Gulf of Thailand; South China Sea and Indonesia.
**Habitat:** Marine; 29–60 m depths.
**Description:** Carapace is much broader than long and is generally smooth. It is finely pitted under magnification, and granulate near antero-lateral borders. Front is straight and is faintly notched medially. Antero-lateral borders of carapace are with 5 or 6 granular denticles followed by a sharp procurved spine. Chelipeds are heterochelous, smooth and unarmed. Fingers are slender and hooked at ambulatory legs are slender and fifth pereiopods are natatorial, with a distinct swimming paddle.
**Biology:** Not known.

### *Libystes lepidus* (Miyake & Takeda, 1970)
Image not available.
**Common name:** Not designated.
**Distribution:** Japan; Indonesia.
**Habitat:** Not known.
**Description:** Carapace is transversely elliptical; its dorsal surface is smooth and convex near the lateral borders. Front is about one third of carapace breadth, with a narrow but distinct edge. Antero-lateral borders are with 5 obscure low teeth. Chelipeds are subequal and their anterior border of merus is with minute granules. Carpus and palm are unarmed, smooth and shining. Palm is not swollen, but it is as stout as the merus. Borders of ambulatory legs are with strong hair brushes. Fifth pereiopods are gressorial.
**Biology:** Not known.

### *Libystes nitidus* (Milne Edwards, 1867)

**Common name**: Not designated.
**Distribution:** Tropical Indo-West-Pacific, from Madagascar to Hawaii, including Japan and Australia; China, Samoa, Fiji, Indonesia, Sri Lanka, Mauritius, Amirante Islands and East African coast.
**Habitat:** Intertidal zone, under dead corals; sandy and stony bottoms or coral reefs, about 10 m deep; intertidal (in dead coral) to 145 m.
**Description:** Carapace is elliptical. Antero-lateral border is entire, without teeth. Merus of third maxilliped is not produced at antero-external angle. Coloration is pink, with brown fingertips. Size of specimen is 4.3 × 2.6 mm.
**Biology:** Unknown.

### *Libystes paucidentatus* (Stephenson & Campbell, 1960)
Image not available.
**Common name:** Not designated.
**Distribution:** Australia
**Habitat:** Not reported
**Description:** Carapace front is narrow and consisting of two slightly rounded lobes. Inner orbital lobes are indistinguishable. There are 5 anterolateral teeth which are all blunt. Among them first has a flat protuberance. Fourth is normally the smallest, and fifth is a projecting procurved spine. Carapace is almost twice as broad as long (from 1·8 to 1·9 times); it is microscopically punctate over most of its surface and granular near anterolateral borders. With a blunt eminence near each posterior angle and a slight concavity of the lateral epibranchial region to give the carapace a quadrilateral appearance. Color of the specimen is dull brown.
**Biology:** Unknown.

### *Libystes villosus* (Rathbun, 1924)

**Common name:** Not designated.
**Distribution:** Japan, South China Sea, Hawaiian Islands, Samoa.
**Habitat:** Not known.
**Description:** Carapace is clearly broader than long and is strongly convex longitudinally. Its surface is smooth and a ridge of fine granules is extending forward from posterolateral angle. Front is very shallowly bilobate and median indentation is hardly visible. Antero-lateral borders are entire, without traces of teeth or lobes. Chelipeds are clearly heterochelous and smooth. Anterior border of merus is minutely granulate. Carpus is unarmed at all. Palm is much stouter than merus. Fingers are a little shorter than palm. Borders of ambulatory legs are fringed with shaggy hair. Fifth pereiopods are gressorial. Color of specimen is dull brown.
**Biology:** Not known.

### *Richerellus moosai* (Crosnier, 2003)

**Common name:** Not designated.
**Distribution:** New Caledonia
**Habitat:** Lagoons at depths of 15–20 and 50 m on muddy or coarse sandy bottoms with many blocks and occasionally gorgonians or coral.
**Description:** Carapace is wider than long. It is slightly convex longitudinally and transversally; smooth to naked eye and is without pronounced regions or transverse rows of granules. Branchial and cardiac regions are slightly swollen. Front is about one-third carapace width and is slightly sinuous, with a small notch medially. Anterolateral margins of the carapace are divided into 4 teeth. Posterior margin of carapace is about three-quarters carapace width. Chelipeds are elongate and unequal. Carpus is with 2 spiny teeth; one is strongest, at antero-internal extremity of upper surface and

other is at antero-external angle of lower surface. Chelae are with no ornamentation on palm except for a very faint carina on external border of upper surface. Fingers are about four-tenths length of large chela and half length of small chela. Dactylus (mobile finger) is with a very faint, partial longitudinal groove. Cutting margins of fingers are with contiguous triangular teeth of variable sizes. Second to fourth pereiopods are elongate (slightly shorter than chelipeds), smooth, glabrous and slender. Male abdomen is triangular with a strong transversal carina on first and second segments. Greatest part of first segment is hidden. Third to fifth segments are fused and suture between them is almost invisible. Telson is triangular, as long as wide at base with lateral borders straight and its distal part is rounded. Body is orange brown with anterior half of carapace slightly darker than posterior one, and with a straight narrow clear median line. On pereiopods there are large transversal bands separated by short clearer ones. Fingers of chelae are distally whitish. Size of specimen is 11.5 × 16.8 mm.

**Biology:** Unknown.

## 3.9 SUBFAMILY CAPHYRINAE

### 3.9.1 CHARACTERISTICS OF THE SUBFAMILY CAPHYRINAE

Carapace is relatively narrow and subcircular to subovate. Dorsal surface is either smooth or traversed on either side by ridges. Antero-lateral margins are cut into four or five teeth. Chelipeds and legs are short. Last pair of legs is either claw-shaped or lanceolate.

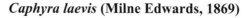

*Caphyra laevis* **(Milne Edwards, 1869)**

**Common name:** Xenia soft coral crab
**Distribution:** Tropical West Pacific, eastern and southwestern Indian Ocean.
**Habitat:** Bare reef, sand with coral-rubble and algae; associated with *soft corals*; and depth range (m): 6–16.
**Description:** Male possesses four anterolateral teeth on each side, of which the first is much the largest, the second and fourth are subequal, and the third slightly the smallest. Meropodite of the chelipeds are denticulated. It has only two slightly prominent lines which are directed inward from the lateral spine and do not meet in the middle of the cephalothorax. This species grows up to 2 cm carapace width.
**Biology:** These crabs use camouflage effectively. This species which is living in association with xeniid alcyonarians, in particular *Heteroxenia* species is white like its host.

**Beautifully-camouflaged *Caphyra laevis* crab**

*Caphyra fulva* **(Stephenson & Campbell, 1960)**
Image not available.
**Common name:** Not designated.
**Distribution:** Red Sea (new record), Madagascar, Indonesia (Kei Islands, Sulu Archipelago), Philippines, and East Australia.
**Habitat:** Coral colonies or in baited traps in coral habitats.
**Description:** Front is four-lobed, the median notch is distinct, but the lobes on each side are almost confluent. Inner orbital lobes are small,

clearly separated from the frontal lobes. Anterolateral teeth are typically 4, sharp, with points directed anteriorly, decreasing in size from front to rear. Carapace is microscopically punctuate with microscopic hairs occasionally. It is slightly broader than long with the broadest point somewhat behind the last anterolateral teeth. Chelipeds are microscopically granular. Anterior border of arm bears 3–5 spines increasing in size distally. Under surface in most specimens bears 5–6 sharp spines. Hand is rounded and without carinae except for a single one on the inner side of the upper surface. Legs are elongate and slender with hairy dactyls. Background color is pale cream with well-developed longitudinal reticulated areas of brown typically well developed. Size of males ranges from 2.44 × 2.62 to 3.54 × 3.72 mm and females 2.27 × 2.83 mm.

**Biology:** This species lives in association with the alcyonarian hosts, *Xenia umbellifera* and *Heteroxenia fiuscescens*.

### *Caphyra minabensis* (Sakai, 1983)

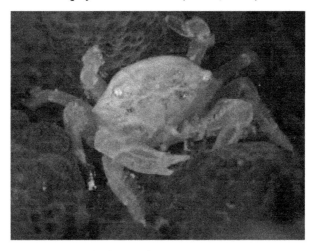

**Common name** Not designated.
**Distribution:** Japan
**Habitat**: Reef slope and green coral reef area.
**Description:** Carapace is smooth and glossy. Front is typically bilobed and median notch is usually distinct. Each of the principal lobes is subdivided into two by a shallow indentation. There are two very sharp and forwardly directed antero-lateral teeth. Behind the second tooth, carapace

is divergent and broad. Cheliped palm is with two longitudinal carinae on upper surface. Fifth pair of legs are curved back dorsally over the carapace. Color of body is white and transparent.

**Biology:** This species is symbiotic with soft corals.

### *Caphyra rotundifrons* (Milne Edwards, 1869)

**Common name:** Turtle-weed crab

**Distribution:** New Caledonia, Madagascar, Seychelles, Mauritius, Japan, Taiwan, China, Mariana Islands, Australia, Fiji and Samoa.

**Habitat:** Associated with turtle weed (filamentous green algae), Chlorodesmis fastigiata which is found growing in rocky pools or on reef flats; depth range, 5–20 m.

**Description:** Carapace is not circular in outline and is not markedly convex. It is smooth to naked eye and microscopically granular upon magnification. Epibranchial ridges are distinct and mesogastrics are obscure. Front is typically bilobed and median notch is usually distinct. There are 4 very sharp and forwardly directed antero-lateral teeth which are decreasing in size fore to aft. Fourth tooth is sometimes rudimentary. Chelipeds are with long hairs on upper and outer faces. Merus is with 3 teeth and 2 tubercles on anterior border. Fifth pair of legs are curved back dorsally over the carapace. Dactylus is slightly shorter than propodus. These crabs have hooks on the ends of their legs, which enable them to cling to the weed, and prevent them from being washed away as the tide sweeps in. Size of the crab is 2 cm.

**Biology:** This species is tightly associated with the clumps of the dark green seaweed called turtle weed and feeds solely on this alga. These crabs use camouflage effectively and has the same color (green) as its host. Further, a defensive metabolite (cytotoxic terpene) of this alga viz. chlorodesmin which is deterrent to several herbivorous fishes acts as a feeding stimulant to this species of crab.

### *Caphyra yookadai* (Sakai, 1933)

**Common name:** Not designated.
**Distribution:** Japan and Australia.
**Habitat:** Shallow waters.
**Description:** Carapace is convex with central area which is smooth and shining. Margins are microscopically granular. Epibranchial ridges are very distinct. Mesogastrics are faint in female and distinct in male. Front is with 4 lobes. Median notch is distinct and the pair of lobes on either side are almost confluent. There are 4 blunt antero-lateral lobes which are sometimes reduced to 3 or even 2. Chelipeds are short and granular. Merus is unidentate on anterior border and its lower border is smooth. Fifth pair of legs are curved back dorsally over the carapace. Dactylus is much shorter than propodus.
**Biology:** Not known.

### *Coelocarcinus foliatus* (Edmondson, 1930)

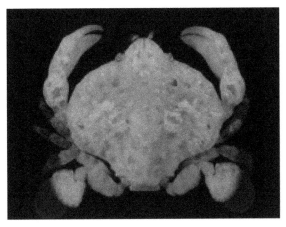

**Common name:** Not designated.
**Distribution:** Madagascar, Japan and Guam.
**Habitat:** Associated with dead coral blocks in shallow water.
**Description:** Carapace is subcircular, about 9 mm in both length and breadth. It is produced into a blunt lobe in front. Anterolateral border consists of four blunt, confluent lobes. Both propodus and dactylus of fifth leg are circular segments.
**Biology:** Not known.

### *Lissocarcinus arkati* (Kemp, 1923)

**Common name:** Red urchin swimming crab

**Distribution:** India, Madagascar, Japan, Hong Kong, South China Sea, Philippines, Indonesia and Australia.

**Habitat:** Coastal; 10–65 m depth

**Description:** Carapace is only slightly broader than long and is not very convex. Surface has numerous setose transverse ridges. Front is little protruding and median notch is not very clear. Antero-lateral borders are with 5 teeth. Cheliped merus is with 2 teeth on anterior border and its posterior border is smooth. Carpus has strong internal and 2 small external teeth. Palm is with transverse setose ridges on upper face. Its external face is densely granulate and lower face is smooth. Colortion of body is entirely red. Size of males and females ranges from 5.2 to 13.0 mm and from 6.5 to 11.9 mm, respectively.

**Biology:** Not known.

### *Lissocarcinus laevis* (Miers, 1886)

Male

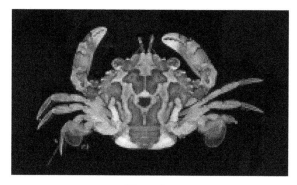

Female

**Common name:** Harlequin swimming crab

**Distribution:** Wide range in the Indo-Pacific: from the Red Sea to Hawaii and Marquesas.

**Habitat:** Sandy bottom with stones and corals; 30–85 m depth

**Description:** Carapace is distinctly broader than long and is very convex. Surface is smooth and glabrous. Front is bilobed and median notch is not clear. Both lobes are sinuous. Antero-lateral borders are with 5 teeth. First is the smallest, $2^{nd}$–$4^{th}$ are subequal and rounded and 5th is small and quite acute. Cheliped merus is without teeth; Carpus is with a strong internal tooth. Palm is smooth except inner border of upper face. Carapace is brown with paired large white spots on either antero and posterolateral areas. Broad white band is seen in middle of anterior part of carapace. White spots are seen between anterolateral teeth. Chelipeds are brown with white transversal bands. Walking legs are proximally brown. Size of males and females is 8.4 mm and 10.3 mm, respectively.

**Biology:** This species lives in association with sea anemones. It hangs onto the edge of a tube anemone. This symbiosis is common throughout the Indo-Pacific region.

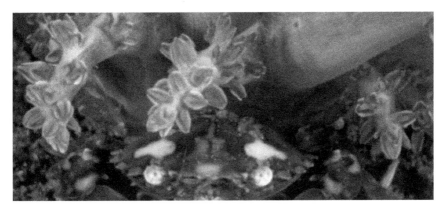

**Harlequin swimming crab with sea anemone**

## *Lissocarcinus orbicularis* (Dana, 1852)

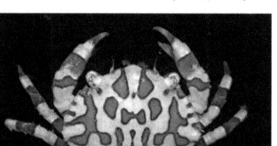

**Common name:** Sea cucumber crab; red-spotted white crab.
**Distribution:** Indo-Pacific.
**Habitat:** Shallow waters; from muddy sand to sand with occasional stones and corals.
**Description:** Carapace is only slightly broader than long and is very convex. Surface is smooth and glabrous, except for epibranchial crest. Front is little protruding and median notch is reduced. Antero-lateral borders are with 5 teeth. Among them, 2nd to 5$^{th}$ are with rounded tips. Cheliped merus is without teeth. Carpus is with an internal tooth and a row of granules on external face. Palm has 2 crests on upper border and both are ending in a rounded tooth. Carapace is light-yellow with regularly placed brown spots. 2 pairs of large spots are on branchial areas, cardiac and 1 postcardiac spots are seen close to posterior edge. Legs have yellow with brown bands. Broad bands are seen on proximal part of chela, meri, and propodi of walking legs. Broad spots are also seen on movable fingers of chelipeds and dactyli of walking legs. While size of males ranges from 3.0 to 16.1 mm, female has a size of 7.6 mm.
**Biology:** These crabs are commensals and are found on the integument and in the buccal/cloacal cavity of several species of holothuroids such as *Halodeima atra, Holothuria pervicax, Thelenota ananas* and *Holothuria argus*. This species is characterized by a weak

Profile of Portunid Crabs (Family: Portunidae)

sexual dimorphism (females are bigger than males) and the presence of pereiopods which are morphologically adapted to fixation on the host integument. Gravid females are observed at each month indicating that the crab reproduces all the year (Caulier et al., 2012).

## KEYWORDS

- biology
- common name
- distribution
- fisheries
- habitat
- species description
- species profile

# CHAPTER 4

# AQUACULTURE OF PORTUNID CRABS

## CONTENTS

4.1 Seed Production and Larval Rearing in Portunid Crabs .............. 273
4.2 Farming of Portunid Crabs: An Introduction .............................. 279
Keywords ......................................................................................... 306

Portunid crabs have great potential in aquaculture. Species of Portunidae targeted for aquaculture include *Scylla serrata, Portunus pelagicus, Portunus trituberculatus, Portunus sanguinolentus* and *Charybdis feriata* However, among these species, the most commonly cultured species is the giant mud crab, *Scylla serrata* and this is largely due to its preference for estuarine habitats, less aggressive behavior and higher value. *Scylla serrata* is successfully cultivated in many Southeast Asian countries and Australia.

## 4.1 SEED PRODUCTION AND LARVAL REARING IN PORTUNID CRABS

### 4.1.1 SEED PRODUCTION AND LARVAL REARING IN SCYLLA SERRATA

**Breeding:** Pond-grown female crabs with matured ovaries which are dark orange in color are held in a concrete tank with sand substrate and PVC pipes (20 cm diameter × 30 cm length) as shelters. They are fed on mussels, squid and fish at 10–15% of body weight daily and a SEAFDEC-formulated diet 2 at 2%. Live marine annelids are offered to these crabs once every 1–2 weeks as a supplement. A water depth of about 30 cm is maintained in the aforesaid tanks. The seawater used for the crab breeders and larvae is pre-treated in a reservoir with 10–20 ppm calcium hypochlorite and then neutralized with sodium thiosulfate after 12–24 h. The water in the tank is changed daily before feeding. Eggs released by the female are found attached to the pleopod hairs of the abdominal flap. Berried (ovigerous) females are transferred individually to 300-liter or 500-liter tank with aerated sea water at 32 ppt. In each spawning, a crab of 350–525 g size produces 0.8–4 million zoeae. Hatching occurs 7–14 days after spawning at temperatures of 26.5–31°C.

**Larval rearing:** Zoeae are stocked at a density of 50 individuals per liter in circular concrete tanks (4 m diameter × 1 m height) and are fed with rotifers at a density of 10–15 rotifers/ml. Nauplii of brine shrimps are also given at 0.5–3/ml to zoea 3 and larger larvae. The zoeae are reared at a salinity of 32–34 ppt and water temperature of 26–30.5°C, and a natural photoperiod of 11–13 hours light and 11–13 hours dark. The rearing water is replaced at a daily rate of 30% starting on day 3 and increasing to 80% as larvae grow bigger.

**Nursery (Tanks):** Megalopae are nursed in concrete tanks or in net cages set in brackishwater ponds. The stocking density of 3–5 days old megalopa may be 1000–2000/ton of water. Black nets are placed at the bottom of these tanks as substrates and some are suspended in the water column. Food of megalopa consists of newly hatched and adult brine shrimps. As soon as the megalopae molt to crab stage, they are fed minced trash fish, or mussel twice daily *ad libitum*. About 30–50% of the volume of the rearing water (26–30 ppt) is replaced daily during the first 5 days and every two days thereafter. Ponds provide a wider surface area for the dispersion of megalopa.

**Nursery (Cages):** Net cages (mesh size 1 mm; bottom surface area 20 m²) are set in ponds for the rearing megalopa. Bamboo poles support the cages and the bottom of the net is buried 3–5 cm into the pond soil. Megalopae to be transferred to net cages are packed in plastic bags at a density of 200–300/liter. Megalopae are stocked at 30/m$^2$ and are fed with adult brine shrimp on the first day in net cages. Food is then changed to minced trash fish and mussel placed in feeding trays. Water depth is maintained at 60–80 cm. About 30% of the water is replaced 3–4 times a month. Strategies to reduce cannibalism include size-grading, trimming of claws, removal of chelipeds 5 and 6, and provision of sufficient shelters. However, chelipeds are not removed from crabs larger than 2.5 cm in carapace width because growth may be affected. Trimming the claws and removal of chelipeds are tedious and are practical only for a small population of crabs.

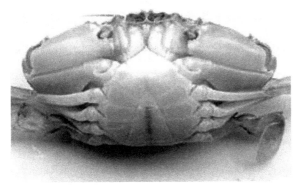

**Mud crab juveniles with claws trimmed**

**Mud crab with removed chelipeds**

276  Biology and Culture of Portunid Crabs of World Seas

The survival from zoea 1 to 3–4 days old megalopa has been found to be 3–7%. The survival from megalopa to juvenile (1–3 g body weight) after 30 days in hatchery tanks or pond cages was found to be 30–50% (Quinitio et al., 2002).

## 4.1.2 SEED PRODUCTION AND LARVAL REARING IN PORTUNUS PELAGICUS

**Collection of broodstock**: *Portunus pelagicus* is a continuous breeder, so the berried crabs are available throughout the year. Healthy ovigerous females with characteristic yellow/orange colored eggs are collected from sea and kept in 1.5 liters capacity fiberglass tanks at a salinity of 32–1 ppt, pH 8.2–0.1 and temperature 28–1°C with continuous aeration. Only filtered seawater is used for the entire rearing operation and 50% of water exchange must be made daily. Usually, the berried crabs do not feed and hence feeding is not required.

**Broodstock development in captivity:** Brood stock crabs can be raised in captivity using either juvenile crabs collected from the wild or using reared crablets produced. Five to ten ton capacity round FRP tanks may be used for the brood stock development. An in-situ biological filter bed of 5–10 cm height was set. Daily 15–20% water exchange is made and once in a week 100% exchange is done. Fecal matter and unused feed are siphoned out in the morning hours before the water exchange. Salinity, temperature, pH, dissolved oxygen, total ammonia and nitrate are monitored and kept optimal regularly. Tank water temperature is maintained between 28–30°C; dissolved oxygen between 5–7 mg/l. and pH between 8.0–8.2. Daily the animals are fed *ad libitum* with clam meat/ shrimp/ squid meat in the morning and evening hours. In such a maturation system with good management practices, crabs will attain maturity within few weeks/months depending on the initial size of the crab. The male crab attains maturity by its 12th molt and female crab by 14th molt. The average size (CW) of the mature male and female crab has been reported to be 82.3 mm and 120.4 mm, respectively. The incubation period ranges between 8 and 10 days mainly depending on the size of the berry and rearing water temperature. The zoeae produced from the captive broodstock are healthy and active like from the wild-berried mothers.

Aquaculture of Portunid Crabs

**Hatching of zoeae:** Female crab with deep grey egg mass is transferred into the hatching tank with known volume of seawater (around 500 liters) during evening hours. Only one berried mother is introduced in a single hatching tank. The total weight and carapace width of the introduce crab are measured. The tank must be cleaned and water exchange should be given until hatching. Anticipating the hatching during the following night mixed phytoplankton dominated with *Chaetoceros* spp. (10,000 cell/ml) and rotifers (5 nos./ml) are added in the hatching tank. Hatching takes place during early morning hours. After full hatching, mother crab is removed from the tank and weight of the crab is taken.

**Larval rearing:** 1–5 ton capacity round/oval fiber glass tanks are generally used for rearing larvae. The newly hatched active zoeae are stocked in the larval rearing tanks at a stocking density of 50,000 no/t. During the entire larval rearing period, every morning 30–40% of the culture tank water is exchanged. For all the zoeal stages vigorous aeration is given, while for megalopa stage, it is marginally reduced.

Desired Range of Various Environmental Parameters in Larval Rearing Tanks

| Parameter | Range |
| --- | --- |
| Salinity | 30–33 ppt |
| Temperature | 27–31°C |
| pH | 8.0–8.5 |
| Dissolved oxygen | 4–8 ml/l |
| Total ammonia | <0.1 ppm |
| Nitrite | <0.05 ppm |

**Feeding:** A combination of algae + rotifers can be given for the first zoeal stage. Among the different phytoplankton feeds used the diatom, *Chaetoceros* has been found to be the best for the first zoea. For the rest of the zoeal stages, a combination of rotifers and brine shrimps; and for megalopa, cladocerans/brine shrimps and prawn-egg custard may give the best results. From Zoea II onwards, *Chaetoceros* is not supplied to the larvae.

**Nursery phase:** The baby crabs are stocked either in rectangular, open outdoor tanks (provided with sand bed and additional substrata) or in earthen ponds, at the rate of 400–500/m². The depth of the water column must be maintained at 80–100 cm. For the first week, feeding rate and schedule are followed as in the case of first crab instar. In the second week of nursery phase, cooked clam meat/small shrimp can be given at 20% of their body weight/day, in addition to the egg custard. 20% water exchange is given on every alternative days by removing water from the bottom layers. The baby crabs attain an average size of 10 mm carapace width at the end of the nursery phase and are ready to stock in a crab farm (Jose, http://eprints.cmfri.org.in/9732/1/Josileen.pdf).

### 4.1.3 SEED PRODUCTION AND LARVAL REARING IN CALLINECTES SAPIDUS

**Broodstock:** Mature females (160–250 g, 15–18 cm carapace width) are held in 2-m³ round, recirculating systems-flat-bottom tanks (22-28°C; 30 ppt; constant photoperiod of 14 h light: 10 h dark). During operation, 7% of the water volume is replaced daily with clean, artificially produced saltwater. The tank bottoms contain 8–12 round plates (30 cm diameter; 4 cm height), which are filled with sand and are easily removable. The plates are provided to allow the females to release their eggs and collect them to an egg mass that is then attached to their abdomen. The crabs are fed once a day with frozen squid, or with shrimp or fish pellets, and are also observed daily for the appearance of egg mass. Approximately, 5 days prior to her anticipated hatching time (indicated by the color of the egg mass), the ovigerous female is transferred to an 800-liter hatching tank. The outflow of this tank is equipped with a 125-μm mesh filter for retaining the larvae. One day prior to hatching, rotifers are introduced into the hatching tank at a density of 40–50 individuals/ml.

**Larval rearing:** Larvae are attracted to the surface of the hatching tank by phototaxis and are removed in 4-l beakers. Hatchery tanks (22-28°C 30 ppt) are stocked at densities of 38–138 individuals/liter. Tank outflows are protected with a 125-μm mesh filter. A flow-through open system is allowed for 50% water exchange/tank/day. Larvae are fed with rotifers.

**Rearing experiments and mass production of blue crab *Callinectes sapidus* juveniles:** Four culture cycles were conducted with *Callinectes sapidus* seed obtained from photoperiod-manipulated

Aquaculture of Portunid Crabs

broodstock. A feeding protocol for early life stages, as well as for juvenile crabs, was established based on microalgae, rotifers, *Artemia* nauplii, and off-the-shelf diets. The captive rearing process was divided into two phases: (1) zoea 1–zoea 8/megalopa and (2) zoea 8/megalopa to 15–30 mm crab juveniles. Each of the phases was accomplished in both open system and recirculated system. Different larval (zoea 1) stocking densities were tested. Within the examined range (40–110 individuals/l), no negative correlation was found for high density. Maximum survival to the zoea 8/megalopa stage was 74% at 95 larvae/l and the average was 30%. Cannibalism was found to be the main cause for blue crab mortality during the second rearing phase. In an effort to reduce the prevalence of cannibalism, experiments providing different shelter substrates, performing size grading, and decreasing stocking density were conducted. During this phase (zoea 8/megalopa until C2–C4), 57% survival was obtained in mass production tanks, using a shelter substrate of snow fence nets and stocking tanks with 2.5 individuals/l. Increasing stocking density by 16-fold (39 individuals/l) and doubling the shelter density resulted in production of about 3000 juvenile crabs/m³, although the survival rate dropped to 7.5% (or 7.6-fold). Cumulatively in the four culturing cycles, 40,000 juveniles were produced, of which 25,000 were individually tagged and experimentally released to the wild (Zmora et al., 2005).

## 4.2 FARMING OF PORTUNID CRABS: AN INTRODUCTION

The grow-out of crabs is undertaken in various systems. The two major system types are: (a) open; which includes ponds and mangrove enclosures (pens) where crabs are maintained at varying densities, and (b) closed; where crabs are held in individual containers, for example, soft shell crab, or restrained in some way, for example, fattening enclosures. Culture techniques have been defined as extensive, intensive and cage culture. In extensive system seed crabs are stocked at 1 crab/5–10 m² with wild seed, 1 crab/2–5 m² for smaller hatchery seed, whilst an intensive stocking rate is considered to be 1–1.5 crabs/ m². Cage culture is a fattening exercise where large crabs (200–400 g) are held in high densities of 35 crabs/ m$^2$ and fed until marketable. Pens in mangrove forests and extensive shallow mangrove silviculture ponds that do not require supplemental feeding. Rice fields are flooded with brackishwater that contains crab and/or shrimp seed, which become an important technique for farmers to supplement their income. More intensive systems that stock ponds at higher densities using purchased crablets commonly use trash fish or shellfish as feed and can produce 1–2 tonnes/ha/crop. In open production systems, especially with mud crabs, cannibalistic behavior of crabs is a major impediment to their high density production. Cannibalism can be minimized by using relatively low stocking densities and by utilizing some form of enclosures or shelter to provide refuge. Antagonistic behavior between crabs can also result in loss of limbs by mud crabs. To minimize the impact of cannibalism on survival, a useful management strategy is to routinely undertake partial harvests of crabs of a commercial size in growout systems, leaving sub-harvest size crabs to grow to harvest size with a reduced incidence of predation, in more space and with less competition for feed (Shelley, 2008).

## 4.2.1  HARD-SHELL CRAB FARMING

In this type of farming, juvenile crabs or crablets collected from wild or produced in laboratory are used as seedstock in farms such as ponds and pens. Culture practices of this type include: (i) monospecies and monosex culture, and (ii) polyculture of crabs with fish, shrimps and seaweeds.

### 4.2.1.1  Monospecies and Monosex Culture of Portunid Crabs

#### 4.2.1.1.1  Culture of Scylla spp.

**Construction of ponds:** The basic design of most earthen ponds used to culture mud crabs is the same as that used to culture marine shrimp. The width of a bank's foot should be 3–4 m depending on the height of the bank. The top of the bank should be 1–2 m wide and at least 0.5 m higher than the highest tide. The foot of the banks is often made of

Aquaculture of Portunid Crabs

bamboo nets for stability. The banks are firmly sealed by solid soil or clay to avoid leakage or slippage. Depending on local features, trees or blocks of woods are planted to prevent the destructive effects of waves, which can cause erosion and collapse. Around the inside of the pond, a canal with a width of about 3–5 m and depth of about 0.5–0.7 m is dug. The excavated soil is used to build the bank. Brushes are often added to the canal to serve as shelters for crabs. There should be 1 or 2 outlets, depending on the area of the pond. One outlet is placed at the lowest point of the canal in order to drain away completely all water in the pond when necessary for harvesting, reforming or cleaning. The outlet's diameter depends on the pond area but is often 0.8–1.5 m. At the inlet, 2–3 valves are installed to control the flow. These openings have mesh to prevent the crabs from escaping when the pond is either emptied or filled. The inlet and outlet are normally made of concrete, prefabricated concrete, prefabricated-concrete pipes, bricks or wood. In large extensive ponds, the parts of the bank near the outlet need to be fenced. These fences extend 20–50 m from the edges of the outlet. Large ponds often have the same structure as the natural environment (with plants, mounds in the middle and space). In small ponds (1–3 ha), creation of mounds, or planting of trees may be useful for shade. Use of bamboo or other kinds of fences around the bank may prevent the escape of crabs from the pond. These fences are 0.7 m or more in height, deeply driven into the inside edge of the pond bank. Instead of building ponds, bamboo fences may also be installed enclosing large areas to raise crabs, shrimps and fish. The height of the fence must be 0.5–1 m higher than the highest tide. On top of the fence, a net can be placed to prevent the escape of crabs. It is convenient raising crabs in fenced water areas because the living conditions are much the same as natural conditions.

**Stock control netting:** Unlike marine shrimp, mud crabs can leave water and spend considerable periods of time on land. As a result, if a barrier of some type did not surround a mud crab aquaculture pond, stock would be able to walk out of the pond. To counter this mud crab behavior, netting typically surrounds mud crab culture ponds. Netting height may vary from 20 to 50 cm in height above the top of the pond. The netting is typically supported by posts and may be topped with plastic. The plastic topping is added as mud crabs are good climbers and they can climb up netting, but are unable to climb up clear plastic sheeting. Netting around the top of the pond is easier to maintain dry raised feeding platforms or mounds.

**Mud crab ponds with net enclosures**

**Size of ponds for monosex/species culture**: Ponds for this type of culture are usually 500–5000 m² in area and the biggest ponds are limited to 2 ha. The pond shape depends on the topography. Generally, ponds are rectangular with a width equal to 40% of the length, with inlet and outlet on opposite ends.

**Preparation of ponds:** Prior to stocking crablets, ponds should be dried out for several weeks and any repairs undertaken. This assists in ensuring any unwanted species are removed from the pond that may be competitors for feed fed to crabs or predate on crablets. Turning over the soil in the bottom of the pond, or tilling, can help in preparing the pond for the next crop. This helps in the breakdown of organic residues and release of nutrients. Tilling can be combined with the addition of lime to pond floors. Liming can be used to improve the pH of pond sediments, accelerate decomposition of organic matter and improve fertilizer response. brackishwater or saltwater being used to fill a mud crab pond should be filtered through a small mesh "sock" or "bag" of approximately 120 μm mesh to reduce the risk of other species entering the pond. Ponds should be filled with brackish to fully saline water (10–35 ppt).

**Seed supply (procurement, packing and transport of seed crabs (crablets) and stoking**: Seed crabs for culture are normally wild-caught and are in the following sizes: 10–15 g (small); 25–40 g (medium) and

Aquaculture of Portunid Crabs

70–100 g (large). Collected seed crabs get their pincers firmly tied and are transported quickly to the culture site. Because smaller crabs molt frequently they are best transported in oxygenated plastic bags containing 2 liters of cool seawater at a density of 1 000 crablets for 0.4–0.6 cm CW, 500–750 crablets for 0.7–1.0 cm, and 250–500 crablets for 1.1–1.5 cm for a maximum of 8 h. Crablets of 1.5–2.0 cm CW may be transported in a box measuring 45L × 35W × 10H cm with wet cloth or sand at a density of 500–700 for maximum of 6 h (http://www.fao.org/fishery/culturedspecies/Scylla_serrata/en). If packed in water, crablets may be cooled to 22–24°C to prevent molting on the way to the farm and to lower oxygen consumption. At the farm, crablets should be put in basins (or similar containers) with a small amount of water from the pond for which they are intended to acclimatize. Once acclimatized to the temperature of the water and its salinity, they can be released into the pond. Normally, seed crabs of the same size are used for each pond. While the small seed crabs are generally grown for 6–7 months, the medium and large sized seed crabs are grown for 4–5 months and 3–4 months respectively. The culture density for seed crabs is 3–5 crabs/m$^2$ for small, 2–4 crabs/m$^2$ for medium and 1–2 crabs/m$^2$ for large crabs. Each pond is filled with a sufficient number of crabs within one or two days. For larger ponds, distributing them from several different points around the pond to assist in distributing them evenly around the pond is recommended

**Stocking for monospecies culture:** Mud crablets for monospecies culture are stocked at relatively low densities (0.5–1.5 crabs/m$^2$) which depend both the size of individual mud crabs and their tendency to cannibalistic behavior. The monoculture of mud crabs can be undertaken in rotation with shrimp culture. It is always better to stock monoculture ponds with crablets of as near the same size as possible, to minimize the risk of larger crabs feeding on smaller ones.

**Stocking for monosex monoculture:** It has been reported that male crabs attain a significantly higher final weight than female crabs. Monoculture ponds may be stocked with advanced crablets. This type of practice may minimize aggressive behavior between crabs associated with sexually maturity. Another potential advantage of this culture is that survival of monosex mud crabs is significantly higher than among mixed sex crabs.

**Monospecies culture of *Scylla serrata* with seaweed:** Male monoculture of *Scylla serrata* with seaweed *Gracilaria* has been found to be economically viable (Triño et al., 1999). Juvenile male crabs stocked at 0.5/m² in 150 m² enclosures in earthen ponds with *Gracilaria* as shelter and fed a mixed diet of 75% fresh brown mussel flesh showed highest survival and lowest production costs as male crabs attained significantly better final weight and specific growth rate than female crabs.

**Pond water management for mud crab culture:** During development, young mud crabs normally live in brackishwaterof 15–25 ppt. However, crabs can withstand wide changes in salinity and can live and develop in salinities from 5 to 36 ppt. The water needs to be clean especially when the mud crabs are kept at high density and fed with raw and fresh food. In areas with daily tidal movement, 30–50% of the pond water may replaced every day, and the water fully replaced once a week. For this, a part or all the water is emptied prior to tidal rise and the outlet is closed. When the tide rises towards the top of the tide, water is taken in from the middle-layer or lower-layers. After harvesting, it is also important to clean the pond. If the pH of the water is less than 6, the pond is emptied and lime powder is spread evenly over the bottom, canal and the inner sides of the pond at 0.07–0.1 kg/m². The bottom of the pond is then exposed for 2–3 days and the pond is filled with fresh water three to four times to empty all the contaminated water.

Optimal Water Quality Parameters for Mud Crab Ponds

| Parameter | Optimal range |
|---|---|
| Dissolved oxygen | >5 ppm |
| pH | 7.8 |
| Temperature | 25–35°C |
| Salinity | 10–25 ppt |
| Total ammonia nitrogen | <3 ppm |
| Alkalinity | 120 ppm |
| Hardness | >2–000 ppm |
| Hydrogen sulphide | <0.1 ppm |

**Feeds:** Mud crabs can be successfully raised on a variety of diets including low-value/trash fish, slaughter wastes, fish wastes, horse

Aquaculture of Portunid Crabs 285

mussels, brown mussels, brackishwater snails, shrimp heads, golden snails (*Pomacea canaliculata*), telescope snails (*Telescopium telescopium*), small bivalves (*Potamorcorbula* spp.), animal hides, entrails and kitchen leftovers.

**Formulated feeds:** It has been shown that *Scylla serrata* gains weight faster and molt more frequently on diets containing higher protein (up to 55% of diet) and lipid (up to 15% of diet) levels. The same study demonstrated formulated feeds can produce feed conversion rations (FCR) of 1.2 to 2.1:1 for juvenile crabs. By increasing the protein content of mud crab feeds from 25 to 45%, there was also a progressive increase in the protein content of crabs.

**Feeding:** Feeding rates utilized are often quoted as a percentage of the body weight of mud crabs in a pond. Feeding frequency is typically once or twice a day, with feeding recommended every day to minimize the risk of cannibalism, which is considered more likely if the crabs are hungry. As mud crabs often seem most active late in the afternoon and early evening, one feed is commonly provided at that time. Mono-cultured crabs live mainly on daily supplied food, as the quantity of natural food in the pond is insufficient. On the other hand, in poly-culture, during the initial months, crabs often use food already present in the ponds as their main food source. However, in the last months of the raising period when the crabs are well grown and have a greater demand for food, the growing crabs should be supplied additional nutrition. Crab food is usually raw and fresh and consists of crushed fish, small crabs, oysters, molluscs, shrimp or fish heads. The quantity of food supplied daily is 4–6% of the estimated total weight. They should be fed once a day between 17 and 19 hrs. Food is spread widely over the pond. Crabs should be fed every day (Shelley and Lovatelli, 2011).

Daily Feeding Rates for *Scylla* spp. During Culture (% body weight)
(Shelley and Lovatelli, 2011)

| Species | Initial | Final |
|---|---|---|
| *S. paramamosain* | 10–15 | 3–5 |
| *S. serrata, S. tranquebarica, & S. olivacea* | 5–10 | 5–10 |

**Yield of mud crabs in monoculture:** Mud crabs cultured singly in bamboo screen compartments of 200 sq. m size with two stocking densi-

286 Biology and Culture of Portunid Crabs of World Seas

Harvest Details of Monoculture Experiment At Different Stocking Densities
(Venugopal et al., 2012)

| Stocking rate (per m²) | 0.5 | 0.75 | 1 |
|---|---|---|---|
| Production (kg/ ha/crop) | 663 | 592 | 489 |

ties viz. 5000/ha and 10,000/ha showed a gross production of 336.2 kg/ ha and 445.3 kg/ha, respectively, in these densities (Lijauco et al., 1980). Venugopal et al. (2012) reported the following production in monoculture of mud crab at three stocking densities viz. $0.5/m^2$, $0.75/m^2$ and $1/m^2$. Their results suggest that the seed crab stocking density of $0.5/m^2$ is profitable.

Monosex experiments conducted in mud crab using all- males and all-females showed higher production in all-males and the crop production in these experiments was 427 kg/ha/crop and 308 kg/ha/crop, respectively (Venugopal et al., 2012).

**Pens for mud crabs in mangroves:** Mangrove forests are the native habitat of mud crabs for most of their juvenile and adult life. Pens may be constructed entirely of wood or plant fencing (e.g., bamboo, or palm trunks), or of netting with wooden supports. The netting or spacing between the fencing needs to be small enough to ensure that the smallest sized crab placed in the pen cannot escape, typically 1–2 cm mesh. Pens may have an inner nursery structure that can be used to hold very small stock for a limited period of time, until they are large enough to be retained in the main pen structure. The inner pen structure in such a design would be constructed of fencing or netting with a smaller opening to retain the smaller stock. This can be considered to be either part of the nursery system or preliminary grow-out stage. The inner pen structure enables smaller stock to be more intensively managed, until they are of sufficient size to be let into the larger pen. The height of the pens must be higher than maximum king tides, so that stock cannot simply swim out of the pen on such tides. Tide tables and local knowledge should be used to establish the appropriate height to construct the pens. Similarly, to ponds, it is preferable that their upper edge be lined with plastic or a similar material that prevents mud crabs from climbing out of the pen.

**Mangrove pen construction:** The walls of mud crab mangrove pens need to be buried in the mud (30–60 cm) to minimize the risk of mud crabs

Aquaculture of Portunid Crabs

burrowing under the walls of the pens. Pens are commonly constructed with vertical support posts every 3 m, and horizontal bracing structures to support the walls. To maximize the longevity of pen structures, posts should be made from wood that is most resistant to marine borers or treated to reduce their impact. Posts covered in plastic or a plastic sleeve to protect the wood, or posts constructed from a synthetic material, will provide the maximum longevity.

**Preparation of mud crab pens prior to stocking:** Prior to stocking with crablets, mud crab pens must be cleared of potential predators, large crabs and any unwanted species. Farmer's need to use baited traps to remove all mud crabs from pens before restocking. Once a good number of baited traps have been set for several days in a pen and no crabs trapped, it can be considered free of mud crabs and ready for restocking.

**Stocking in mangrove pens:** The size at which crablets are stocked into mangrove pens depends on the size of netting used and also if a nursery is installed within the pen. In the case of a pen with a nursery, small crablets of less than 1 cm CW can be stocked. Once the crablets are larger than 2 cm they can be released into the main pen. The density at stocking that is used should produce 1–1.5 tonnes/ha. A stocking density of 5,000 to 10,000 crablets per hectare has been suggested as an appropriate stocking density for mud crab pens. Survival rates from 20 to 86% have been reported from mud crabs stocked in mangrove pens at various initial sizes and stocking densities.

**Feeding in mangrove pens:** It has been found that if a mangrove pen is stocked with wild-collected crablets, these can survive for one month without supplementary feed. However, if stock is from a hatchery and is used to being fed, supplementary feeding is required from the time of stocking. If, while monitoring for food consumption using feed trays, it is found that all the food provided is not being consumed, the amount of feed per feeding event should be reduced. Feeding is usually undertaken just before, or during an incoming tide, as this is when crabs emerge from their burrows in the mud to feed.

**Maintenance of mangrove pens:** The perimeter fencing of mud crab pens must be checked routinely to ensure that it has not been damaged or compromised in any way so that stock are able to escape. Farmers must be vigilant to prevent and minimize losses from poachers.

288                    Biology and Culture of Portunid Crabs of World Seas

**Stocking of crab seed in pens:** Crab seeds are collected from the surrounding mangrove swamps and stocked at a density of 2 crabs/m$^2$. Each seed is measured for its carapace length and width, sex and weight. Size of crab seeds for stocking ranges from 20 to 39.9 mm carapace length and weight is < 19.9 g. However, crabs seeds of other size ranges and weight classes were also stocked. Before stocking in the pen, the seed crabs are allowed to acclimatize in cool water from the pond. The ratio of females to males is 1:1.75.

**Feeds and feeding:** Growing crabs are fed with trash fish (fish heads, fish skin, broken fish, fish entrails and squid heads) once a day during high tide.

**Mud crab harvesting:** Harvesting of mud crabs start normally after 4 months using a beach seine net and setting of baited traps. The targeted crabs are those weighing 300 g and above. Growth estimates are evaluated by calculating specific growth rate (SGR) and total weight gained (TWG) for the different weight groups of crabs harvested, as follows:

$$SGR = \frac{\log e \, (FBW - IBW)}{No. \text{ of culture days}} \times 100$$

where FBW = final body weight; IBW = initial body weight; and

$$TWG = \frac{FBW - IBW}{IBW}$$

**Pond construction:** The pond was fenced with palmyrah rachis on top of the earthen bund, to prevent the escape of crabs. Natural shelters in the form of soil mounds with mangroves were provided which remained submerged during high tide and got exposed during low tide. Few stoneware pipes were also provided inside the pond as shelters for crabs, especially during molting.

**Stocking:** A total of 500 crablets were stocked. The crabs were stocked at the rate of 8/m$^2$, at 54.6:45.4 ratio of male to female. At the time of stocking, the carapace width of the males ranged from 75 to 117 mm and females from 75 to 120 mm. The total average weight was 133 g.

**Water quality management:** Depth of water in the pond was 1 m. According to the tidal changes, water was exchanged daily through the sluice gate and the sluice screen was cleaned every alternate day for easy exchange of water. Water quality parameters, mainly temperature

Aquaculture of Portunid Crabs 289

(water and atmosphere), salinity, pH, dissolved oxygen and turbidity were recorded, once a week. Temperature during the period ranged from 27 to 34°C, salinity from 19 to 47 ppt, pH from 7.7 to 8.5, dissolved oxygen from 3.8 to 10.7 ml/1 and turbidity from 3 to 10.7.

**Food and feeding:** Chicken waste was fed for four days a week, trash fish twice a week and both once per day at the rate of 5% of body weight of crabs. The feed was placed on the side of the pond and crabs were found to come out and take away the feed. They were fed during morning hours around a fixed time.

**Harvest:** The crabs were harvested, after a culture period of 273 days. During high tide, the crabs, which gathered near the sluice gate were caught with scoop net and the rest handpicked after draining the pond water at low tide, using a 5 Hp pump.

**Results on growth:** At the end of the culture period, the carapace width of males ranged from 110 to 136 mm, overall growth increment being 40 mm. In females, the carapace width ranged from 110 to 145 mm with an overall increase of 60 mm. Percent survival was 23.8%. The average weight increased from 133 g at the time of stocking to 428 g at harvest. The overall growth rate in terms of weight was 1.089 g/day. In males at harvest, the carapace width showed a growth rate of 0.146 mm/day and in females, 0.219 mm/day (Pillai et al., 2002).

### 4.2.1.1.2  *Culture of* Portunus pelagicus

Maheswarudu et al. (2008) have reported on the culture of Portunus pelagicus. Earthen ponds were preferred for the grow-out culture of this species. Pond preparation was carried out as in shrimp farming to ensure the best environmental conditions for the growth and survival of the growing crabs. For grow-out culture, first crab instars of this species were stocked in a 0.06 ha earthen pond at a stocking density of 2.6 nos./m2. During the culture period, the first crab instars/crablets/crabs were fed with shrimp feeds, gradually increasing the size of the feed as the crab showed progress in growth. Water depth in the pond was maintained at 0.7–1.0 m up to 60 days and thereafter 70% water exchange was provided once every week. Ambient water parameters such as salinity, temperature, pH and dissolved oxygen were recorded at fortnightly intervals. The ranges of salinity and temperature were 28 to 39 ppt and 25.2 to 28.5°C, respectively. pH and dissolved oxygen ranges were 8.2–8.8 and 3.9–4.1 ml/l, respectively. After 135 days of culture period, the average carapace width and weight of the crabs were 116 mm and 111.8 g, respectively.

Results of a small scale farming of *Scylla paramamosain* in central Vietnam (Petersen et al., 2011)

| | |
|---|---|
| Average number of mud crab crops per year | 1.5 |
| Average length of grow-out period (months) | 3.4 |
| Weight of seed crab (g/crablet) | 13 |
| Length of seed crab (cm/crablet) | 2.9 |
| Average stocking density (crablets/m3) | 0.40 |
| Number of crablets stocked | 6,000 |
| Survival rate (%) | 53 |
| Weight of mud crabs at harvest (kg) | 0.30 |
| Average total harvest biomass (kg/crop) | 960 |

### 4.2.1.1.3 *Culture of* Callinectes sapidus

**Culture of blue crab in ponds:** Two Ponds, each 0.25 acres with an average depth of 4.5 feet were selected for the experiments. Ponds were filled with well water, averaging 0.4 to 0.5 ppt salinity. In the first pond, artificial seagrass squares were placed around the perimeter of the pond, approximately 3 feet from the pond edge. The artificial seagrass was a substrate for the juvenile crabs to colonize. Artificial substrate was added to the second pond. However, additional substrate was provided by barnyard grass (*Echinochloa* spp.) that was present when the ponds were dry.

**Blue crab pond with artificial seagrass squares**

Aquaculture of Portunid Crabs 291

**Seed stock:** Hatchery-reared juvenile blue crabs of 11.5 mm CW.

**Food and feeding:** Crabs were fed a commercial shrimp diet starting at rate of 13.5% of the biomass, which was gradually reduced to 2.2% by the end of the study.

**Sampling and harvest:** Periodic sampling monitored crab growth, and a final harvest was conducted after 96 days. All samples were taken using peeler pots. The final harvest was carried out by draining the pond and netting all remaining crabs. From the two ponds, a total of 1,960 crabs were harvested between 62 and 96 days after stocking, with an overall survival rate of 20.4%. The majority of crabs came from the pond with barnyard grass (1,253). Average crab size for both ponds was 106.3 mm CW, and growth ranged from 1.0 to 1.9 mm CW per day during the study. The bulk of the crabs harvested ranged from 81 to 110 mm CW (http://nsgl.gso.uri.edu/ncu/ncug12001).

Results of two experimental ponds in the production of blue crab, *Callinectes sapidus* (http://nsgl.gso.uri.edu/ncu/ncug12001)

| | Pond 1 | Pond 2 |
|---|---|---|
| Area (acres) | 0.51 | 1.41 |
| No. of crabs stocked | 220 | 581 |
| Crab density (No./acre) | 431 | 412 |
| No. of crabs harvested | 30 | 71 |
| Total survival (%) | 14 | 12 |
| Avg. initial size, mm (CW) | 35 | 16 |
| Avg. final size, mm (CW) | 136 | 155 |

## Culture of blue crab in cages

Yadon (http://hdl.handle.net/1969.3/20730) has reported on the culture of blue crabs in cages. Juvenile crabs were separated by sex and carapace width (CW) into two size classes (small <30 mm, large 30–100 mm carapace width). These crablets were randomly assigned to one of three chambers within a 163 × 43 × 43 cm mesh enclosure. While, the larger cages were used to monitor the growth of crabs from 30–100 mm carapace width (CW) juveniles <30 mm, were monitored in small cages. The small cages were 101 × 23 × 23 cm in size and constructed in the same manner as the larger cages. Five large and small cages, containing a total of 30 crabs, in as equal a sex ratio as possible, were developed at two sites each year. No food was added to the cages. For the growing crabs in the enclosures, the mesh was assumed to be sufficiently large to permit prey to enter, providing food. Carapace width was measured and each crab was checked daily for signs of molting. Water temperature and salinity were recorded daily. The presence of shell remains or a crab with an incompletely hardened new shell was used to determine a molt. After the new shell of a molted crab hardened significantly, the new CW was measured. A total of 60 crabs were monitored for 113 days and their size varied from 14.13 to 80.07 mm CW.

### 4.2.1.2 Bispecies Culture of Portunid Crabs

The production performance of *Scylla paramamosain* and *Scylla olivacea* reared together in ponds and provided with different supplementary diets (crustaceans or trash fish) were evaluated with that of an unfed control group relying only on natural food available within the pond (Christensen et al., 2004).

**Pond design:** Rectangular brackishwater ponds were modified into semi-intensive crab ponds for the experiments. Each of the nine ponds used measured 40 m × 12.5 m with a depth of 0.8 m. Fences were constructed on the dykes around each pond to prevent crabs from escaping. The fences were made from a 1.2-m wide fine mesh (1-mm) black nylon net supported by bamboo sticks angled by 201 towards the center of each pond. The lower ends of the nets were embedded 10 cm along the base of the enclosures. Water was exchanged at high tides twice monthly. The grow-out trial was operated for 130 days.

**Treatments:** Both mud crab species, *Scylla paramamosain* and *Scylla olivacea* of both sexes, were allocated to each pond Their average carapace width was 3.18 cm and their average body weight 8.69 g. Each pond group contained 250 crabs representing a stocking density in the ponds equivalent to 0.5 crab /$m^2$ in 500-$m^2$ earthen ponds. Three feed treatments were selected. They are Trash fish, consisting mainly of *Tilapia mossambica* which breed freely in the ponds and canals; Mangrove sesarmid crabs; and No supplementary feeding, relying on natural food available within the pond.

Aquaculture of Portunid Crabs 293

Feeding rates (% wet bodyweight for *Scylla paramamosain* and *S. olivacea* reared together in experimental ponds (Christensen et al., 2004)

| Days | 0–30 days | 30–60 days | 60–90 days | More than 90 |
|---|---|---|---|---|
| **Feeding rate** | **15** | **10** | **7** | **5** |

**Sampling and harvest:** Sixty crabs were measured from each pond population at stocking. Sampling of each pond was conducted at days 30, 60, 90 and 115 by using a hook and line plus a scoop net.

**Environmental conditions:** Water quality was monitored once weekly throughout the experimental period. During the first month of culture, pond water salinity values ranged between 5 and 10 ppt, and the salinity increased steadily throughout the experiment. From day 55 to day 110, salinity remained within the optimum range, but during the last 20 days of culture it reached 32–39 ppt. Temperature remained fairly constant around 30–34°C. pH ranged from 8.0 to 8.5. During the first 120 days of culture, the mean dissolved oxygen levels remained above the recommended minimum of 4.0 mg $O_2$/liter. During the last 10 days, the dissolved oxygen fell below 4.0 mg $O_2$/liter. Turbidity values were highest in the first half of the experiment (average range: 23–33 cm) and decreased during the second half (average range: 19–24 cm). Water level ranged from 50 to 85 cm with an average depth of 70 cm.

### 4.2.1.3 Polyculture of portunid crabs

### 4.2.1.3.1 Polyculture of Mud Crabs in Brackishwater Ponds

The ponds for polyculture range in size from one to tens of hectares, located in brackish-water coastal areas or saline flooded areas. The best sites for polyculture ponds have little wind or waves, with a low current and slope in order to avoid building high banks. The bottom of the pond has a deep layer of mud (up to 30 cm) or sandy-mud, or loamy soil mixed with sand. It is possible to have trees and mounds but they usually cover less than 30% of the area of surface water. In these seminatural ponds, there is an abundant source of food.

**Preparation of brackishwater ponds for polyculture:** Prior to stocking of fish or crabs in polyculture ponds, Tea-seed (saponin) and Sodium cyanide are applied to eradicate obnoxious pests and nuisance organisms. Ponds are then partially filled and chemicals are added to encourage the growth of natural food. Urea and lime are also used. Natural food includes lab-lab which is a mixed benthic algal microorganism and animal community. It often includes several species of bacteria, blue-green algae, diatoms, protozoans, copepods, amphipods, ostracods, nematodes, polychaetes, molluscs, cladocerans, isopods and other organisms. Ammonium Phosphate is also used to encourage the natural food growth of lumut.

**Associated species:** Mud crab is cultured with milkfish, tiger prawns (*Penaeus monodon*) or white leg shrimps (*Penaeus vannamei*), and tilapias.

**Stocking:** Stocking size of mud crabs for grow-out in polyculture ponds start from early stage (10 mg) up to 125 g and their stocking rates are commonly 0.5 to 1.5 crabs/m$^2$ in experimental trials. When mud crabs are cultured at higher densities, mortality and cannibalism significantly reduce the harvest. Single size and juvenile are normally stocked. These crabs are cultured for four months.

**Feeding:** Feeds are collected from the nearby riverbeds and distributed to shrimps and crabs in addition to the natural production in the water. Common items are sea snails (*Cerithadia* sp.) and mussels (*Modiolus* sp.)

**Production in volume and value:** The mud crab annual production has been estimated at 192 kg/ha/yr. Overall production of the fishpond has been estimated between 1,357 kg/ha/yr. and 1,576 kg/ha/yr. depending on the species and survival rate. The higher production in volume is normally obtained with the white leg shrimp (http://aquatrop.cirad.fr/FichiersComplementaires/Mud%20crab_Final%20report.pdf).

### 4.2.1.3.2 Polyculture of Mud Crabs and Milkfish

Mud crabs (*Scylla serrata*) were cultured in combination with milkfish (*Chanos chanos*) in bamboo screen compartments of each 200 sq. m each at the following two stocking densities.

Experiment 1: Milkfish, 2500/ha and mud crab, 5000/ha
Experiment 2: Milkfish, 2500/ha and mud crab, 10000/ha

Aquaculture of Portunid Crabs

The values of gross production in the first experiment was 372.9 kg/ha and 499.7 kg/ha, respectively. In the second experiment the values were 395 kg/ha and 529 kg/ha respectively. Net production of crab was found to be higher in the polyculture than in the monoculture in both stocking densities. The reverse was observed in the case of milkfish. In these experiments, crabs were observed to molt at an interval of about two weeks coinciding with the occurrence of the neap tides. After molting, discarded shells were seen strewn about at the pond bottom. During this time, the crabs could hardly be seen for about 2 to 3 days. Normally, the crabs could be observed swimming against the current flowing in from the inlets, or occasionally basking atop the trellises in the early morning hours and shortly before feeding time in the afternoon (Lijauco et al., 1980). Venugopal et al. (2012) reported a mud crab production of 292 kg/ha/crop and milkfish production of 462.5 kg/ha/crop in their polyculture experiments conducted for a period of 5 months. In their experiments, the juvenile crabs of *Scylla serrata* were stocked at $0.5/m^2$ along with the milkfish, *Chanos chanos* fingerlings of 20–30 g at a stocking density of $0.25/m^2$.

### 4.2.1.3.3 Polyculture of Soft-Shell Mud Crabs and Tilapia

| Culture species | Red tilapia (*Oreochromis niloticus*), black tilapia (*Oreochromis placidus*) and soft-shell mud crabs |
|---|---|
| Stocking size of male and female crabs | 30–100 g |
| Stocking density | Not reported |
| Feed | No feeding |
| Culture period | 6–10 months |
| Size of crab at harvest | 250 g– 500 g |
| Total production | Not reported |

### 4.2.1.3.4 Polyculture of Mud Crab, Fish and Shrimp or Sea Weed

Mud crabs may be polycultured successfully with fish species such as milkfish (*Chanos chanos*), siganids, mullets and tilapia or shrimps such as grass shrimp, *Litopenaeus vannamei* and tiger shrimp (*Penaeus monodon*) or sea weed, *Gracilaria* spp. Due to the cannibalistic nature of mud crabs, polyculture is commonly practiced in earthen brackish-water ponds (1–10 ha). The culture of mud crabs with tiger shrimp or fish (siganids or mullets) has been found to provide a better economic option. Crabs are commonly stocked at 500–800/ha, together with milkfish (*Chanos chanos*) at 1,000–2,000/ha and black tiger prawns (*Penaeus monodon*) at 20 000–50 000/ha. After pond preparation, 'lab-lab' (a complex group of minute plants and animals that form a brownish, greenish or yellowish mat on the pond) is grown first with a low water level. When the lablab growth is abundant, the water level is increased gradually to 80 cm. Shrimp postlarvae ($PL_{15-18}$) are stocked, followed by crabs after about a week and milkfish fingerlings (2.5–3.0 cm) after 15–30 days. Food items consist mainly of natural food (lablab and filamentous algae; higher aquatic macrophytes such as *Hydrilla verticillata*, and *Ruppia* spp.; and fish and molluscs). To reduce supplementary feed inputs, the growth of natural food such as lablab, filamentous algae and other macrophytes is maintained but controlled. Small snails (*Thiara* spp.) are even introduced into the ponds and allowed to reproduce to serve as food for crabs and shrimp. Commercially available formulated diets are given to the shrimp and milkfish only when the natural food is overgrazed. About 30% of the water is changed every spring tide; the amount may be increased as the culture period progresses.

Yields per crop are 200–350 kg/ha for mud crabs, 250–600 kg/ha for milkfish and 250–1,300 kg/ha for black tiger prawns. Two crops are achieved annually. Polyculture of crabs (*Scylla* sp.) with milkfish conducted in India has shown a production of about a ton/ha of crabs and 0.7 tonnes/ha of milkfish (Troell, 2009; http://www.fao.org/fishery/culturedspecies/Scylla_serrata/en).

**Polyculture of fish and shrimps/crabs in Indonesia:** Traditional polyculture pond farming sytem ( tambak) in Indonesia mixes milkfish (*Chanos chanos*) with different species of shrimp (*Penaeus vannamei, Penaeus stylirostris, Penaeus monodon*) and wildfish (i.e., mullet (*Mugil* sp.) and barramundi, (*Lates calcarifer*). Compared to intensive shrimp farming, these traditional systems need less inputs, as they are supplied by natural tidal inundations with larvae and most of their foods and nutrients.

Aquaculture of Portunid Crabs 297

### 4.2.1.3.5 Polyculture of Mud Crabs and Shrimp

A study on the feasibility of bi-culture of mud crab (*Scylla serrata*) and tiger shrimp (*Penaeus monodon*) in brackishwater earthen ponds (0.1 ha each) was carried out for a period of five months. Shrimp and crab were stocked at the rate of $2/m^2$ and $1/m^2$, respectively (Treatment 1) and at $2/m^2$ and $0.5/m^2$, respectively (Treatment 2). The growing crabs were fed with chopped trash tilapia while shrimp were fed with Saudi-Bangla shrimp feed twice daily. Among these two experiments, the treatment 1 showed promising results as the mud crab showed a production of 568.80 kg/ha and a highest total production of 871.29 kg/ha was in this treatment (Begum et al., 2007).

### 4.2.1.3.6 Polyculture of Blue Swimming Crab and Red Seaweed

Polyculture experiments using blue swimming crabs (*Portunus pelagicus*) (initial weight 43.0 g.; CW 9.1 cm) and red seaweeds (*Kappaphycus alvarezii*) were carried out in 1 $m^3$ cages for 45 days. For crab production experiment, 10 crabs were placed in a cage with varying seaweed stocking density: 10 + 500 (g red seaweed), 10 + 750 and 10 + 1,000. Treatment without red seaweeds (10 + 0) served as the control. For seaweed production, two culture schemes (hanging and bottom method) were experimented with the same stocking density used in crabs. However, set-up with 1,000 g red seaweed and without crab (0 + 1,000) served as control. For crabs, highest net production was obtained in the 10 + 750 treatment. For seaweed production experiment, the bottom culture scheme showed significantly higher net production. Further, red seaweeds cultured in the 10 + 750 showed highest net production among the stocking density combination evaluated. The study suggests polyculture of blue swimming crab and red seaweeds using the bottom culture scheme at 10 + 750 as the optimum stocking density (Suerte, 2015).

### 4.2.2 FATTENING OF PORTUNID CRABS

Crab fattening is fattening the soft crabs or post molt crabs ("empty crabs," "thin crabs," or "water crabs") in confined spaces to fill up

298 Biology and Culture of Portunid Crabs of World Seas

their meat within a shorter period of time. Just-molted mud crabs caught from lagoon and shallow coastal waters are fattened under captivity, e.g., in cages, floating cages, ponds or tanks, with a variety of food such as fish offal, fish heads, cockles, etc., for 2–3 weeks. The growing crabs in fattening systems are fed 7–10% of their total body weight per day, sometimes even higher (10–15%). To reduce cannibalism and damage to other crabs, crabs for fattening systems typically have their claws tied shut. Water quality is a critical factor in crab fattening. Crabs kept in high densities and in close proximity to each other, oxygen demand will be higher than in low density grow-out systems. Oxygen levels should therefore be kept at acceptable levels (>5 mg/l). If cages are located in the shallow waters in the sea, a good flow of water is essential to maintain good oxygen levels. In pond culture systems, access to good-quality marine and freshwater sources is important. While mud crabs can survive a wide salinity range in culture (5–40 ppt), optimal growth appears to be in the range of 10–25 ppt. Optimal growth is at a temperature of 30°C, with good growth from 25 to 35°C (Sathiadhas, 2006; Anon., 2013).

### 4.2.2.1 Fattening of *Scylla* spp.

**Mud crab fattening in cages in mangrove areas:** A bamboo cage ($0.25 \times 0.7 \times 2$ m³) is constructed with a green nylon net (12 mm mesh size) for side walling, bottom flooring and top movable cover. Three main divisions are made which are further divided into eight cells. A cage thus would have 24 cells. Each cage is provided with floats and rings in the corners. The cage is set or staked at the fringes of the mangrove area, such that, at the lowest tide, ¾ of the cage is still submerged. Cage is covered with coconut fronds as crab shelter. One crab is stocked per cell (lean crabs, weighing at least 100 g or 300 g (if female giant crab); males weighing 200 g (lean crabs) or 350 g (giant crabs). These crabs are fed with trash fish or mixed diet of 75% brown mussel meat and 25% trash fish at 10% of the crab biomass per day. Selective harvest is made and restocking of harvested cages may also be done. Fattening can normally take 15–30 days.

**Bamboo box for mud crab fattening**

**Bamboo boxes mud crab individual fattening**

**Fattening "empty crabs" in ponds, enclosures or cages:** In empty crabs, the male crabs have a thin, soft carapace with little flesh, and female crabs have little ovary tissue. Such crabs maybe fattened for marketable size. Thin, male crabs are fed for 25–35 days (fattening period) so that their shell hardens, muscle flesh develops, or in the case of females the ovaries develop.

**Fattening systems:** Fattening systems include ponds, enclosures or cages. For areas where ponds or enclosures are not suitable, fattening cages are used. The cage is usually made of bamboo and a popular size is 2–3 m wide, 3–4 m long, and 1.0–1.2 m high. A wide opening in the top of the cage (0.6–1.2 m), covered with bamboo, is used for access and feeding. It must be tightly closed and locked. The cage is kept afloat by buoys, the top about 0.2–0.3 m out of the water, and is anchored by cables tied to stakes on the bank. It is ideal to set where strong water flows.

**Stocking density:** Thin crabs are fattened in small ponds (200–500 $m^2$), enclosures (100–300 $m^2$) or cages. The density for fattening is 0.5–1.0 $kg/m^2$ for ponds and enclosures. However, this density is increased to 10–25 $kg/m^2$ for cages.

**Food and feeding:** Thin crabs need large amounts of food, which is usually small fish, clams, solens, fiddler crabs etc. The quantity of food should be 5–8% of the biomass of crabs.

**Management of water quality:** During fattening, if the pond is heavily contaminated, all the water is emptied and the crabs are collected. Then the bottom is cleaned and excessive food is removed. This can be carried out in cool weather taking in new water at high tide. In pond culture systems, access to good-quality marine and freshwater sources is important. While crabs can survive a wide salinity range in culture (5–40 ppt), optimal growth appears to be in the range of 10–25 ppt. Oxygen levels should be kept at acceptable levels (>5 mg/ liter). Water temperature can affect crab survival. Optimal growth is at a temperature of 30°C, with good growth from 25 to 35°C. (http://www.islf.org/index.php?option=com_content&view=article&id=64:crab-fattening-a-potential-marine-aquaculture-industry-for-small-scale-fishermen&catid=34:featuresarticles).

### Mud crab fattening in ponds of Philippines

Since lean mud crabs (*Scylla serrata*) have very low demand in the domestic market, these crabs are fattened for 15 to 20 days to attain the

Aquaculture of Portunid Crabs 301

highly priced meat of the large pincers of males and the bright red roe of gravid females. Crab fattening in cages is economically viable but can only accommodate a limited number of stocks. It also requires high maintenance requirements in terms of labor. Hence, crab fattening in ponds is a better alternative to accommodate a large volume of crabs at one time. The present study was conducted to determine the growth, survival and production of both undersized and lean marketable sized mud crab, *Scylla serrata* fattened in ponds.

**Materials and Methods:** Two independent experiments on mud crab (*Scylla serrata*) fattening were conducted simultaneously in 150 m² ponds for 30 days. In Experiment I, monosex cultures of undersized male (286 ± 1.2 g) vs. undersized female (267 ± 0.9 g) were stocked at 0.5m², and in Experiment II monosex cultures of lean marketable sized male (338 ± 3.1 g) or female (338 ± 2.8 g) vs. mixed sex (338 ± 3.4 g) were stocked at 0.25 m². The crabs were fed daily a mixed diet of 75% brown mussel (Modiolus metcalfei) flesh and 25% fish bycatch (consisting of *Leiognathus* sp.) at a rate of 10% of the crab biomass daily until final harvest. The daily ration was given equally between 0730 h and 1700 h by broadcasting the feeds. The body weight, carapace length (CL) and width (CW), specific growth rate (SGR), survival, and production of crabs were calculated from the total harvest.

**Results:** Crabs in Experiment I attained a mean final body weight of 496 g for males and 432 g for females. Crabs in Experiment II gave a mean final body weight of 520 g for males, 484 g for females and 517 g for mixed sex (Triño and Rrodriguez, 2001).

**Fattening of mud crabs in cages and brackishwater ponds in Bangladesh:** Begum et al. (2009) made a study to compare the survival, production and economics of mud crab fattening in cage and in encircled earthen brackishwater pond. Thirty cages of 1 m (L) × 1 m (W) × 0.3 m (H) partitioned into 16 compartments (each 25 × 25 × 30 cm) were set in a 40 m² pond and another pond with same area was encircled with bamboo fence. Mud crab fattening in cage and in encircled earthen area were considered as Treatment-1 and Treatment-2, respectively. Single adult non-gravid female crab (204.42 g) was stocked into each compartment of the cages and 80 crabs (204.42 g) were also stocked into earthen pond @ 2 individual/m². The crabs were fed with chopped tilapia @ 8% of body weight twice daily.

Survival rate of crab was found 93.75 and 86.12, respectively, in cages and encircled earthen area. Significantly higher total production of crab from cages (3.30 kg/m²) was recorded than the encircled earthen area (0.37 kg/m²). Comparative benefit-cost analysis showed that bamboo cage fattening attained higher net profit than crab fattening in encircled earthen area from 12 crops (12–16 days per crop) fattening period. The present study revealed that mud crab fattening using bamboo cage might be better than encircled earthen area with fencing especially in Bangladesh.

**Mud crab fattening in Sri Lanka and Bangladesh:** The fishermen catch just-molted crabs mud crabs from lagoon and shallow coastal waters and fatten them under captivity.They use different types and sizes of cages for crab fattening. Some use ponds constructed along the beach equipped with gates to let in seawater and drain the water as required. The most commonly used cages are 2 m × 2 m × 0.5–0.75 m for easy handling. Inthis study,crab cages were setup into the shallow water fronts.A total of 20 cages were set-up for the study. Each cage was seeded with 10–15 crablings (soft-shellcrabs) and the crabs were fish-fed with a variety of food such as fish offal, fish heads, cockle setc.The feeding was done two times each day for 2–3 weeks at the rate of 10% of the biomass of the crabs.The experiment shows that from sales of two batches of fattened crabs,the profit was around Rs.6,388 per cage (Anon.,2013). Similar studies made in Bangladesh in 30 bamboo cages of 1m (L) x 1m (W) x 0.3m (H) size yielded a net profit of 91630 Bangladeshi Taka for 12 crops (Begum et al., 2009)

### 4.2.2.2 Fattening of Portunus pelagicus

For fattening studies in *Portunus pelagicus*, two types of experiments were conducted. First experiment was designed to know the optimum salinity and second experiment was to know the suitable live feeds for fattening. They were kept in different salinities (20, 25, 30 ppt). Weight gain of the crabs cultured in 30 ppt (58.3 g) was significantly higher than 20 (49.6 g) and 25 ppt (52.3 g) and also the shell hardens in shorter duration (11.6 days). Hence 30 ppt salinity was selected as optimum salinity for further experiment. After knowing the salinity, 3 different live feeds viz., oyster, clam, trash fish were offered individually. Fourth feed was offered in combined manner. Weight

Aquaculture of Portunid Crabs 303

gain was higher when the water crabs were fed with mixed feed (59.00 g). The shell hardening duration (11.6 days) also shorter when compared to other feeds. From the present study 30 ppt salinity was optimum and mixed feed is suitable feed for fattening of *P. pelagicus* as evidenced by shorter duration and higher weight gain (Soundarapandian and Raja, 2008).

### 4.2.2.3  *Fattening of* Charybdis natator

The weight gain of the water crabs of *Charybdis natator* cultured in 30 ppt was (58.0 g) significantly higher than 20 (46.7 g) and 25 (50.4 g) ppt. In this salinity, the shell was becoming harder in shorter duration (10.6 days). The water crabs of this species offered with trash fish showed a significantly higher weight gain (50.40 g). The crabs were hardened significantly in very short duration when they were fed with trash fish (14 days). The survival of the crabs was 100% irrespective of the feeds used (Soundarapandian et al., 2013).

### 4.2.3  *SOFT-SHELL CRAB FARMING*

Soft-shell crabs are a specialized commercial crab product. In soft-shell crab (water crab) farming, crabs of varying sizes are caught and put into soft shell shedding facilities. Such crabs are commonly placed in individual containers and monitored until they molt. Soft-shell crab farming may also be undertaken in small ponds. In this method, the shell of the crab is not allowed to harden after molting. After a series of molting, the crabs will be marketed when they reach a suitable size. Soft-shell mud crab farming is practiced now in a number of Asian countries such as Thailand, Myanmar, Vietnam, Malaysia, Indonesia, and Philippines. The soft-shell mangrove crab can be eaten whole (carapace and all limbs) when cooked. Soft-shell crabs are made available whole year round in these countries. Because of its profitability, there is an increasing interest to engage in this type of mangrove crab farming business venture. Income is generated daily as fresh soft-shell mangrove crabs are sold to local restaurants and frozen crabs are exported to Hong Kong, Singapore, South Korea, Japan, Taiwan, Europe and the United States.

**Molting process:** The physiology of molting crabs has been well studied. A crab responds to the lack of growth space within its shell by entering the pre-molt stage, also called the "peeler" or "buster" stage. At this stage, the crab can absorb more water and ions into its circulatory system, dramatically swelling its body to crack its hard shell. Once it backs out of the old shell, or molts, the crab begins to harden its new carapace or shell using calcium from the water. A crab molts 20 to 25 times during its life. The molting process is the most difficult and stressful time in the life of the crab and the time it is most vulnerable to cannibalization from other crabs. Even small mechanical, chemical or physiological problems during this time will result in death for a crab, so it is absolutely necessary to give peeler crabs the highest level of care before they are placed in the shedding system and throughout the shedding process. Peelers obtained by different collection methods will yield crabs in various stages of the pre-molt cycle. Pre-molt stages are indicated by subtle (yet obvious to the trained eye) physical changes to certain body parts, primarily the lower two segments of the swimming legs ("paddles"). Stages are distinguished by the color of the line between the old and new shell at the edge of the swimming paddles. This line turns from white to pink, then to red just before molting. Segregating pre-molt crabs based on these stages is standard practice. Earlier stage crabs are much more likely to cannibalize late-stage molting crabs (Gaudé and Anderson, 2011).

**Soft-shell mud crab farming:** Immature wild-sourced *Scylla* spp. crabs (60–130 g) in the intermolt stage are held either in communal or individual rearing systems. In communal rearing, crabs are held in tanks, bamboo enclosures or polyethylene nets set in ponds. The chelipeds are removed to avoid cannibalism when the crabs molt. Removal of chelipeds may also induce the molting of the crabs. The most common system is individual rearing, in which intact crabs are held in perforated plastic boxes with base measurement of 215 × 155 × 90 cm and detachable covers of 215 × 155 × 25 cm. Several boxes are positioned in pontoons or floating platforms. One pontoon may have 27 sections that can hold 648 boxes. To maximize the utilization of the space in the pond, herbivorous or omnivorous fish are also stocked. Soft-shell

Aquaculture of Portunid Crabs

crabs are fed fish and other cheap protein sources at 8 percent of the biomass or to satiation every two or three days. Replacement of 50–80 percent of the total water volume is carried out for 2–3 days every spring tide (Lwin, https://www.was.org/meetings/mobile/MG_Paper. aspx?i=34840).

**Soft-shell crab farming in ponds and floating cages:** Ponds for soft crabs are rectangular with an area from 100–200 $m^2$. The bottom of the pond is covered with a 20–25 cm layer of mud or sandy-mud. Crablets of 30–60 g are stocked in these ponds. Prior to stocking of these crablets, both pincers and the three pairs of legs are removed from each crab. The pleopods (oars) are kept so the crab can swim. The stocking density is 100–120 kg/100 $m^2$. The daily quantity of food used is 2–4% of the total biomass of the crabs. Crabs are fed twice a day, e.g., early morning and evening. In the first few days, the crabs tend to eat a great amount of food but from the ninth or the tenth day on, the crabs' food consumption capacity is reduced slightly. By the 11th or 12th day when its pincers become big enough, the crab passes into a premolt stage. When all the crabs pass into this stage, the pond is harvested by complete draining (Dat, 1999; Keenan and Blackshaw, 1999).

Crabs with developing pincers and legs are selected for soft-shell production in cages. They are placed in a special floating cage, which consists of a bamboo frame 1.5 × 1.0 × 0.25 m covered with curtain. Such a cage is stocked with 3–7 kg and placed into a cool pond with a good supply of fresh water. They are not fed, but are examined every two hours. Crabs that have just thrown off their shell are left in this cage from 20–40 minutes, then harvested and arranged on trays in a lateral position, resting against each other. The basket or tray is covered with a thin piece of cloth or a layer of young grass, kept in a cool shady position to avoid sun and wind, and carefully transported to local markets (Dat, 1999; Keenan and Blackshaw, 1999).

**Farming of soft-shell blue crabs *(Callinectes sapidus)* in Virginia:** Wire enclosures (pens) are staked out in the shallows of the tidal zone and these pens are filled with hard blue crabs which are fed and watched closely for molting. This method is difficult to manage as numerous crabs are lost to cannibalism or died as a result of wide variations in temperature, salinity or water quality (Oesterling, http://

nsgl.gso.uri.edu/vsgcp/vsgcpc00001/1996/6-shellfish_and_fish_pro-duction.pdf).

Economic aspects of fattening mud crabs (Anon., 2013)

| Expenditure (in Rs) | | |
| --- | --- | --- |
| Wooden cages | | 13,250 |
| Cost of watery crabs | | 18,000 |
| Feed cost | | 4410 |
| Cage repair and maintenance | | 700 |
| Harvesting charges | | 1000 |
| *Total* | | *37,360* |
| **Income (in Rs)** | | |
| Crab sales 1 | 38 kg @ Rs.1500/kg | 57,000 |
| Crab sales 2 | 8.2 kg @ Rs.1500/kg | 12,300 |
| *Total* | | *69,300* |
| *Profit per cage* | | *Rs.6388* |

## KEYWORDS

- **bispecies culture**
- **fattening**
- **hard-shell crab farming**
- **larval rearing**
- **monosex culture**
- **monospecies culture**
- **polyculture**
- **seed production**
- **soft-shell crab farming**

# CHAPTER 5

# NUTRITIONAL VALUES OF PORTUNID CRABS

## CONTENTS

5.1 Health Benefits of Crabmeat .......................................................... 307

Keywords ............................................................................................ 328

The crabmeat is nutritionally good as it is high in vitamins and high quality proteins and amino acids. It is also rich in minerals such as calcium, copper, zinc, phosphorus and iron, while having lower levels of fat and carbohydrates. The taste, texture and nutritional benefits of crabmeat make it a versatile delicacy used in a wide variety of dishes like salads, soups, pates, starters, or served as a main course. Crabs are a sweet and succulent delicacy that is extremely popular universally.

## 5.1 HEALTH BENEFITS OF CRABMEAT

**Good for diabetics:** Along with all other shellfish, crabs are rich in chromium, which helps insulin to metabolize sugar, and thereby lowers the blood glucose levels in the body.

**Anti-cancer properties:** All crabs have plentiful amounts of selenium. Selenium is an anti-oxidant, and cancels out the carcinogenic effects of cadmium, mercury and arsenic, which can cause tumors. Higher levels of selenium in the blood lead to lower rates of cancer.

**Heavy Cholesterol:** Blue crabs have 28 mg of cholesterol per ounce. While crabmeat is low in saturated fat, the presence of chromium helps to increase the level of HDL (good cholesterol) in the body, and thus reduces the risk of strokes and coronary and circulatory heart disease. In fact, crabs contain sterol, which restrict the absorption of other cholesterol, eaten during a meal.

**Omega-3 fatty acids:** Crabs are also rich in omega-3 fatty acids, which it gets from phytoplanktons and algae. Omega-3 fatty acids, help in reducing the stickiness of blood platelets, thus making red blood cells more flexible and ensuring smoother flow. Omega-3 helps to reduce the level of tri-glycerides and LDL (low-density lipo-proteins), which choke up artery walls as deposits.

**Source of copper:** Crabmeat contains nearly 30 times the copper found in cod and 56 times that found in salmon, chicken and beef. 100 g of crab provides 62% of daily recommended value for adult men and women (https://www.salcombefinest.com/news/top-5-reasons-why-eating-crab-is-good-for-you).

The nutritional values of portunid crab species for which research has already been carried out are given below.

### *Callinectes amnicola*

| Proximate composition | (%) |
| --- | --- |
| Protein | 20.1 |
| Carbohydrate | 2.9 |
| Fat | 0.0 |
| Ash | 1.8 |
| Moisture | 74.5 |
| Fiber | 0.1 |

| Mineral composition | (mg/100 gm) |
| --- | --- |
| Fe2 | 98 |
| Ca2 | 402 |
| Na | 441 |
| K | 361 |
| NO3 | 52 |
| P | 50 |
| Mg | 180 |
| Cu2 | 8 |
| Flesh | |

*Source*: Jimmy and Arazu (2012).

## Callinectes bocourti

| Proximate composition | (%) |
|---|---|
| Moisture | 79.8 |
| Proteins | 17.7 |
| Ash | 2.2 |
| Lipids | 2.8 |

| Fatty acids | (%) |
|---|---|
| Myristic | 2.6 |
| Palmitic | 24.3 |
| Palmitoleic | 6.0 |
| Heptadecanoic | 1.6 |
| Stearic | 9.4 |
| Oleic | 13.9 |
| Elaidic | 3.3 |
| Linoleic | 6.9 |
| A linolenic | 5.2 |
| Eicosatrienoic | 4.6 |
| Eicosapentenoic | 10.9 |
| Docosahexenoic | 2.3 |
| Unidentified | 8.9 |
| Saturated | 37.9 |
| Monounsaturated | 23.2 |
| Polyunsaturated | 29.9 |

*Source*: Lira et al. (2007).

## Callinectes danae

| Proximate composition | (%) |
|---|---|
| Moisture | 79.0 |
| Crude protein | 15.5 |
| Lipids | 0.5 |
| Ash | 2.5 |

| Minerals | (%) |
|---|---|
| Calcium | 0.1 |
| Phosphorus | 0.2 |

*Source*: da Costa et al. (2012).

## *Callinectes exaperatus*

| Proximate composition | (%) |
|---|---|
| Moisture | 83.6 |
| Crude protein | 12.7 |
| Lipids | 0.4 |
| Ash | 2.4 |
| **Minerals** | **(%)** |
| Calcium | 0.1 |
| Phosphorus | 0.2 |

*Source*: da Costa et al. (2012).

## *Callinectes latimanus*

| Proximate composition | (g/100 g edible portion) (dry weight) |
|---|---|
| Crude protein | 19.1 |
| Carbohydrate | 46.2 |
| Crude fat | 0.9 |
| Total ash | 28.6 |
| Moisture | 5.3 |
| **Minerals** | **(mg/100 g dw)** |
| Fe | 8.1 |
| Cu | 0.6 |
| Co | 0.1 |
| Mn | 1.4 |
| Zn | 1.8 |
| Ca | 14.9 |
| Mg | 962 |
| K | 259 |
| Na | 987 |
| P | 2,904 |
| **Minerals** | **(mg/kg)** |
| Calcium | 3844.0 |
| Potassium | 1489.0 |
| Magnesium | 3842.8 |
| Sodium | 2056.0 |
| Iron | 9.3 |

# Nutritional Values of Portunid Crabs

| | |
|---|---|
| Manganese | 34.4 |
| Zinc | 7.9 |
| Copper | 84.8 |

*Source*: Elegbede and Fashina-Bombata (2013).

### *Callinectes sapidus*

| Proximate composition (%) | Female | Male |
|---|---|---|
| Protein | 22.5 | 21.4 |
| Lipid | 1.0 | 1.1 |
| Water | 74.3 | 75.5 |
| TMS* | 2.0 | 1.9 |

* TMS: Total mineral substance.
*Source*: Ayas and Özoğul (2011).

| Fatty acids ( % of total fatty acids) | Female | Male |
|---|---|---|
| SFA | 4.8 | 23.3 |
| MUFA | 29.6 | 26.6 |
| PUFA | 39.2 | 42.8 |
| Unidentified | 6.5 | 7.3 |
| **Minerals ( µg/g)** | **Female** | **Male** |
| Cu | 29.5 | 21.6 |
| Zn | 127.8 | 113.4 |
| Fe | 24.1 | 22.6 |

*Source*: Ayas and Özoğul (2011).

### Amino acid composition (g amino acid/100 g edible portion)

| Amino acids | Claw meat | Breast meat |
|---|---|---|
| Aspartic acid | 1.61 | 1.60 |
| Threonine | 1.05 | 0.97 |
| Serine | 0.87 | 0.88 |
| Glutamic acid | 2.60 | 2.58 |
| Glycine | 1.24 | 1.18 |
| Alanine | 0.70 | 0.78 |
| Valine | 0.93 | 0.89 |
| Methionine | 0.59 | 0.55 |

| | | |
|---|---|---|
| Isoleucine | 0.94 | 0.10 |
| Leucine | 1.31 | 1.41 |
| Tyrosine | 0.67 | 0.66 |
| Phenylalanine | 0.74 | 0.75 |
| Histidine | 0.45 | 0.42 |
| Lysine | 1.31 | 1.28 |
| Arginine | 1.58 | 1.03 |

*Source*: Küçükgülmez and Çelik (2008).

### *Carcinus maenas*

| Proximate composition | (%) |
|---|---|
| Moisture | 68.0 |
| Ash | 16.6 |
| Protein | 12.3 |
| Fiber | 2.9 |
| Fat | 0.2 |

*Source*: Fulton and Fairchild (2013).

| Amino acid content | (%) |
|---|---|
| Methionine | 0.17 |
| Cystine | 0.06 |
| Lysine | 0.36 |
| Phenylalanine | 0.30 |
| Leucine | 0.46 |
| Isoleucine | 0.33 |
| Threonine | 0.28 |
| Valine | 0.74 |
| Histidine | 0.15 |
| Arginine | 0.46 |
| Glycine | 0.66 |
| Aspartic Acid | 0.72 |
| Serine | 0.23 |
| Glutamic Acid | 1.26 |
| Proline | 0.47 |
| Hydroxyproline | 0.01 |
| Alanine | 0.47 |

| | |
|---|---|
| Tyrosine | 0.30 |
| Tryptophan | 0.02 |

*Source*: Fulton and Fairchild (2013).

| Minerals | (%) |
|---|---|
| Calcium | 5.7 |
| Zinc | 3.8 |
| Potassium | 0.2 |
| Hg | <5.0 |

*Source*: Fulton and Fairchild (2013).

**Fatty acid composition** (% of total oils)

| Saturated | |
|---|---|
| 12:0 | 0.28 |
| 14:0 | 1.90 |
| 15:0 | 1.04 |
| 16:0 | 15.56 |
| 17:0 | 0.90 |
| 18:0 | 3.56 |
| 20:0 | 0.05 |
| **Unsaturated** | |
| 14:1 | 0.24 |
| 16:1 | 8.06 |
| 16:2 | 0.26 |
| 17:1 | 0.95 |
| 18:1ω9 | 15.14 |
| 18:1ω7 | 4.88 |
| 18:2ω6 | 3.06 |
| 18:3ω6 | 0.19 |
| 18:3ω3 | 0.79 |
| 18:4ω3 | 0.28 |
| 20:1ω11 | 2.08 |
| 20:1ω9 | 3.65 |
| 20:1ω7 | 2.23 |
| 20:2ω6 | 1.57 |

| 20:3ω3 | 0.18 |
|---|---|
| 20:4ω6 | 2.58 |
| 20:5ω3 | 8.73 |
| 22:1ω11 | 2.99 |
| 22:1ω9 | 0.28 |
| 22:4ω6 | 0.51 |
| 22:5ω3 | 1.28 |
| 22:6ω3 | 7.69 |
| 24:1 | 0.37 |
| Other | 8.72 |
| **Total %ω3** | **18.95** |
| **Total %ω6** | **7.91** |
| **Total % Saturated** | **23.29** |
| **Total % Unsaturated** | **67.98** |

*Source*: Fulton and Fairchild (2013).

### *Charybdis feriatus*

| **Proximate composition** | **(%)** |
|---|---|
| Protein | 19.11 |
| Fat | 0.07 |
| Carbohydrate | 0.00 |
| Moisture | 83.52 |
| Ash | 2.16 |

*Source*: Wisespongpand et al.,
http://www.annualconference.ku.ac.th/cd53/04_012_P49.pdf

| **Proximate composition** | **(%, dry weight)** |
|---|---|
| Moisture | 78.7 |
| Crude protein | 84.8 |
| Crude fat | 3.9 |
| Ash | 4.2 |

*Source*: Lu et al. (2014).

| **Amino acids** | **(%, dry weight)** |
|---|---|
| Taurine | 1.23 |
| Aspartic acid | 7.61 |
| Threonine | 3.31 |
| Serine | 3.07 |

# Nutritional Values of Portunid Crabs

| | |
|---|---|
| Glutamic acid | 12.42 |
| Glycine | 5.51 |
| Alanine | 4.41 |
| Valine | 3.31 |
| Methionine | 1.50 |
| Isoleucine | 3.09 |
| Leucine | 5.79 |
| Tyrosine | 3.05 |
| Phenylalanine | 3.15 |
| Histidine | 2.06 |
| Lysine | 6.41 |
| Arginine | 7.31 |
| Proline | 2.92 |
| Cystine | 3.02 |
| Tryptophan | 0.62 |

*Source*: Lu et al. (2014).

| Fatty acid composition | (% of total lipid) |
|---|---|
| C15:0 | 0.88 |
| C16:0 | 24.73 |
| C17:0 | 0.45 |
| C18:0 | 11.80 |
| C16:1 | 6.09 |
| C18:1n9 | 22.11 |
| C20:1n9 | 0.29 |
| C18:2n6 | 1.24 |
| C20:2n6 | 0.36 |
| C20:3n3 | 9.19 |
| C22:5n3 | 0.85 |
| SFA | 37.87 |
| MUFA | 28.49 |
| PUFA | 33.64 |

SFA: Saturated fatty acids;
MUFA: Monounsaturated fatty acids;
PUFA: Polyunsaturated fatty acids
*Source*: Lu et al. (2014).

### *Charybdis helleri*

| Proximate composition | (%) |
|---|---|
| Protein | 17.4 |
| Fat | 0.1 |
| Carbohydrate | 0.0 |
| Moisture | 81.5 |
| Ash | 2.3 |

Source: Wisespongpand et al.,
http://www.annualconference.ku.ac.th/cd53/04_012_P49.pdf

### *Charybdis lucifera*

| Proximate composition | (%) |
|---|---|
| Protein | 15.6 |
| Fat | 0.2 |
| Carbohydrate | 0.6 |
| Moisture | 81.2 |
| Ash | 2.4 |

Source: Wisespongpand et al.,
http://www.annualconference.ku.ac.th/cd53/04_012_P49.pdf

### Essential amino acids (g/100g)

| Amino acids | Male | Female |
|---|---|---|
| Histidine | 1.13 | 1.98 |
| Isoleucine | 1.57 | 1.95 |
| Leucine | 0.38 | 0.71 |
| Lysine | 0.28 | 0.36 |
| Methionine | 1.06 | 1.65 |
| Phenylalanine | 0.32 | 0.41 |
| Threonine | 0.52 | 0.62 |
| Tryptophan | 1.00 | 1.62 |
| Valine | 0.12 | 0.22 |

## Non essential amino acids (g/100g)

| Amino acids | Male | Female |
|---|---|---|
| Alanine | 0.04 | 0.15 |
| Arginine | 0.80 | 0.95 |
| Aspartic acid | 1.65 | 1.97 |
| Aspargine | 0.11 | 0.15 |
| Cystine | 0.90 | 2.00 |
| Glutamic acid | 2.23 | 3.62 |
| Glutamine | 0.36 | 0.41 |
| Glycine | 0.39 | 0.52 |
| Proline | 0.82 | 0.99 |
| Serine | 0.10 | 0.13 |
| Tyrosine | 0.86 | 0.99 |

*Source*: Nachimuthu et al. (2015).

## Minerals (mg/100g)

| Minerals | Male | Female |
|---|---|---|
| Calcium | 11.2 | 12.5 |
| Magnesium | 18.3 | 22.2 |
| Iron | 1.0 | 1.4 |
| Sodium | 37.5 | 46.7 |
| Potassium | 25.9 | 36.8 |
| Zinc | 0.3 | 0.5 |
| Manganese | 1.3 | 1.6 |

*Source*: Kumari et al. (2015).

## *Charybdis natator*

| Proximate composition | (%) |
|---|---|
| Protein | 16.2 |
| Fat | 0.1 |
| Carbohydrate | 0.1 |
| Moisture | 81.4 |
| Ash | 2. 1 |

*Source*: Wisespongpand et al.,
http://www.annualconference.ku.ac.th/cd53/04_012_P49.pdf

**Mineral composition** (mg/100 g)

| Minerals | Male | Female |
|---|---|---|
| Calcium | 12.6 | 11.5 |
| Magnesium | 20.0 | 18.3 |
| Iron | 1.3 | 1.1 |
| Sodium | 45.7 | 36.5 |
| Potassium | 35.8 | 24.9 |
| Zinc | 0.5 | 0.3 |
| Manganese | 1.5 | 1.3 |

*Source*: Soundarapandian et al. (2013).

**Essential amino acids** (g/100 g)

| Amino acids | Male | Female |
|---|---|---|
| Histidine | 1.88 | 1.45 |
| Isoleucine | 1.95 | 1.75 |
| Leucine | - | 1.00 |
| Lysine | 0.31 | 0.28 |
| Methionine | 1.21 | 1.06 |
| Phenylalanine | 0.46 | 0.35 |
| Threonine | 0.51 | 0.48 |
| Tryptophan | 1.08 | 1.01 |
| Valine | 0.14 | 0.13 |

- Absent

*Source*: Soundarapandian et al. (2014).

**Non essential amino acids** (g/100 g)

| Amino acids | Male | Female |
|---|---|---|
| Alanine | 0.14 | 0.05 |
| Arginine | 0.91 | 0.80 |
| Aspartic acid | 1.98 | 1.75 |
| Aspargine | 0.14 | 0.10 |
| Cystine | 1.98 | 0.85 |
| Glutamic acid | 3.01 | 2.00 |
| Glutamine | 0.37 | 0.25 |

# Nutritional Values of Portunid Crabs

| | | |
|---|---|---|
| Glycine | 0.45 | 0.39 |
| Proline | 1.06 | 1.01 |
| Serine | 0.10 | 0.09 |
| Tyrosine | 0.98 | 0.85 |

*Source*: Soundarapandian et al. (2014).

### *Charybdis smithii*
**Proximate composition** (% dry weight)

| Protein | Carbohydrate | Lipid |
|---|---|---|
| 63.0 | 2.8 | 6.9 |

*Source*: Balasubramanian and Suseelan (2001)

### *Ovalipes punctatus*

| Proximate composition | (%) |
|---|---|
| Crude protein | 17.9 |
| Crude fat | 1.4 |
| Moisture | 73.3 |
| C rude ash | 4.7 |

*Source*: Jing et al.,
http://en.cnki.com.cn/Article_en/cjfdtotal-spgy201206033.htm

### *Podophthalmus vigil*
**Proximate composition** (%)

| Carapace | Protein | Carbohy-drate | Fat | Waterwidth mm |
|---|---|---|---|---|
| 51-60 | 19.6 | 0.3 | 4.3 | 74.5 |
| 61-70 | 18.0 | 0.5 | 5.1 | 72.1 |
| 71-80 | 20.2 | 0.6 | 6.9 | 70.2 |
| 81-90 | 19.0 | 0.5 | 5.7 | 72.3 |
| 91-100 | 15.8 | 0.6 | 9.7 | 69.5 |

*Source*: Radhakrishnan and Natarajan (1979).

**Mineral composition** (mg/100g)

| Mineral | Male | Female |
|---|---|---|
| Calcium | 11.5 | 14.6 |
| Magnesium | 1.0 | 2.5 |
| Iron | 7.5 | 13.3 |
| Sodium | 11.5 | 15.9 |
| Potassium | 8.5 | 10.5 |
| Phosphorus | - | 3.1 |
| Zinc | - | 1.7 |

*Source*: Soundarapandian et al. (2014).

**Fatty acid composition** (%)

| | |
|---|---|
| **Saturated fatty acids** | |
| Palmitic acid | 0.8 |
| Stearic acid | 0.3 |
| **Monounsaturated fatty acids** | |
| Oleic acid | 1.0 |
| **Polyunsaturated fatty acids** | |
| Linoleic acid | 1.1 |
| Alpha-linoleic acid | 0.7 |

**Amino acid composition** ( g/100g ( dry weight)

| Essential amino acids | |
|---|---|
| Arginine | 1.8 |
| Histidine | 1.0 |
| Isoleucine | 0.6 |
| Leucine | 0.8 |
| Lysine | 1.0 |
| Methionine | 1.6 |
| Phenylalanine | 0.6 |
| Threonine | 1.3 |
| Valine | 0.4 |

# Nutritional Values of Portunid Crabs

| Non-essential amino acids | |
|---|---|
| Alanine | 1.5 |
| Aspartic acid | 0.3 |
| Aspargine | 0.4 |
| Cystine | 0.9 |
| Glutamic acid | 0.8 |
| Glycine | 1.9 |
| Proline | 1.0 |
| Serine | 1.7 |
| Tyrosine | 1.8 |

*Source*: Sudhakar et al. (2011).

## *Portunus haani*

| Proximate composition | (%) |
|---|---|
| Protein | 16.8 |
| Fat | 0.2 |
| Carbohydrate | 0.0 |
| Moisture | 81.2 |
| Ash | 2.0 |

*Source*: Wisespongpand et al.,
http://www.annualconference.ku.ac.th/cd53/04_012_P49.pdf

## *Portunus pelagicus*

| Proximate composition | (%) |
|---|---|
| Protein | 16.8 |
| Fat | 0.7 |
| Carbohydrate | 0.0 |
| Moisture | 81.0 |
| Ash | 2.2 |

*Source*: Wisespongpand et al.,
http://www.annualconference.ku.ac.th/cd53/04_012_P49.pdf

| Proximate composition | (%) |
|---|---|
| Moisture | 89.4 |
| Ash | 13.0 |
| protein | 75.7 |
| Fat | 2.2 |

*Source*: Premarathna et al., 2015

| Proximate composition | (%) |
|---|---|
| Protein | 23.2 |
| Lipid | 1.2 |
| Moisture | 73.6 |

*Source*: Ayas and Ozogul (2011).

**Fatty acids** (% of total fatty acid)

| MUFA | |
|---|---|
| Palmitoleic acid | 2.39 |
| Hepatodecanoic acid | 0.01 |
| Arachidic acid | 0.90 |
| **PUFA** | |
| Linoelaidic acid | 0.20 |
| Linolenic acid | 0.25 |
| Eicocadienoic acid | 4.02 |
| Arachidonic acid | 0.22 |
| Eicosapentaenoic acid | 2.07 |

MUFA: Mounsaturated fatty acids;
PUFA: Polyunsaturated fatty acids;
Source: Rameshkumar et al. (2009).

# Nutritional Values of Portunid Crabs

**Amino acids** (%)

| Amino acid | Wet wt basis | Dry wt basis |
|---|---|---|
| Isoleucine | 0.6 | 2.7 |
| Leucine | 1.1 | 5.2 |
| Lysine | 1.1 | 5.0 |
| Methionine | 0.4 | 1.9 |
| Phenylalanine | 0.6 | 2.7 |
| Tyrosine | 0.5 | 2.5 |
| Threonine | 0.6 | 2.7 |
| Valin | 0.6 | 2.9 |
| Arginine | 1.5 | 6.9 |
| Aspartate | 1.4 | 6.4 |
| Serin | 0.6 | 2.8 |
| Glutamate | 2.4 | 10.9 |
| Glycine` | 0.5 | 2.1 |
| Alanine | 0.9 | 4.2 |
| Histidine | 0.4 | 1.6 |

Source: Jacoeb et al. (2012).

**Minerals** (mg/kg dry weight)

| K | | Na | | Ca | | P | | S | |
|---|---|---|---|---|---|---|---|---|---|
| Male | Female | Male | Female | Male | Female | Male | Female | Male | Female |
| 233 | 239 | 262 | 270 | 655 | 649 | 822 | 811 | 939 | 999 |

*Source*: Abdel- Salam and Hamdi (2015).

### *Portunus sanguinolentus*

**Proximate composition of hard and soft shell crabs**

| Crab | Protein(%) | Carbohydrate (%) | Lipid(%) |
|---|---|---|---|
| Hard shell | 32.6 | 1.2 | 2.4 |
| Soft shell | 17.2 | 0.7 | 1.5 |

*Source*: Sudhakar et al. (2009).

**Essential amino acid contents of hard and soft shell crabs** (%)

| Amino acid | Hard shell crab | Soft shell crab |
| --- | --- | --- |
| Threonine | 4.88 | 5.26 |
| Valine | 6.48 | 6.95 |
| Arginine | 8.39 | 9.24 |
| Isoleucine | 5.41 | 3.14 |
| Leucine | 8.39 | 9.24 |
| Lysine | 6.96 | 6.66 |
| Phenylalanine | 6.15 | N.D |
| Histidine | 4.44 | 3.13 |

ND: No data.
Source: Sudhakar et al. (2009).

**Nonessential amino acids of hard and soft shell crabs** (%)

| Amino acid | Hard shell crab | Soft Shell crab |
| --- | --- | --- |
| Glutamic acid | 11.54 | 12.38 |
| Tyrosine | 1.91 | 2.05 |
| Taurine | 4.88 | N.D |
| Alanine | 2.94 | 3.14 |
| Aspargine | 12.88 | 12.30 |
| Glycine | 4.44 | 4.80 |
| Proline | 5.54 | 4.68 |
| Serine | 9.65 | 10.37 |

ND: No data.
Source: Sudhakar et al., 2009

**Minerals of hard and soft shell crabs** (mg/100g)

| Mineral | Hard shell crab | Soft shell crab |
| --- | --- | --- |
| Calcium | 2.03 | 1.02 |
| Sodium | 0.52 | 0.52 |
| Potassium | 0.48 | 0.50 |
| Zinc | 0.44 | 0.46 |
| Magnesium | 0.51 | 0.52 |

*Source*: Sudhakar et al. (2009).

**Fatty acids** ( % of total fatty acids)

| Saturated fatty acids | | |
|---|---|---|
| **Fatty acid** | **Lyophilized*** | **Frozen** |
| C14:0 | 0.83 | 0.62 |
| C15:0 | 0.56 | 0.27 |
| C16:0 | 15.64 | 19.54 |
| C17:0 | 0.87 | 0.67 |
| C18:0 | 10.45 | 12.19 |
| **Mono-unsaturated fatty acids** | | |
| C14: 1n7 | 2.40 | 3.46 |
| C16: 1n7 | 8.66 | 7.82 |
| C16: 1n5 | 0.62 | 2.91 |
| C17:1 | 0.44 | 0.48 |
| C18: 1n9 | 19.61 | 17.19 |
| C18: 1n7 | 4.12 | 3.54 |
| C20: 1n9 | 1.73 | 1.91 |
| C20: 1n7 | 0.88 | 0.83 |
| **Poly-unsaturated fatty acids** | | |
| C18: 2n6 | 11.24 | 9.44 |
| C18: 3n4 | 1.95 | 1.25 |
| C18: 3n3 | 0.21 | 0.19 |
| C18: 4n3 | 0.15 | 0.26 |
| C20: 2n6 | 0.82 | 1.73 |
| C20: 3n6 | 0.63 | 0.75 |
| C20: 4n6 | 2.83 | 4.16 |
| C20: 5n3 | 4.80 | 5.54 |
| C22: 2n6 | 0.13 | 0.07 |
| C22: 5n3 | 0.43 | 0.65 |
| C22: 6n3 | 3.10 | 3.43 |

Lyophilised* – Freeze-drying

*Source*: Kala et al. (2015).

## Portunus trituberculatus

| Proximate composition | (%) |
|---|---|
| Crude protein | 74.0 |
| Crude fat | 57.3 |

*Source*: Liang et al. (2009).

## Scylla olivacea

| Proximate composition | (%) |
|---|---|
| Protein | 14.2 |
| Fat | 0.1 |
| Carbohydrate | 0.1 |
| Moisture | 83.7 |
| Ash | 2.2 |

*Source*: Wisespongpand et al.,
http://www.annualconference.ku.ac.th/cd53/04_012_P49.pdf

## Scyla serrata

| Proximate composition | (mg/g) |
|---|---|
| Protein | 5.9 |
| Carbohydrate | 39.1 |
| Lipid | 3.1 |

*Source*: Viswam (2015).

| Proximate composition | (%) |
|---|---|
| Moisture | 79.2 |
| Protein | 17.5 |
| Fat | 0.2 |
| Ash | 1.4 |
| Carbohydrate | 2.7 |

*Source*: Santhanam (2015).

# Nutritional Values of Portunid Crabs 327

**Fatty acids** ( % by weight of total fatty acids)

| Mounsaturated fatty acids | |
|---|---|
| Palmitoleic acid | 4.0 |
| Hepatodecanoic acid | 0.1 |
| Arachidic acid | 0.1 |

| Polyunsaturated fatty acids | |
|---|---|
| Linoelaidic acid | 0.0 |
| Linolenic acid | 0.4 |
| Eicocadienoic acid | 2.5 |
| Arachidonic acid | 1.0 |
| Eicosapentaenoic acid | 4.8 |

*Source*: Rameshkumar et al. (2009).

| Amino acids | (% of total free amino acids) |
|---|---|
| Aspartic acid | 3 .09 |
| Threonine | 0.85 |
| Serine | 0.43 |
| Glutamic acid | 0.72 |
| Proline | 0.45 |
| Cystine | 0.68 |
| Valine | 1 .09 |
| Methionine | 0.34 |
| Isoleucine | 0.46 |
| Leucine | 1 .03 |
| Phenylalanine | 0.55 |
| Tyrosine | 0.25 |
| Histidine | 2.28 |
| Lysine | 4.12 |
| Arginine | 1 .20 |
| Tryptophan | 0.17 |

*Source*: George and Gopakumar (1987).

## KEYWORDS

- amino acid content
- fatty acid content
- health benefits
- mineral content
- nutritional values
- proximate composition

# CHAPTER 6

# BIOMEDICAL, FOOD, AND INDUSTRIAL APPLICATIONS OF PORTUNID CRAB WASTES

## CONTENTS

6.1   Biomedical Applications of Chitin and Chitosan .......................... 330

6.2   Food Applications of Chitin, Chitosan and Their Derivatives
in the Food Industry ................................................................. 339

6.3   Industrial Applications of Portunid Crab Wastes ........................ 344

Keywords ............................................................................................ 348

Crab wastes (exoskeleton) are one of the main sources for the production of value-added products such as chitin and chitosan. The latter along with their derivatives have been reported to show several biomedical, food and industrial applications.

**Chitin:** Crab waste contains 25–35% of chitin on a dry weight basis. The shells are typically washed in dilute acid and rinsed in water. The proteins are then removed by washing with concentrated alkaline solution and then rinsed with water. The material is dried to produce powdered chitin. Chitin is a naturally occurring, non-toxic, biodegradable polymer. A wide range of potential uses have been reported for chitosan. It serves as an as an antimicrobial agent; in edible and biodegradable film; as a food additive; and in biomedical and pharmaceutical applications.

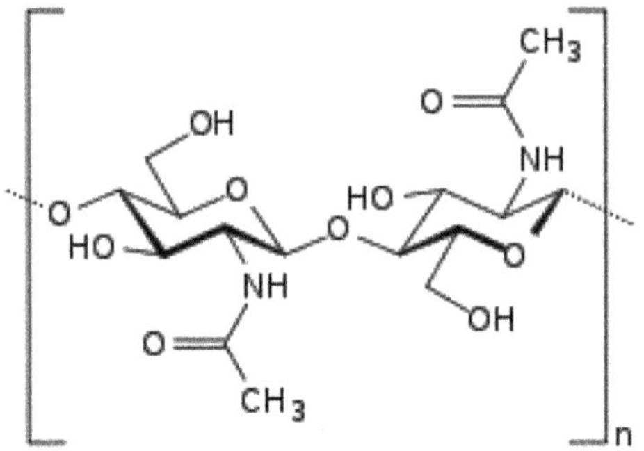

**Chitin**

**Chitosan:** Chitosan is a derivative of chitin in a readily usable form. Chitin is treated with an alkaline solution to produce chitosan. It is then dried into flakes or powdered form. A wide range of potential uses have been reported for chitosan. For example, food ingredient, medical treatment, cosmetics, and nutraceuticals.

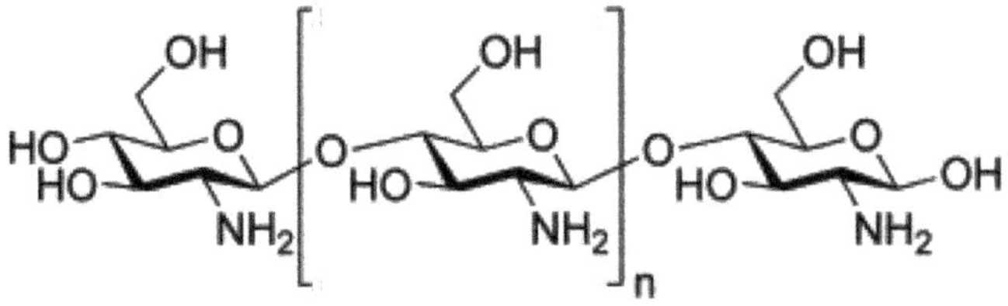

**Chitosan**

**Yield of chitin and chitosan:** The extraction of the chitin and chitosan of various body parts of mud crab *Scylla tranquebarica* showed that the leg shell yield is more than that of carapace shell and claw shell. The chitosan yielded representing 6.59%, 4.12% and 8.42%, respectively in the shell of carapace, claw and legs of *Scylla tranquebarica* (Thirunavukkarasu and Shanmugam, http://dx.doi.org/10.18000/ijabeg.10048).

## 6.1 BIOMEDICAL APPLICATIONS OF CHITIN AND CHITOSAN

The biomedical applications of chitin, chitosan and their derivatives are largely due to their antibacterial, antifungal, antioxidant, antidiabetic, anti-inflammatory, anticancer, and hypocholesterolemic properties.

Biomedical, Food, and Industrial Applications 331

### 6.1.1 APPLICATIONS OF CHITOSAN IN HUMAN HEALTH

**Oral drug delivery:** The potential of chitosan films containing 'diazepam' as an oral drug delivery has been investigated in rabbits. The results indicated that a film composed of a 1:0.5 drug-chitosan mixture might be an effective dosage form that is equivalent to the commercial tablet dosage form. The ability of chitosan to form film may permit its use in the formulation of film dosage form as an alternative to pharmaceutical tablets.

**Stability of drug:** Chitosan polymer has been used to increase the stability of drug in which, the drug is complexed with chitosan to make slurry which is released for 45 min until dough mass appears. This dough mass is passed through a sieve (No. 16) to get granules which are completely stable at different conditions.

**Osteo-properties:** Chitosan has been found to be useful in promoting tissue growth, repairing tissues and accelerating wound healing and bone regeneration. Chitosan as a biopolymer exhibits osteo-conductive and antimicrobial properties which make it attractive for use as a bioactive.

**Dental medicine:** Chitosan has a variety of biological activities and serves as an important ingredient in healthy foods which are effective for improvement and/or care of various disorders such as arthritis, cancer, diabetes hepatitis, etc. It is worthy of mention that chitosan dressing has dramatically recovered 90% of burnt skin besides preventing scar formation in a 3-year-old boy. In dental medicine, chitosan dressing has yielded desired results in oral mucous wounds and is used as a tampon following radical treatment of maxillary sinusitis. Further, chitosan can also serve as an absorbing membrane for periodontal surgery.

**Bone formation:** Chitosan composite microgranules could be fabricated as bone substitutes for the purpose of obtaining high bone forming efficiency.

**Cholesterol lowering:** Incorporation of chitosan in the diet may reduce cholesterol significantly.

**Permeation enhancer:** Chitosan due to its cationic nature is capable of opening tight junctions in cell membranes. This property has leaded the use of chitosan as a permeation enhancer for hydrophilic drugs that may otherwise have poor bioavailability such as peptides.

**Mucoadhesive recipient:** Bioadhesivity is often used as an approach to enhance residence time of a drug in the GI tract thereby, increasing the oral bioavailability. A comparison between chitosan and other commonly used polymeric excipients indicates that the cationic polymer has higher bioadhesivity compared to other natural polymers such as cellulose, xanthium gum and starch.

**Ophthalmic drug delivery:** Chitosan has been found to be a unique material for the design of ocular drug delivery vehicles owing to its bioadhesion and permeability-enhancing properties and interesting physicchemical characteristics. Due to their elastic properties, chitosan hydrogels offer better acceptability with respect to solid or semisolid formulation for ophthalmic delivery such as suspensions or ointments. Ophthalmic chitosan gels improve adhesion to the mucin which coats the conjunctiva and the corneal surface of the, eye an increase precorneal drug residence times showing down drug elimination by the lachrymal flow. In addition, its penetration enhancement has more targeted effect and allows lower doses of the drugs. In contrast, chitosan-based colloidal systems were found to work as transmucosal drug carriers either facilitating the transport of drugs to the inner eye (chitosan-coated colloidal system containing indomethacin) or their accumulation into the corneal/conjunctival epithelia (chitosan nanoparticulate containing cyclosporine). The microparticulate drug-carrier (microspheres) seems to be a promising means of topical administration of acyclovir to the eye. The duration of efficacy of the ofloxacin was measured by using high MW (1930Kd) chitosan.

**Gene delivery:** Chitosan has been used as a carrier of DNA for gene delivery applications. Also chitosan could be a useful oral gene carrier because of its adhesive and transport properties in the GI tract. Plasmid DNA containing cytomegalo virus promoter sequence and a luciferase reporter gene could be delivered in vivo by chitosan and depolymerized chitosan oligomers to express luciferase gene in the intestinal tract (Bansal et al., 2011; Ramesh et al., 2012).

### 6.1.2 CHITOSAN AND ITS DERIVATIVES IN PREVENTION AND TREATMENT OF AGE-RELATED DISEASES

Degenerative age-related diseases such as cardiovascular and cerebrovascular disease, diabetes, osteoporosis and cancer are the common diseases

Biomedical, Food, and Industrial Applications     333

in older persons, and these diseases are also diet-oriented. Older people often suffer from impaired immunity due to the deficiency of trace elements such as zinc, iron, selenium and copper; and vitamins A, B, C and E. Functional foods and nutraceuticals with antioxidant, anti-inflammatory, anti-diabetic and anticancer properties may prevent age-related and diet-related diseases.

**Oxidative stress:** Chitosan has been reported to have a direct antioxidant activity in systemic circulation by lowering the indices of oxidative stress in both in vitro and in vivo studies. Dietary chitosan supplementation attenuates isoprenaline-induced oxidative stress in rat myocardium and the antiaging effect of dietary chitosan supplementation on glutathione-dependent antioxidant system in young and aged rats has been reported.

**Inflammation:** Oxidative stress and inflammation are involved in the pathology of age-related diseases such as cardiovascular diseases, cancer, neurodegenerative diseases, rheumatoid arthritis, and diabetes. Chronic inflammation is considered as a major risk factor for age-related diseases.

**Diabetes Mellitus:** Diabetes Mellitus is considered as a risk factor for incidence of age-related dementia, Alzheimer's disease, and cardiovascular diseases Chitosan and its derivatives have the potential for use in several antidiabetic therapeutic applications. Chitosan has shown blood glucose-lowering and lipid-lowering effects in neonatal streptozotocin-induced diabetic mice. Chitosan lactate has an antidiabetic effect also in obese diabetic KK-Ay mice. Chitosan may possess a potential for alleviating type-1 diabetic hyperglycemia through the decrease in liver gluconeogenesis and increase in skeletal muscle glucose uptake and use. Long-term administration of chitosan has been reported to reduce insulin resistance through suppression of lipid accumulation in liver and adipose tissues and amelioration of chronic inflammation in diabetic rats.

**Hypercholesterolemia**: Effect of media-milled chitosan on the decrease of serum triacylglycerol, total cholesterol and LDL cholesterol is higher compared to chitosan. It has been also demonstrated recently that total cholesterol content in mice blood fed during 12 weeks with γ-irradiated chitosan (30–100 kGy) was significantly lower than that of the control.

**Cancer:** Chitin and chitosan oligosaccharides have been evaluated as functional foods against cancer. Curcumin-loaded chitosan-coated nanoparticles can be used for the local treatment of oral cavity cancer. Oral administration of chitosan based nanoformulated green tea polyphenol EGCG effectively inhibits prostate cancer cell growth. EGCG has been encapsulated also into chitosan-coated nanoliposomes and anticancer effects are expected in the treatment of breast cancer. The curcumin/5-fluorouracil loaded thiolated chitosan nanoparticles showed enhanced anticancer effects on colon cancer cells in vitro.

**Neurodegenerative diseases:** Water-soluble chitosan inhibits the production of pro-inflammatory cytokine in human astrocytoma cells activated by A$\beta$ and IL-1$\beta$ and may reduce the pathological events associated with Alzheimer's disease (Kerch, 2015).

### 6.1.3 ANTIOXIDANT AND ANTIMICROBIAL ACTIVITIES OF CHITOSAN

**Antioxidant and antimicrobial activities of astaxanthin of *Portunus sanguinolentus* and *Callinectes sapidus:*** Astaxanthin pigment has been isolated from the powdered shells of portunid crab species viz. *Portunus sanguinolentus* and *Callinectes sapidus* and its antioxidant and antimicrobial activities were evaluated. The yield of astaxanthin from *Portunus sanguinolentus* and *Callinectes sapidus* was found to be 29.01 µg/g and 38.98 µg/g respectively. Total antioxidant activity of astaxanthin in these species was found to be 50% and 83%, respectively (standard astaxanthin showed 58%). The antimicrobial activity of isolated and standard astaxanthin was also studied against the isolated organism (*E.coli*) from spoiled milk and rotten meat. The astaxanthin pigment was found to be more effective against isolated strains at concentration of 50 µg. The astaxanthin of the portunid crabs may therefore be of great use in food and pharmaceutical industries (Suganya and Asheeba, 2015)

**Antioxidant activity of haemolymph of hard and soft shell crabs of *Charybdis lucifera:*** Antioxidant activity of soft shelled and hard shelled crabs of the species *Charybdis lucifera* has been studied. The soft shelled crabs showed maximum phenolic content of 48% which is tentatively higher than that of hard shelled crabs. The total antioxidant potential of soft shelled crab exhibited maximum antioxidant potential of 49% and

Biomedical, Food, and Industrial Applications 335

minimum effect of 32% was recorded in hard shelled crab. The results suggest that the soft-shelled crab may be used for the preparation of antioxidant (Soundarapandian et al., 2014).

**Antimicrobial activity of chitin, chitosan and their derivatives against bacteria:** Chitosan concentrations of 0.005% were found to be sufficient to elicit complete inactivation of *Staphylococcus aureus*. A much higher concentration of chitosan (1±1.5%) is required for complete inactivation of *Staphylococcus aureus* after two days of incubation at pH 5.5 or 6.5 in the medium. *Bacillus cereus* required chitosan concentrations of (0.02% for bactericidal effect, while *Escherichia coli* and *Proteus vulgaris* showed minimal growth at 0.005%, and complete inhibition at 0.0075%). Complete inactivation of *Escherichia coli* after a 2-day incubation period was with concentrations of 0.5 or 1%, at pH 5.5. Higher concentrations (0.1%) were required to inhibit *Escherichia coli*. Only 0.0075% chitosan was needed to inhibit the growth of *Escherichia coli*. Chitosan glutamate and chitosan lactate were also bactericidal against both gram-positive and gram-negative bacteria. A concentration of 0.2 mg/mL chitosan lactate appeared most effective against *Escherichia coli* (Shahidi et al., 1999).

**Antimicrobial activity of chitin, chitosan and their derivatives against fungi:** Chitosan has been found to reduce the in vitro growth of numerous fungi with the exception Zygomycetes. Chitosan has a dual function, that is direct interference of fungal growth and activation of several defense processes Chitosan (with 7.2% NH2) reduced markedly the radial growth of *Botrytis cinerea* and *Rhzopus stolonifer,* with a greater effect at higher concentrations. After 14 days of storage, chitosan coating at 15 mg/mL reduced decay of strawberries caused by the same fungi by more than 60%, and also observed that coated fruits ripened normally and did not show any apparent sign of phytotoxicity. The growth of *Aspergillus niger* was found inhibited by the addition of chitosan (0.1±5 mg/mL) to the medium (pH 5.4), Chitosan coating for the inhibition of Sclerotinia rot of carrot showed that the incidence of rotting was significantly reduced (from 88 to 28%) by coating carrot roots with 2 or 4% chitosan Coating fruits and vegetables with chitosan or its derivatives may have some positive advantages for long term storage of these foods (Shahidi et al., 1999).

336          Biology and Culture of Portunid Crabs of World Seas

**Antibacterial activity in the haemocytes of the shore crab, *Carcinus maenas:*** The presence of antibacterial activity in the haemocytes of the shore crab, *Carcinus maenas* (L.) (Crustacea: Decapoda), was investigated using a selection of Gram-positive and Gram-negative bacteria. Preliminary investigations into the relationship between this activity and the prophenoloxidase activating system (proPO) were also carried out. Antibacterial activity against both Gram-positive and Gram-negative organisms were found to reside exclusively in the granular haemocytes and eight of the twelve bacteria tested were susceptible to this effect. Additional studies, using *Psychrobacter immobilis* (= *Moraxella* sp.), revealed that the factor (or factors) responsible was 90% effective within 60 min and was also heat stable, independent of divalent cations, and non-lytic in character. Although antibacterial activity resides in the same cell population that carries the pro-PO system, there appears to be no relationship between antibacterial activity and phenoloxidase itself. Other components of the proPO system may however be involved (Chisholm and Smith, 1992).

### Antiviral drugs from the chitin of crab wastes

Recently researchers have developed a technique in which chitin has been used to produce a currently very-expensive source of antiviral drugs. Many of the presently-used antiviral drugs are derived from N-acetylneuraminic acid (NANA). The substance can be synthesized or obtained from natural sources, but in either case, it is very costly – at about 2,000 euro (US$2,626) a gram, it is approximately 50 times more valuable than gold. To produce the cheaper NANA, scientists have introduced bacterial genes into a very common fungus called *Trichoderma*. Normally, the fungus feeds on chitin and breaks it down into monomer amino sugars. With the addition of the new genes, however, a couple of extra steps are added to the process, with the chitin ending up as N-acetylneuraminic acid. The reportedly eco-friendly process has been patented, and the researchers hope to see industrial-level production of cheap NANA beginning soon (Coxworth, http://www.gizmag.com/chitin-cheaper-antiviral-drugs/21457/).

Biomedical, Food, and Industrial Applications 337

### *Hemostasis effects of chitosan and its derivatives*

The effects of chitin, chitosan, and their derivatives on in vitro human blood coagulation and platelet activation were comparatively studied. The coagulation was assessed by the measure of the whole blood clotting time (WHBCT) and plasma recalcification time (PRT). The tested materials were chitin, chitosan, partially N-acetylated chitosan (PNAC), *N*, *O*-carboxymethylchitosan (NOCC), *N*-sulphated chitosan, *N*-(2-hydroxy)propyl-3-trimethylammonium chitosan chloride, and SPONGOSTAN® standard (a positive control). The results revealed that the WHBCTs of whole blood mixed with chitin, chitosan, NOCC, or SPONGOSTAN® standard were significantly decreased with respect to that of the pure whole blood (a blank control) ($P < 0.05$), while the WHBCT value of whole blood mixed with PNAC was not significantly reduced (Janvikul et al., 2006).

### *Clot-inducing celox from the chitosan of crab shell wastes*

Researchers have created a novel substance Celox, a clot-inducing granular hemostatic agent using crab shell wastes. Celox halts minor bleeding in seconds and can stop lethal femoral artery bleeding with 3 minutes compression. Even in hypothermic conditions, Celox creates its gel-like clot in 30 seconds. It is also able to function when anticoagulants (like warfarin or heparin) are in our system. Celox works by bonding with red blood cells and gelling with fluids to produce a sticky pseudo clot. This clot sticks well to moist tissue to plug the bleeding site. Celox does not set off the normal clotting cascade, it only clots the blood it comes directly into contact with. (Behrman, http://gizmodo.com/5806944/celox-how-crab-shells-will-stop-you-from-bleeding-to-death/).

### *Hemostatic wound dressing based on crab wastes*

Materials for dressings must have hemostatic mechanisms built in to aid in the formation of blood coagulation. "The hemostatic agents are plant (cellulose)-based, gelatin-based, collagen-based, fibrin-based, thrombin-based, chitin/chitosan-based or mineral-based." Chitosan has been shown to be more effective than chitin in controlling severe hemorrhage. It is used often and has shown to have no side effects when used. Therefore, chitosan finds applications in the manufacturing of hemostatic wound dressings. This material is effective in hemostasis (Anon. Hemostatic Wood Dressingfshttps://biotextiles2013.wordpress.com/hemostatic_wound_dressings/).

Based on the hemostatic properties reported from studies of chitin, chitosan and their derivatives, commercial chitin- and chitosan-based hemostatic dressings have flooded the market and are used in medicine.

## D-Glucosamine

D-glucosamine

D-glucosamine is a structural unit of chitosan, and is produced commercially by the hydrolysis of chitosan in hydrochloric acid. D-glucosamine could be oxidized to D-glucosaminic acid which is one of the carbohydrate units used to manufacture various biotic substances. It has various physiological functions.

## Biological activities of D-Glusosaminic acid

In recent years, research on D-glucosaminic acid has increased because of its industrial, agricultural, food and medical applications. D-glucosaminic acid has recently been identified as a promising sweetener and condiment. D-glucosaminic acid has been found to be a biocompatible, nontoxic ligand chelated with many metals for potential medical applications.

Biomedical, Food, and Industrial Applications 339

D-glucosaminic acid iron (III) complexes are potential pharmaceuticals in human and veterinary iron therapy since their stabilities are high enough to prevent metal ion hydrolysis in biological systems. The platinum (IV)/D-glucosaminic acid complex has received considerable attention for cancer therapy because of lower toxicity and the possibility of oral administration. Moreover, D-glucosaminic acid has also been selected to design a new kind of chromium drug candidate for anti-diabetic purposes. Two chromium (III) 1:1 and 2:3 (Cr: glucosaminate) complexes of glucosaminic acid were synthesized by neutralization and exchange reaction. The effect of the complexes on decreasing blood sugar was investigated on type-2 diabetes model rats induced by tetraoxypyrimidine. The results indicated that the effect on decreasing blood sugar was comparable to that of picolinate chromium complex (Cr (pic)3) currently used worldwide (Zhang et al., 2010).

### Haemocytes of *Carcinus aestuarii* in immune responses

Haemolymph of *Carcinus aestuarii* has been reported to possess a high quantity of total proteins, functional properties of which need to be better investigated in future studies. It is also reported that *Carcinus aestuarii* haemocytes are not very active phagocytic cells, but they are more active in terms of both hydrolytic and oxidative enzyme activities and superoxide anion production (Matozzo and Marin, 2010).

## 6.2 FOOD APPLICATIONS OF CHITIN, CHITOSAN AND THEIR DERIVATIVES IN THE FOOD INDUSTRY

**Nutritional effects of chitin and chitosan in foods:** Multiple actions of chitin and chitosan in food systems relate to their effects as dietary fiber and as functional ingredients. The United States Food and Drug Administration (USFDA) has approved chitosan as a feed additive 1983. Chitosan is also used in the food industry as a food quality enhancer in certain countries. Japan produces dietary cookies, potato chips and noodles enriched with chitosan because of its hypocholesterolemic effect. Furthermore, vinegar products containing chitosan are manufactured and sold in Japan, again because of their cholesterol lowering ability. Recently, the nutritional significance of chitinous polymers has been demonstrated in animals and the effectiveness of chitin and chitosan as feed additives has also been indicated. Normal growth patterns were observed with hens and broilers fed <1.4 g of chitosan/kg of body weight per day for up to 239 days and with rabbits fed <0.8 g of chitosan/kg of body weight per day for the same period. Furthermore, the serum cholesterol and triacylglycerol values of rabbits, hens and broilers were kept low by feeding 2% chitosan. In a similar study, increased high-density lipoprotein (HDL) concentrations were observed after feeding chitosan containing diet to broiler chicken. This could be attributed to the enhanced reverse cholesterol transport in response to intestinal losses of dietary fats. The effect of chitin, chitosan and cellulose as dietary supplements on the growth of cultured fishes such as red sea bream, Japanese eel, and yellow tail has been investigated by several researchers. The growth rate of all fishes fed with a 10% chitin supplement was the highest, thus indicating its applicability in feed. Feed efficiency in the red sea bream and Japanese eel fed a 10% chitin supplemented diet was also the highest.

### Applications of chitinous materials of crab shell wastes in the food industry

**Bioconversion of chitin to single cell protein**: Bioconversion of crab shell wastes into single cell protein, which is a suitable feed supplement for animals and aquatic organisms, has been studied by several researchers. This process can play an economical role not only in shellfish processing, but also in integrated aquaculture systems. Indirectly, it serves as a method of waste management which may reduce large quantities of shellfish waste from processing plants. There are four steps in production of single cell protein from crab shell wastes. These are; (I) drying, size reduction and chemical purification of shellfish processing waste; (II) extraction of chitinase enzyme from purified waste; (III) chitin hydrolysis; and (IV) fermentation in submerged culture (product generation stage). The appropriate microorganism for extracellular chitinase production has been found to be *Serratia marcescens* QMB 1466. It is also found that a temperature of 30°C and an initial pH of 7.5 in the medium were suitable for chitinase production of the above species (Tom and Carroad, 1981).

**Supplementation of chitin in the production of ethanol from cane molasses:** Studies have been made in accelerating the rate of ethanol production by supplementation with 0.2% carbohydrates such as acacia gum, chitin, xylan, pullulan, cellobiose, dextrin, inulin and agar. When two yeast strains viz. *Saccharomyces cerevisiae* NCIM 3526 and *Saccharomyces uvarum* NCIM 3509 were used for all fermentation reactions, it was found that, among carbohydrate supplements used, chitin was found to be most effective in accelerating the rate of ethanol production from cane molasses. Approximately 5.38% ethanol was formed after 36 h at 37°C from cane molasses containing 16% reducing sugar with chitin supplements (0.2%) in the fermentation medium. In the absence of any supplement, more than 72 h was needed to produce the same amount of ethanol. Chitin supplementation can therefore reduce the fermentation time to one-third and hence the cost of ethanol production can considerably be reduced. It is also demonstrated that the rate of ethanol formation was enhanced in the presence of chitin, acacia gum, xylan, dextrin or cellobiose in the broth culture media. In the same study, these researchers further observed that the rate of ethanol production was higher for all the culture strains used in the presence of chitin than simply with yeast extract supplements or controls with no supplements (Patil and Patil, 1989).

**Applications of chitin and chitosan in food for human health**: Research on preparation and physiological activities of chitin and chitosan oligomers has attracted much attention in the food and pharmaceutical fields due to their versatile antitumor activity, immuno-enhancing effects, protective effects against some infectious pathogens in mice, antifungal activity, and antimicrobial activity. Studies have shown the antitumorigenic properties of chitin and chitosan oligomers in the inhibition of the growth of tumor cells via an immuno-enhancing effect. Chitin and chitosan oligomers, (GlcNAc)6 and (GlcN)6, have shown a tumor growth-inhibitory effect in allogenic and syngeneic mouse system, including sarcoma 180 solid tumor and MM46 solid tumor. Two oligosaccharides, (GlcNAc)6 and (GlcN)6, have been reported to exhibit growth-inhibitory effect against Meth-A solid tumor transplanted into BALB/c mice. A significant antimetastatic effect has been reported for (GlcNAc)6 in mice bearing Lewis lung carcinoma. Chitin and chitchitosan oligomers were also found responsible for enhancing protective effects against infection with some pathogens in mice (Shahidi et al., 1999).

**Antioxidative properties of chitosan and chitosan derivatives on muscle foods:** Muscle food products are highly susceptible to rancidity development caused by oxidation of their highly unsaturated lipids. Effectiveness of chitosan treatment on oxidative stability of beef has been studied. Addition of chitosan at 1% resulted in a decrease of 70% in the 2-thiobarbituric acid (TBA) values of meat after 3 days of storage at 4°C. The effect of N-carboxymethylchitosan to prevent the warmed-over favor (WOF) in uncured meat has been studied and the results show that N-carboxymethylchitosan is effective in controlling WOF over a wide range of temperature. Use of 5000 ppm N-carboxymethylchitosan in ground beef resulted in a 93% inhibition of TBA and 99% reduction in the hexanal content in the products. Further, N,O-carboxymethylchitosan (NOCC) and its lactate, acetate and pyrrolidine carboxylate salts were found effective in controlling the oxidation and favor deterioration of cooked meat over a nine-day storage at refrigerated temperatures. The mean inhibitory effect of NOCC and its aforementioned derivatives at 500 ppm was 46.7, 69.9, 43.4 and 66.3%, respectively, as reflected in their TBA values. The mechanism by which this inhibition takes place is thought to be related to chelation of free iron which is released from hemoproteins of meat during heat processing. This would in turn inhibit the catalytic activity of iron ions. Further, addition of 3000 ppm N-carboxymethylchitosan to cooked pork was found to be sufficient to prevent the oxidative rancidity of the product (Shahidi et al., 1999).

**Chitin and chitosan in the edible film industry:** The use of edible films and coatings to extend shelf life and improve the quality of fresh, frozen and fabricated foods has been examined during the past few years due to their ecofriendly and biodegradable nature. These outer layers/films can provide supplementary and sometimes essential means of controlling physiological, morphological and physicochemical changes in food products. High density polyethylene film, a common packaging material used to protect foods, has disadvantages like fermentation due to the depletion of oxygen and condensation of water due to fluctuation of storage temperature, which promotes fungal growth. There are many mechanisms involved in extending shelf life of food by coating films. Due

Biomedical, Food, and Industrial Applications 343

to their film-forming properties, chitin and chitosan have been successfully used as food wraps. The use of N, O-carboxymethylchitin films to preserve fruits over long periods has been approved in several countries. Due to its ability to form semipermeable film, chitosan coating can be expected to modify the internal atmosphere as well as decrease the transpiration loss and delay the ripening of fruits. Rigid chitosan films can be formed using crosslinking agents such as glutaraldehyde, divalent metal ions, polyelectrolytes, or even anionic polysaccharides. The preparation of chitosan and chitosan laminated films with other polysaccharides has been reported by various authors; these include chitosan films, chitosan/pectin laminated films and chitosan/methylcellulose films. Chitosan films are tough, long-lasting, flexible and very difficult to tear. Most of these mechanical properties are comparable to many medium-strength commercial polymers. It is also reported that chitosan films have moderate water permeability values and could be used to increase the storage life of fresh produce and foodstuffs with higher water activity values (Shahidi et al. 1999).

**Effect of chitosan coating on storability and quality of fresh fruits:** Extension of the storage life and better control of decay of peaches, Japanese pears and kiwifruits by application of chitosan film has been reported. Similarly, cucumbers, and bell peppers, strawberries, and tomatoes could be stored for long periods after coating with chitosan. These results may be attributed to decreased respiration rates, inhibition of fungal development and delaying of ripening due to the reduction of ethylene and carbon dioxide evolution (Shahidi et al. 1999).

**Antimicrobial properties of chitosan and chitosan-laminated films:** Chitosan and chitosan-laminated films contain antimicrobial agents. Preservatives released from this packaging film deposit on the food surface inhibit the microbial growth. The presence of preservatives in chitosan films also reduces the intermolecular electrostatic repulsion in the chitosan molecules and facilitates formation of intramolecular hydrogen bonds (Shahidi et al., 1999).

**Application of chitosan for clarification and deacidification of fruit juices (fining agent):** Chitosan salts, which carry a strong positive charge may be used to control acidity in fruit juices. They are also good clarifying agents for grapefruit juice either with or without pectinase treatment. With the addition of chitosan at 0.8 kg/m³, apple juice has yielded zero turbidity products. Further, chitosan has a good affinity towards polyphenolic compounds such as catechins, proanthocyanidins, cinnamic acid and their derivatives that can change the initial straw-yellow color of white wines into deep golden-yellow color due to their oxidative products. By adding chitosan to grapefruit juice at a concentration of 0.015 g/mL, total acid content was found reduced by about 52.6% due to decreasing the amount of citric acid, tartaric acid, l-malic acid, oxalic acid and ascorbic acid, by 56.6, 41.2, 38.8, 36.8 and 6.5%, respectively (Shahidi et al. 1999).

## 6.3  INDUSTRIAL APPLICATIONS OF PORTUNID CRAB WASTES

### 6.3.1  USE OF CHITIN AND CHITOSAN IN WASTEWATER TREATMENT

**Use of chitosan in wastewater treatment in metal-plating plants:** The crab shell wastes provide a solution to the hazardous-waste disposal problem faced by the metal-plating industry. The shell of the red crab has been reported to clean wastewater from electroplating operations. Chitosan of crab shells bonds to the metals in the wastewater of plating plants, and the metals can then be recovered and used again. A plating plant that spends $2,019 a month using standard treatment technology could cut those costs to $790 by using crab shells. In the laboratory, crab shells have been used, cleaned and reused as many as 10 times without losing their effectiveness. A plating plant might need a few hundred pounds of the crab shells a year for waste-water treatment. When the shells' use is exhausted, they can be cleaned a final time and crushed into animal feed for roughage (Hamilton, http://www.nytimes.com/1987/09/20/nyregion/crab-shells-seen-as-industry-waste-solution.html).

Chitin and chitosan as coagulation and flocculation substances in food processing wastewater: Aluminum sulphate (alum), ferrous sulphate, ferric chloride and ferric chloro-sulphate are commonly used as coagulants. However, a possible link of Alzheimer's disease with conventional aluminum based coagulants has become an issue in wastewater treatment. Hence, special attention has been made towards using biodegradable polymer, chitosan in treatment, which is more environmental friendly. Chitosan, with its partial positive charge, can effectively function as a polycationic coagulant in wastewater treatment. As a coagulating agent for waste treatment systems, chitosan is particularly effective in removing proteins from wastes; the coagulated by-products could serve as a source of protein in animal feed. Chitosan has been found to reduce the suspended solids of various food processing wastes to different contents. There are two stages that result in destabilization of a colloidal system; coagulation and flocculation. The former is the process where the forces holding the particles in suspension are neutralized, whereas flocculation is the process in which destabilized suspended particles are brought together to form larger aggregates.

**Chitosan as coagulation and flocculation substances in treating textile industry wastewater:** Experiments were carried out on textile industry wastewater by varying the operating parameters, such as chitosan dosage, pH and mixing time in order to study their effect in flocculation process by using chitosan. The results obtained proved that chitosan had successfully flocculated the anionic suspended particles and reduce the levels of Chemical Oxygen Demand (COD) and turbidity in textile industry wastewater. The optimum conditions for this study were at 30 mg/l of chitosan, pH 4 and 20 minutes of mixing time with 250 rpm of mixing rate for 1 minute, 30 rpm of mixing rate for 20 minutes and 30 minutes of settling time. Moreover, chitosan showed the highest performance under these conditions with 72.5% of COD reduction and 94.9% of turbidity reduction. Chitosan is therefore found to be an effective coagulant, which can reduce the level of COD and turbidity in textile industry wastewater (Hassan et al., 2011).

**Purification of copper and cadmium contaminated waters by crab shell wastes:** The crab shell wastes can be effective for elimination of copper and cadmium from contaminated waters. Industrial and agricultural processes can increase the concentrations of copper and cadmium in sources. Copper can affect health causing symptoms like nausea while cadmium can be harmful for the kidneys. Bioabsorbents, which are the naturally available waste materials, have been found to purify water of the contaminants without causing much effect on environment and human health. The crushed shells from the crab *Scylla serrata* may be used as biosorbents to treat wastewater and remove its copper and cadmium contents.

Calcium carbonate present in the shells has been found to form strong bonds when exposed to copper and lead. Chitin absorbs the dissolved metals after that. Being a common waste material from the fishing industry, crab shells can be used effectively as an economic heavy metal remover. Crab shells have been found to eliminate up to 95% of copper and 85% of cadmium content in 6 hours, which have lead the researchers to the conclusion that that crab shell wastes can work as heavy metal absorbents for industrial wastewater (http://www.water-technology.net/news/news-crab-shells-effective-in-purifying-contaminated-water-say-malaysian-researchers-4658258).

**Application of chitin and chitosan in the removal of metals from industrial wastewater:** Conventional methods for the removal of metals from industrial wastewater, may be ineffective or expensive. Chelation ion exchange is a technique which can be used to recover metal ions from wastewater. Environmentally safe biopolymers are capable of lowering transition metal ion concentrations to parts per billion levels. Chitosan can be utilized as a tool for the purification of wastewater because of its high sorption capacity. The capacity of chitin and chitosan to form complexes with metal ions has been exploited in Japan for water purification. The use of commercially available chitosan for potable water purification has been approved by the United States Environmental Protection Agency (USEPA) up to a maximum level of 10 mg/liter. The effectiveness of cross-linked N-carboxymethylchitosan in removing lead and cadmium from drinking water has already been demonstrated. A study of the metal binding capacity of chitosan has shown that it has a high binding capacity with metals such as copper and vanadium. It is also reported that unfunctionalized chitosan is effective in removing $Cu2+$, but not $Pb2+$. However, for copper at 10 and 100 ppm, the best adsorbents were found to be carboxymethylchitosan and the ion exchange resin, respectively. The ability of chitosan to remove polychlorinated biphenyls (PCB) from contaminated stream water has already been tested and it was found that chitosan is highly efficient and more effective than activated charcoal for purification of PCB contaminated water.

**Chitosan in the removal of aquatic pollutants:** Chitin and chitosan-derivatives have gained much attention as effective biosorbents due to low cost and high contents of amino and hydroxyl functional groups which

Biomedical, Food, and Industrial Applications 347

show significant adsorption potential for the removal of various aquatic pollutants such as (a) metal cations and metal anions; (b) radionuclides; (c) different classes of dyes; (d) phenol and substituted phenols; (e) different anions and other miscellaneous pollutants. However, still there is a need to find out the practical utility of such developed adsorbents on commercial scale (Bhatnagar and Sillanpää, 2009).

### Production of chitin and L(+)-lactic acid from crab wastes

Shell wastes of the portunid crab, *Callinectes bellicosus* have been employed for simultaneous production of chitin and L(+)-lactic acid by submerged fermentation of *Lactobacillus* sp. B2 using sugar cane molasses as carbon source. Fermentations in stirred tank reactor using selected conditions produced 88% demineralization and 56% deproteinization with 34% yield of chitin and 19.5 g/liter of lactic acid (77% yield). The chitin purified from fermentation showed 95% degree of acetylation and 0.81 and 1 % of residual ash and protein contents, respectively (Flores-Albino et al., 2012).

### Biogas production from crab shell wastes

Crab shells are known to contain non-renewable biological substances. They may be used to produce biogas, a viable energy source (http://www.euronews.com/2014/05/12/from-crab-shells-to-raw-materials/).

### Production of polymers (plastics) from crab shell wastes

Biochemists have developed yeast strains to convert the chitin into fatty oils through fermentation in a period of 5–7 days. The natural oil so obtained from crab shells may be transformed into a pure chemical, the raw material for the production of plastic. Next step is the polymerization. Through this process, a high performance biopolymer is produced in a reactor at temperatures of up to 300°C. It is then chopped into granulates. The biopolylmer is then molded into sample pieces for quality control.

The crab shell wastes may therefore be used for the production of plastics in the near future (http://www.euronews.com/2014/05/12/from-crab-shells-to-raw-materials/).

### Battery materials from crab shell wastes

**Lithium batteries**: The shells of portunid crabs may be used to fabricate materials for the production of batteries. The nanostructures found in these shells could serve as templates to make sulphur and silicon electrode materials for lithium-ion batteries. (Hongbin et al., 2013). The portunid crab shells could serve as an inexpensive, environmentally friendly alternative to those templates. It is learnt that every year, the seafood industries waste about 0.5 million tons of crab shells, and this could be converted into a low-cost, renewable source of these templates (Boerner, http://cen.acs.org/articles/91/web/2013/06/Crab-Shells-Help-Researchers-Make.html; Hongbin et al., 2013).

### KEYWORDS

- **biomedical applications**
- **chitin**
- **chitosan**
- **food applications**
- **human health applications**
- **industrial applications**

# CHAPTER 7

# DISEASES AND PARASITES OF PORTUNID CRABS

## CONTENTS

7.1 Shell Diseases of Portunid Crabs ................................................. 349
Keywords ....................................................................................... 373

Portunid crabs are prone to shell diseases which lead to internal damages causing variation in haemocyte counts and histopathological alteration in internal tissue and organs. The crab shell diseases are associated with infectious agents like bacteria, virus, fungi, dinoflagellates, protistans, ciliates, metazoans, and foulers. Among them, fungi are occasionally involved in shell diseases. Shell diseases are characterized by various types of erosive lesions on carapace. Black spot and white spot lesions are some of the classic and common shell diseases in portunid crabs. Infections in portunid crabs may lead to great economic loss (Joseph and Ravichandran, 2012)

## 7.1 SHELL DISEASES OF PORTUNID CRABS

### 7.1.1 BACTERIAL DISEASES

(a) Bacterial necrosis ('black spot,' 'brown spot,' 'burnt spot,' 'shell disease' or chitinolytic bacterial disease).

Host species: *Scylla serrata, Liocarcinus puber, L. corrugatus, L. depurator and Carcinus maenas*

Causative agent: Chitinolytic bacteria (*Vibrio* spp., *Pseudomonas* spp., *Aeromonas* spp., and *Spirillium* spp.)

**(b) Bacterial shell disease**
Host species: *Scylla serrata.*
Causative agent: Chitinolytic or chitin digesting bacteria viz. *Vibrio vulnificus, V. parahaemolyticus, Vibrio spendidus*

**(c) Filamentous bacterial disease**
Host species: *Scylla serrata*
Causative agent: Filamentous bacteria such as *Leucothrix mucor, Thriothrix* spp. and *Flexibacter* spp.

**(d) Luminescent vibriosis**
Host species: *Scylla serrata*
Causative agent: Rod-shaped bacteria, *Vibrio harveyi*

**(e) Black gill disease**
Host species: *Scylla serrata*
Causative agent: Unidentified bacteria

**(f) Milky disease**
Host species: *Carcinus maenas*
Causative agent: Tiny, bacteria-like particles (alpha- proteo bacteria) in hepatopancreas cells, as well as bacteria apparently escaping from ruptured cells (Eddy et al., 2007).

**(g) Red sternum syndrome**
Host species: *Scylla serrata*
Causative agent: Rod-, curve rod-, or coccus-shape bacteria

### *7.1.2 VIRAL DISEASES*

http://www.inspection.gc.ca/animals/aquatic-animals/diseases/reportable/white-spot-disease/fact-sheet/eng/1336065470439/1336068112774.

Diseases and Parasites of Portunid Crabs 351

**(a) White spot disease**

**Host species**: *Carcinus maenas Callinectes sapidus, Charybdis annulata, Charybdis feriatus, Charybdis granulata, Charybdis lucifera, Liocarcinus puber, Podophthalmus vigil, Portunus pelagicus, Portunus sanguinolentus, Scylla serrata and Thalamita danae.*

**Signs of white spot disease:** White spot disease causes death in juvenile and adult crustaceans. Infected animals die within 3 to 10 days of the appearance of the disease. Affected crustaceans may exhibit any of the following signs:

- Behavior
  - Decreased appetite
  - Gather near the surface and edges of the rearing units
- Appearance
  - White spots in the shell
  - Appear darkened and red or pink in color
  - Covered with debris (not preening to keep clean)
  - Gut appears white in color
  - Shell may be loose

**(b) White spot syndrome virus (WSSV)**

**Host species:** *Scylla serrata*

**Causative agent:** Whispovirus

## 7.1.3 FUNGAL DISEASES

**Host species:** *Scylla serrata*

**Causative species:** *Fusarium* sp., *Lagenidium* sp., *Haliphthoros* sp., *Halocrusticida* sp. and *Atkinsiella* sp.

## 7.1.4 PARASITIC INFECTIONS

**(A) Protozoan parasitic infections**

**Host species:** *Scylla serrata*

**Causative species:** Species of *Vorticella, Zoothamnium, Epistylis* and *Acineta*

**(B) Metazoan parasitic infections**
**i) Platyhelminthes**
**(a) Cestodes**
**Host species**: *Portunus pelagicus* (females)
**Causative agents**: Tetraphyllidean cestode (Shields, 1992)
**Host species:** *Portunus pelagicus* (Juvenile crabs)
**Causative agents:** Lecanicephalid cestode, *Polypocephalus moretonensis* (Shields, 1992)
**Host species:** *Carcinus maenas*
**Causative agents:** *Dolfusiella martini* and *Trimacanthus aetobatidis* (Torchin et al., 2001)
**Host species**: *Portunus pelagicus*
**Causative agents:** *Polypocephalus moretonensis*
**(b) Trematodes**
**Host species:** *Callinectes sapidus*
**Causative agents:** *Microphallus* sp.
**(c) Turbellaria**
**Host species**: *Carcinus maenas*
**Causative agents**: Parasitic crab flatworm, *Fecampia erythrocephala*
**(d) Acanthocephalan**
**Host species:** *Carcinus maenas*
**Causative agents**: *Profilicollis botulus* (Torchin et al., 2001)

**ii) Nemertean worms**
**Host species:** *Callinectes sapidus, Liocarcinus puber, Liocarcinus depurator* and *Carcinus maenas*
**Causative agents**: *Carcinonemertes carcinophila*
**Host species:** *Portunus pelagicus*
**Causative agents:** *Carcinonernertes mitsukurii*
**Host species:** *Carcinus maenas*
**Causative agents**: *Carcinonemertes epialti*

**iii) Nematodes**
**Host species:** *Charybdis japonica, Ovalipes catharus* and *Callinectes aminicola*
**Causative agents:** Ascaridoid worms

Diseases and Parasites of Portunid Crabs    353

**iv) Polychates** (serpulids)
**Host species:** *Liocarcinus puber*
**Causative agents:** *Pomatoceros triqueter* and *Hydroides norvegica*

**v) Entoprocta**
**Host species:** *Liocarcinus puber*
**Causative agents:** *Barentsia matsushimana*

**vi) Bryozoans** (Ectoprocta)
**Host species:** *Ovlaipes catharus*
**Causative agents:** *Triticella capsularis*

**vii) Crustaceans**
**(a) Copepoda**
**Host species:** *Carcinus maenas*
**Causative agents**: *Chioniosphaera cancorum* and *Lecithomyzon menaedi* (Torchin et al., 2001)
**(b) Isopoda**
**Host species:** *Carcinus maenas*
**Causative agents:** *Portunion maenadis* (Torchin et al., 2001)
**(c) Cirripedia**
**Host species**: Several species of portunid crabs
**Causative agents**: *Octolasmis* sp., *Balanus* sp., *Sacculina* sp. and *Loxothylacus* sp.

## 7.1.5 NON-INFECTIOUS AND OTHER DISEASES

i.     Rust spot shell disease
ii.    Injury
iii.   Nutritional
iv.    Albinism
v.     Toxicity
vi.    Deformities

*Diseases and parasites of the commercially important portunid species.*

***Achelous spinicarpus:*** Parasites: Female crabs serve as hosts for sacculinid parasites which are carried under the abdomen, usually one specimen, sometimes two. The parasitized specimens are all small (carapace length 8 to 19 mm).

***Achelous spinimanus:*** Parasites: The infestation rates of stalked barnacle species *Octolasmis lowei* (Cirripedia: Poecilasmatidae) in male and female hosts were similar, 11% of adults and 1.1% of juveniles had *Octolasmis lowei*. Among infested crabs, 91.8% were in the intermolt stage; 55.9% of females were in an initial stage of gonad development and 53.6% of males presented fully developed gonads. The obtained data suggest that *Octolasmis lowei* infests adult crabs preferentially, and it doesn't interfere with the gonad development. However, it seems to negatively affect the weight of infested crabs (Santos, 2002).

***Octolasmis lowei***

***Arenaeus cribarius:*** Epibionts: Male crabs showed a higher infestation rate by *Chelonibia patula* when compared to mature non-ovigerous females. *Octolasmis lowei* infestation was found associated to adult crabs. The ovigerous females of this species showed a higher prevalence of infestation than males and non-ovigerous females. A synchrony between the life cycle of the epibionts and their hosts was evident and promotes the continuity of the former (Costa et al., 2010).

*Chelonibia patula*

***Callinectes amnicola:*** Ectoparasitic barnacles have been recovered from the carapace, chelipeds and walking legs of this species. Only the male crabs were found infected with a prevalence of 13.46% (Omuvwie and Atobatele, 2013).

The overall prevalence of ecto and endoparasites in *Callinectes aminicola* in the Cross River estuary was 12.38%; parasites recovered were *Trichodina* and nematodes scantly distributed on the skin and gut of *C. aminicola* respectively. Prevalence of parasite in relation to carapace width (cm) was highest in width class 5–9.9 cm (61.54%) followed by 10–14.9 cm (38.46%) and 0.00% for 15–19.9 cm (Ekanem et al., 2013).

Callinectes bocourti*:* Bacterial species identified from the cultures of the hemolymph of this species include *Aeromonas hydrophyla, Aeromonas salmonicida, Bacillus sp., Hemophilus paracuniculus, Hemophilus* sp., *Legionella sp., Pasteurella multocida, Pasteurella* sp., *Proteus sp. Pseudomonas cepacia, Pseudomonas mallei, Psuedomonas putrefasciens, Pseudomonas sp. Salmonella* sp., *Serratia rubidaea, Shigella flexeri, Vibrio cholera, Vibrio fisheri, Vibrio fluvialis, Vibrio metschnikovii, Vibrio proteoliticus, Yersinia pseudotuberculosis* and *Zoogloea* sp. Incidence of *Haemophilus* sp., *Pseudomonas* sp. and *Salmonella* sp. were the highest in the hemolymph. A suggested hypothesis explaining high densities of bacteria in crabs' hemolymph is that the immune system of crabs is slow acting; therefore, bacteria which enter into the hemolymph would proliferate for a brief period before being removed by the crabs' immune mechanisms. From a medical viewpoint, several species isolated from the hemolymph are considered human pathogens. *Aeromonas hydrophila* has been associated with septicemia, pneumonia and gastroenteritis. *Pasteurella multocida* can be responsible for various syndromes including localized abscesses and septicemia. *Pseudomonas mallei* is the agent of glanders disease; while *Pseudomonas cepacia* and *Pseudomonas putrefasciens* are considered opportunistic pathogens. *Shigella jiexeri* invades intestinal epithelial cells. *Yersinia pseudotuberculosis* is associated with mesenteric lymphadenitis or terminal ileitis. Given the high density and diversity of bacteria in the hemolymph, the ingestion of uncooked crab may represent a potential threat to human health (Rivera et al., 1999).

***Callinectes danae*** and ***Callinectes ornatus***: Stalked barnacles *Octolasmis lowei* are frequently found attached to decapod crustaceans. Their epibiotic association depends on many factors, which are mainly related to characteristics of the host's biology. All infested hosts were in the intermolt period. The mean size of infested crabs was larger than that observed for non-infested individuals. Internally, stalked barnacles were concentrated on the central gills or walls and floor of branchial chambers, suggesting that these gills provide more favorable conditions for the settlement and development of these epibionts. These results highlight the relationship between epibiont infestation and host biology, as well as the role of decapod crustaceans as a suitable substrate for the development of stalked barnacle *Octolasmis lowei* (Machado et al., 2013)

***Callinectes rathbunae*** **and** ***Callinectes sapidus:*** The rhizocephalan cirripede *Loxothylacus texanus* was found almost exclusively in *Callinectes rathbunae* with a mean prevalence of 7.58%, while less than 1% of all *Callinectes sapidus* were parasitized. *Callinectes rathbunae* constitutes a new host record for this parasite. A study of infection revealed significant variation in prevalence and host size throughout the study period. The sex ratio of parasitized crabs differed from that of the total sample with males being parasitized more often, and the comparison of carapace width-weight relationships revealed lower weights of parasitized crabs (Alvarez et al., 1999).

*Carcinus aestuarii:* Shell Disease Syndrome (SDS) has been reported in wild specimens of *Carcinus aestuarii* from the polluted Volturno River estuary. Collected crabs showed erosive and ulcerative "black spot" lesions on the exoskeleton; histologically the lesions ranged from a mild to extensive and severe damages and intense haemocyte infiltration. Bacterial isolation confirmed the presence of some marine bacterial species with chitinolytic activity. The authors suggest that the detection of SDS in wild crabs living in polluted waters could be used to monitor the health status of this environment (Mancuso et al., 2013).

*Carcinus maenas:* This species has been found infected by Grey Crab Disease for which the protozoan *Paramoeba perniciosa* is the causative species. Parasitic amoebae progressively invade the connective tissues along the midgut, the antennal gland, Y organ and the haemal spaces in the gills. Eventually the infection becomes systemic. Pathological changes caused by large numbers of amoeba include: tissue displacement, probable lysis of some types of tissue including haemocytes, and significant decreases in protein, haemocyanin and glucose. Epizootics involving mortalities as high as 17% of the blue crab population have been reported from Chincoteague Bay in early summer and mortalities in shedding tanks (for production of newly molted soft-shell crab) have been estimated at 20–30% (http://www.dfo-mpo.gc.ca/science/aah-saa/diseases-maladies/paramcb-eng.html)

Bitter Crab Disease (BCD) is a fatal disease of marine crustaceans caused by parasitic dinoflagellates, species of *Hematodinium.* Common names include Bitter Crab Syndrome (BCS) and Pink Crab Disease (PCD). The type species *Hematodinium perezi* was described in swimming (*Liocarcinus depurator*) and green crabs (*Carcinus maenus)* in Europe (Chatton and Poisson 1931; Morado et al., http://www.ices.dk/sites/pub/CM%20Doccuments/CM-2008/D/D0408.pdf. ICES CM 2008/D:04).

A total of 5 endoparasites were recorded from *Carcinus maenas.* The included 2 species of microphallid trematodes, *Microphallus claviformis* and *Maritrema subdolum,* cystacanths of the acanthocephalan *Profilicollis botulus,* and 2 species of parasitic castrators, the cirripede *Sacculina carcini* and the isopod *Portunion maenadis* (Zetlmeisl et al., 2011).

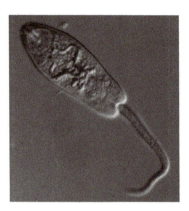

*Maritrema subdolum*

In this species, milky disease is due to the fact that the haemolymph (crab blood) of infected crabs takes on a milky appearance. Crabs with milky disease show no-external symptoms and the only obvious sign is the milky appearance of the haemolymph (crab blood). Infected crabs have higher levels of ammonia and glucose in the blood, but reduced numbers of circulating blood cells and protein. Certain cells of the hepatopancreas and gills of infected crabs appear to be physically affected by the disease, but other tissues appear normal. Nonetheless, the infection is lethal; usually within seven days of when the crabs are first observed to have milky blood. Tiny, bacteria-like particles (alpha-proteo bacteria) in hepatopancreas cells, as well as bacteria apparently escaping from ruptured cells. Sequence analysis of the 16S rRNA gene from the milky disease bacteria indicated that they are a previously undescribed species of a-proteobacteria with little phylogenetic similarity to members of the order Rickettsiales (Eddy et al., 2007).

*Carcinus maenas* **infected with milky disease**

**Trematode infection in *Carcinus maenas*:** Species of trematodes infecting *Carcinus maenas* include *Microphallus claviformis*, *Microphallus similis* and *Maritrema subdolum*

**Turbellarian infection in *Carcinus maenas*:** *Carcinus maenas* is commonly parasitized in the habitats where the worm cocoons are abundant. The turbellarian, *Fecampia erythrocephala* did not infect crabs larger than 11 mm carapace width, and prevalence decreased significantly with crab size. Prevalences reached 11% in areas where cocoons were abundant. Together with the large size of these worms relative to the size of the host crabs and the observations on worm emergence, these life history features indicate that *Fecampia erythrocephala* is a parasitoid of young shore crabs. *Fecampia erythrocephala* cocoon abundance is often high in localized areas and size-prevalence information suggests that worms mature rapidly in these crabs. This suggests that *Fecampia erythrocephala* is an important contributor to crab mortality and to the ecology of shore crabs at these sites (Kuris et al., 2002).

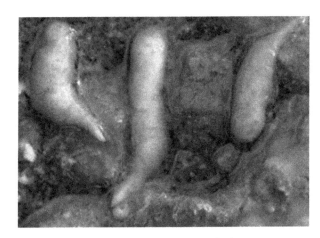

*Fecampia erythrocephala*

**Acanthocephalan infection in *Carcinus maenas*:** *Carcinus maenas* is the 1st host of the acanthocephalan helminth *Profilicollis botulus* which infects eider ducks (*Somateria mollissima*) by ingestion of infected crabs. Juvenile eider ducks suffer some mortality from heavy infections and crabs are infected by eggs of the parasite from duck faeces (http://www.marlin.ac.uk/biotic/browse.php?sp=4286)

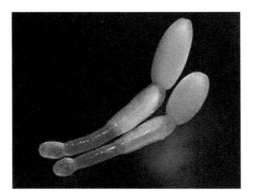

*Profilicollis* sp.

**Sacculina infection in *Carcinus maenas*** : The ecologically important shore crab *(Carcinus maenas)* is commonly infected by the rhizocephalan parasite Sacculina carcini. This infection causes behavioral change, castration and ceased molting, in the host. Further it also feminizes its male host morphologically resulting in reduced cheliped size.

*Carcinus maenas* **infested with** *Sacculina carcini*

*Carcinus mediterraneus*: A viral agent (Reovirus disease: W2 Virus Infection) was isolated and studied in the Mediterranean shore crab *Carcinus mediterraneus*. This agent, developed in the cytoplasm of connective tissue cells of *Carcinus mediterraneus* and produced unusual viral structures, 'rosettes,' consisting of an empty sphere bounded by arrangements of

Diseases and Parasites of Portunid Crabs 361

viral particles. Absence of aggressiveness, increasing weakness and lack of appetite. The connective tissue of many organs (hepatopancreas, digestive tract, gills, hematopoietic organs) of W2-infected animals showed severe damage. The most obvious lesions were observed in the connective tissue surrounding the tubules of the hepatopancreas (Mari and Bonami, 1988). The occurrence of metacercariae of parasitic trematodes (Microphallidae) has also been reported from the body cavity of this species (Vivares, 1970).

**Ectosymbionts *in Charybdis* spp*., Thalamita coeruleipes* and *Ovalipes catharus:* *Charybdis japonica* and *Ovalipes catharus* harbored different ectosymbionts. Serpulid polychaete tubes occurred on the exoskeleton of 85.4% of *C. japonica* examined, but were absent from *Ovalipes catharus*. The bryozoan, *Triticella capsularis* occurred on 97.4% of *Ovalipes catharus* but was not found on *Charybdis japonica*. Few endoparasites were present in either species. An unidentified juvenile ascaridoid nematode occurred in the hindgut of 5.9% of *Charybdis japonica,* but was not found in sympatric populations of *Ovalipes catharus.* A second, unidentified species of ascaridoid nematode occurred in 7.1% of *Ovalipes catharus*. Melanized lesions were observed in the muscle tissue of almost half (46.6%) of the *Charybdis. japonica* examined (Miller et al., 2006).

*Chaybdis feriatus, Charybdis variegata, Charybdis natator, Charybdis miles, Charybdis lucifera, Charybdis hoplites, Charybdis hellerii, Charybdis granulate, Charybdis amboinensis* and *Thalamita coeruleipes* have been found infected by rhizocephalan cirripedes (Elumalai et al., 2014).

*Charybdis callianassa* has been reported to be infected with parasitic barnacle species viz. *Octolasmis angulata* and *Heterosaccus lunatus* (Walker, 2001).

An epizoic stalked barnacle, *Octolasmis angulata*, was identified within the branchial chambers of *Charybdis callianassa*, a swimming crab from Moreton Bay, Queensland, Australia, making this crab a new host for O. angulata. In the present study fifty-two crabs, 30–49 mm carapace width, were dissected, and thirty-three were found to have the epizoite. The number and position of the O. angulata within the branchial chambers were noted. *Octolasmis angulata* is principally found attached to the cuticle of the anterior chamber wall in the epibranchial space, although attachment to the gills does occur. *Charybdis callianassa* is also parasitized by the sacculinid barnacle *Heterosaccus lunatus,* and one such parasitized crab contained eighty-seven *Octolasmis angulata*, the highest number recorded in the present study.

The factors governing *Octolasmis angulata* distribution within the branchial chambers of *Charybdis callianassa* are discussed (Walker, 2001).

**Charybdis longicollis:** *Heterosaccus dollfusi* has been a very common parasite on the crab *Charybdis longicollis* along the Mediterranean coast of Israel. It accompanied from the Red Sea this Lessepsian migrant crab. The parasite is more common on male than female crabs, where it becomes external on hosts of many sizes shortly after a molt and undergoes characteristic growth stages. The parasite usually causes the complete loss of pleopods in both sexes and a feminine broadening of the abdomen in male crabs. Parasite survival is the same in both sexes of the host. More than one parasite per host is frequent, each probably originating from an individual cypris larva rather than from asexual budding. Hosts with single parasites outlive those with multiple infections.(Galil and Lutzen, 1995)

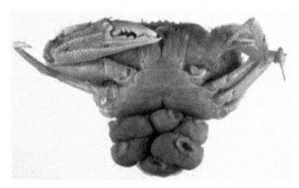

**Host and parasite**

**Charybdis truncata:** The shallow water swimming crab *Charybdis truncata* is occasionally infected by *Sacculina* species in Hong Kong. Histological sectioning showed that the paired male receptacles are globular in shape and separated from each other. The paired receptacle ducts are united in the middle region and again separated near the duct openings which are light microscopically cuticularized. This species was identified as *Sacculina scabra* (Chan, 2004). This species has also been infected by *Loxothylacus nierstrasz*.

**Liocarcinus depurator:** Epibiont ciliates have been recorded from the exoskeleton of the decapod crustacean *Liocarcinus depurator*, a portunid crab collected off the Mediterranean coast of Spain. These ciliates included 11 species, belonging to three subclasses. The chonotrich *Chilodochona* *quennerstedti*, the most abundant species, colonized specific sites on the crab, the mouthparts. The peritrichs, identified as *Zoothamnium* sp. and *Cothurnia* sp., and the suctorians, identified as *Acineta tuberosa, Acineta papillifera, Ephelota gemmipara, Actinocyathula homari, Paracineta limbata, Conchacineta constricta, Corynophrya anisostyla,* and *Acinetides symbiotica*, were generally less site-specific. Feeding activities, burying behavior, and molt state of the host crab, and competition between ciliate species, are mentioned as the most important factors in the occurrence of ciliates on the crustacean body (Fernandez-Leborans et al., 1997).

Bitter Crab Disease (BCD) is a fatal disease of marine crustaceans caused by parasitic dinoflagellates, species of *Hematodinium*. Common names include Bitter Crab Syndrome (BCS) and Pink Crab Disease (PCD). The type species *H. perezi* was described in swimming (*Liocarcinus depurator*) and green crabs (*Carcinus maenus*) in Europe (Chatton and Poisson, 1931; Morado et al., http://www.ices.dk/sites/pub/CM%20 Doccuments/CM-2008/D/D0408.pdf. ICES CM 2008/D:04)

**Liocarcinus puber and Liocarcinus depurator:** Ovigerous *Liocarcinus puber* and *Liocarcinus depurator* were caught in small numbers, and both were infested with the nemertean parasite *Carcinonemertes carcinophila* (Comely and Ansell, 1989). *Liocarcinus puber* has also been reported to be infested with crenate barnacles, *Balanus crenatus*. This species (an epibiont), with comparatively larger size, are found distributed on the carapace, ventral surface of the cephalothorax and the pereiopods; (Fernandez-Leborans and Gabilondo, 2008).

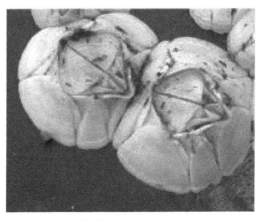

***Balanus crenatus***

***Liocarcinus depurator* and *Liocarcinus puber***: The epibionts of *Liocarcinus depurator* and *Liocarcinus puber* are the hydrozoans *Clytia gracilis* and *Leuckartiara* sp. While *Clytia gracilis* is seen on the second pereiopod, *Leuckartiara* sp., principally on chelipeds, carapace and the fourth right pereiopod (Fernandez-Leborans and Gabilondo, 2005). Reports are also available about the association of this crab species serpulid worms *Spirobranchus triqueter (=Pomatoceros triqueter)* and *Hydroides norvegica (Polychaeta: Serpulidae)* and entoproct, *Barentsia matsushimana.*

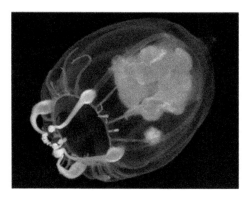

***Leuckartiara* sp.**

***Clytia gracilis***

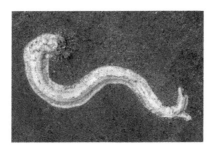

*Spirobranchus triqueter*

*Hydroides norvegica*

***Ovalipes ocellatatus:*** This species has been found affected by milky disease due to *Hematodinium* sp. The bryozoan, Triticella capsularis (Bryozoa, Ctenostomata) has been newly described as an ectosymbiont of this species in central New Zealand. The bryozoan produces the longest zooids known in the genus Triticella, with colonies forming a "fur" up to almost 10 mm thick on large crabs, mostly males. The densest area of colonization is the ventral anterior half of the crab. The bryozoan lives only on this species and probably benefits from its "messy" and voracious feeding habits, opportunities for gene exchange during crab swarming behavior, and dispersal. Although visually striking when dense, the bryozoan growth is only superficial and affects neither the behavior of the crab nor the quality of its meat (Gordon and Wear, 1999).

***Portunus binoculus:*** Females of this species have been reported to carry the sacculinid parasite under their abdomen.

Portunus hastatus: This species serves as the host for the parasites, *Loxothylacus carinatus* and *Sacculina carinata.*

***Portunus hastatoides:*** *This species has been found infected by* rhizocephalan cirripedes (Elumalai et al., 2014).

*Portunus pelagicus:* The rhizocephalan *Sacculina granifera* has a significant effect on the gonad development and growth of *Portunus pelagicus*. *Sacculina* infection may also cause degeneration of the gonads in both male and female crabs and a modification of the secondary sexual characteristics in the male crab resulting in the acquisition of female characteristics. These crabs may also be castrated by the parasite.

Fifteen parasites and symbionts were found in the tissues, in the branchial chambers, and on external surfaces of the sand crab *Portunus pelagicus*. Female crabs possessed more species of parasites and symbionts. Females also had a higher prevalence of the peritrich ciliate *Operculariella* sp., a tetraphyllidean cestode, and the barnacles *Octolasmis* spp. There were positive correlations between intensity of infection and host size (carapace width) for only a lecanicephalid cestode parasite (*Polypocephalus. moretonensis*) and 2 symbionts (the nemertean *Carcinonernertes mitsukurii*, and *Octolasmis* spp.). Not surprisingly, the molt condition of the crab influenced the abundances of the sessile external symbionts. Crabs in the postmolt condition had fewer *Operculariella* sp., *Chelonobia patula*, and *Octolasmis* spp. The abundances of the motile *Carcinonemertes mitsukurii* and the internal parasites and symbiont were not, however, affected by the molt condition of their hosts.

Four species of *Octolasmis* have been identified from this species. They are: *Octolasmis angulata, Octolasmis warwickii, Octolasmis tridens* and *Octolasmis lowei. Octolasmis angulata* and *Octolasmis lowei* showed the highest intensity (16.8). The infestation occur entire body of the P. pelagicus such as gill, abdomen and the carapace area. The distribution of the *Octolasmis* spp. is random among the gill area. The highest abundance of barnacle was recorded in others part of the host body compared to the gill attachment. The attachments of barnacle influenced by certain factor, for example the current or water flow through gill (Jeffries et, al., 1994). Water enters the crab hypo-branchial chamber through opening at the bases of thoracic appendages, this occurs as results of pressure created. It will influence the site selection of *Octolasmis* spp attachment. *Octolasmis lowei* mostly preferred attached at the gill because of the morphology structure (less hard structure at the head area compared to other species that found in this study (Ihwan et al., 2015).

*Octolasmis tridens*

*Octolasmis warwickii*

***Portunus sanguinolentus:*** In this species, parasitization by sacculinids has been reported to induce severe modifications in its morphology, behavior and reproduction. These sacculinids which were found on crabs as external parasites ranged from 56 mm to 88 mm in carapace width (Elumalai et a., 2012).

***Portunus segnis:*** Three specimens of the swimming crab *Portunus segnis* and two specimens of the blue crab *Callinectes sapidus* were found infested by the symbiotic barnacles *Chelonibia patula*. This is the first record of *Chelonobia patula* on the carapace and cheliped of *Portunus segnis* from the Turkish coasts. (Özcan, 2012).

***Chelonibia patula*** **on its host** ***Portunus segnis***

***Portunus trituberculatus:*** An epidemic disease named "milky disease" happened in cultured swimming crab *Portunus trituberculatus* and lead to high mortality. The disease crab is usually thin and has the symptom of muscle white, milky liquid can be seen when open the lid. No predominance bacteria colonies were isolated from the typical milky crab. The study of the histopathology showed that the haemolymph in the disease crab decreased sharply, and was replaced by large amount of parasites, which was observed in muscle, heart and hepatopancreas from the disease crab. The pathological characteristics mainly appeared as cells swelling, necrosis, nucleus condensation and collapse. The opalescent hemolymph and abnormal tissues were observed under electron microscope, masses of the parasite were observed but no virosomes had been found. The parasite is 5–7 μm in size, and oval in shapes, with two flagellums. Preliminary study showed that the dinoflagellate *Hematodinium.* sp was the main pathogen of the milky disease of *Portunus trituberculatus* (Wen-Jun et al., 2007). A pathogenic yeast strain (WCY) *Metschnikowia bicuspidata* which could cause milky disease in this species has also been reported (Wang et al., 2007).

## Important diseases/infections of mud crabs (*Scylla* spp.)

**A) Viral diseases:** Four types of viruses *viz.*, white spot syndrome virus (WSSV), muscle necrosis virus, reovirus and baculovirus have been reported in mud crab *Scylla* spp.

**i) White spot syndrome virus (WSSV):** Mud crabs are known carriers and vectors of WSSV in shrimp culture facilities without showing any signs of the disease. Signs are inapparent and the crab can maintain the experimental WSSV infection for many months, though molting frequency was found to be reduced In India, the natural prevalence of WSSV in crab is about 5.06%, while in culture ponds it is about 30%.

**ii) Muscle necrosis virus:** It is an icosahedral virus, 150 nm in diameter, recorded in cultured mud crab causing a disease characterized by muscle necrosis.

**iii) Reovirus:** Recently, a reovirus designated as mud crab reovirus (MCRV) from cultured mud crab *Scylla serrata* with signs of 'sleeping disease,' high mortality and heavy economic loss has been reported. This cytoplasmic cosahedral non-enveloped virus (70 nm diameter) infects connective tissue cells of the hepatopancreas, gills and intestine in mud crab.

**iv) Baculovirus infection:** A baculovirus infection in the form of intra-nuclear inclusion bodies in hepatopancreas epithelium of juvenile and mature mud crabs (*Scylla serrata*) without much clinical manifestation has been reported.

### B) Bacterial diseases

**i) Bacterial necrosis:** This is a common disease observed in larvae, post-larvae or adults. It is variously termed as ' black spot,' 'brown spot,' 'burnt spot,' 'shell disease' or chitinolytic bacterial disease. This is caused by the invasion of chitinolytic bacteria, which break down the chitin of the exoskeleton, leading to erosion and melanization (dark brown to black pigmentation) at the site of infection. Several chitinolytic bacteria (Gram negative rods) such as *Vibrio* spp., *Pseudomonas* spp., *Aeromonas* spp., and *Spirillium* spp. are involved.

**ii) Filamentous bacterial disease:** Filamentous bacteria such as *Leucothrix mucor, Thriothrix* spp. and *Flexibacter* spp. sometimes cause mortalities by discoloration of gills and associated secondary infections. In berried crab, the eggs may become infected with filamentous bacteria accompanied with other microbial infections causing retarded embryonic development, longer incubation time and egg mortality due to depleted oxygen exchange across egg membrane.

**iii) Luminescent bacterial disease:** Luminescent bacterial disease is a severe, economically important bacterial infection caused by members of the genus *Vibrio* and other related genera Adult animal often shows symptoms of loss of appetite, reduced growth, dark hepatopancreas and morality in large numbers. *Vibrio harveyi* often infect the crab larvae reared in hatchery conditions.

**iv) Other opportunistic bacterial pathogens:** Co-infection with one or more bacterial species and/or other infectious agents like viral, fungal and parasitic infection is a common occurrence. *Vibrio harveyi* and *Vibrio campbellii* were found to be predominant often in WSSV infected mud crabs. Other *Vibrio* species found were *Vibrio vulnificus, Vibrio nereis, Vibrio fischeri* and *Vibrio fluvialis.* Infected crabs were characterized by hard carapace, red chelipeds and joints, pale hepatopancreas, gills, and soft muscles almost immobile before succumbing to death. The haemolymph revealed three stages of the syndrome, namely orange, orange-white and milky-white in color. The disease is suspected to be bacterial in nature due to the isolation of five different types of bacteria from the infected tissues.

**C) Fungal diseases:** Three fungi, *viz., Lagenidium, Atkinsiella* and *Haliphthoros* have been identified as possible agents of mud crab egg mortality.

**D) Parasitic diseases:** These diseases belong to protozoans (*Hematodinium* sp., *Epistylis* sp. and *Acineta* sp.) and metazoans (helminths under microcephallid trematode metacercaria, *Levinseniella* sp., metacestodes such as *Polypocephalus* sp. and dorylaimoid nematode and two cirriped crustaceans; *Octolasmis* sp. and *Chelonibia* sp.). Symbionts belonging to the *Vorticella* sp., oysters, polychaetes, hydrozoans, amphipods and turbellarian were also occasionally found in mud crab.

**i) Protozoan parasites:** Fouling protozoans such as peritrich ciliates (*Zoothamnium, Vorticella* and *Epistylis*) and suctorian ciliate (*Acineta* sp.) both on the gills may be a problem during the hatchery phase of mud crab culture mainly on eggs and larval stages. Once infested, cotton wool like growth attaches to the body as well as appendages and disrupt mobility and feeding. Suctorians feed mainly on other protozo-

# Diseases and Parasites of Portunid Crabs

ans and in high numbers may interfere with respiration, while peritrichs interfere with locomotion, feeding and molting of larvae causing stress and even death. Dinoflagellates of the genus *Hematodinium* sp. that infects the hemolymph and tissues of the crab, cause a disease condition known as 'milky disease' with gross symptoms such as moribund behavior, opaquely discolored carapace, cooked appearance, milky body fluid, unpalatable flavor and high mortality. Among sporozoan protozoa, histozoic microsporidian, *Thelohania* sp. and enteric cephaline gregarines are encountered in mud crab though not associated with any clinical disease.

**ii) Metazoan parasites:** Most common metazoan parasites of mud crabs are cirripedians, either fixed or parasitic in their adult stage. Barnacles or cirripedes are mostly commensals and tend to be predaceous while others like *Sacculina* are exclusively parasitic on crabs. Two rhizocephalan parasites affecting mud crabs are *Sacculina* sp. and *Loxothylacus* sp. These infections seldom lead to mortality, but extensive shell erosion and perforation may lead to entry of opportunistic pathogens Typically, pedunculate barnacles of the genus *Octolasmis* spp. inhabit the branchial chambers of mud crabs, especially on the ventral surfaces of the gills. The stalk bears at its free end, the rest of the body known as the capitulum enclosed in a mantle formed by the carapace. *Balanus* (acorn barnacle) is found attached to rocks below high water marks or bottom side of crab hideouts differing from *Octolasmis* sp. in having no peduncle and the shell is directly attached to the substratum. The shells form a sort of cone shaped case surrounding the body, and the opening is closed by a lid formed of four opercular plates. Several species of balanid cirripedes (*Balanus* and other barnacles) live attached to the carapace, chelipeds of crabs. The presence of parasite causes degeneration of tissues of the crab and hinders the formation of cuticle at the site of attachment during every successive molting, through which the body of *Sacculina* project freely as a fleshy mass. *Sacculina* is a typical example of parasite induced castration in that the infected male crab exhibits a tendency to develop characteristics of females; the abdomen becomes broader like females, the copulatory organs (pleopods) get reduced and become suited for carrying eggs whereas in female crabs the swimmerets (pleopods) becomes reduced.

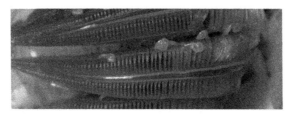

**Gills of *Scylla olivacea* infested with pedunculate barnacles**

**Pedunculate barnacles isolated from the gills of *Scylla olivacea***

**(E) Non-infectious diseases**

**i) Deformities:** Environmental parameters and nutrient inadequacy are implicated in the delay or failure of molting. A range of deformities have been observed in grow-out culture. These include molting failure, missing legs, abdominal flaps, abnormalities of chelate legs, claws *etc.* when hard shelled crabs attack the freshly molted crab, abnormal outgrowth, or injury due to various farming operations. All these factors enhance the entry of opportunistic pathogens.

**ii) Albinism:** Partial albinism on carapace and legs has been observed in pond reared juveniles of *S. tranquebarica.*

**F) Other diseases:** A new shell disease of non-infectious nature and uncertain etiology in *Scylla serrata* has been reported. This disease is characterized by irregularly shaped circular lesions commonly called 'rust spot shell disease' and unique histopathological alteration. Crabs also show symptoms arising out of any stress such as overcrowding, extreme temperature or pH, low dissolved oxygen, ammonia, etc. Blackening of gills may sometimes be found as a manifestation of several other disease syndromes, precipitation of dissolved chemicals, turbidity, Vitamin C deficiency, etc. General discoloration of gills may occur due to melanization of tissue and necrosis, which may be visible through the side of carapace (Jithendran et al., 2010).

***Thalamita cooperi*** and ***Thalamita coeruleipes:*** These species are found infested with rhizocephalan cirriped parasites viz. *Sacculina anomala* and *Sacculina* spp.

***Thlamita sima***: A new entoniscid isopod species *Cancrion australiensis* (isopoda: Entoniscidae) has been reported from this species (Shields and Earley, 1993).

## KEYWORDS

- bitter crab disease
- black gill disease
- milky disease
- red sternum syndrome
- rust spot shell disease
- shell diseases
- white spot disease

# REFERENCES

Abdel-Salam, H. A., Hamdi, S. A. H. Evaluation of the edible muscles of four species of crustaceans from three regions of Egypt and Saudi Arabia. *Global Advanced Research Journal of Agricultural Science* (2015). *4,* 105–112.

Abellóa, P. Reproductive biology of *Macropipus tuberculatus* (Roux, 1830) (Brachyura: Portunidae) in the Northwestern Mediterranean. *Ophelia* (1989). *30,* 47–53.

Abramowitz, A. A. Color changes in cancroid crabs of Bermuda. Proceedings of the National Academy of Sciences, Zoology (1935). *21,* 678–681.

Adeyeye, E. I., Oyarekua, M. A., Adesina, A. J. Proximate, mineral, amino acid composition and mineral safety index of *Callinectes latimanus. International Journal of Development Research* (2014). *4,* 2641–2649.

Adeyeye, E. I. Comparative Study of the Lipid and Fatty Acid Composition of Two Shell Fish: Lagoon and Fresh Water Crabs. *Malaya Journal of Biosciences,* (2015). *2,* 57–74.

Alvarez, F., Gracia, A., Robles, R., Calderon, J. Parasitization of *Callinectes rathbunae* and *Callinectes sapidus* by the Rhizocephalan Barnacle Loxothylacus texanus in Alvarado Lagoon, Veracruz, Mexico. *Gulf Research Reports,* (1999). *11,* 15–21.

Anand, T. P., Chellaram, C., Shanthini, C. F., Karthika, G., Vijayalakshmi, C. Antioxidant properties of natural dietary common seafoods from Pulicat coast. *Journal of Chemical and Pharmaceutical Research,* (2014). *6,* 611–613.

Andrade, L. S., Antunes, M., Lima, P. A., Furlan, M., Frameschi, I. F., Fransozo, A. Reproductive features of the swimming crab *Callinectes danae* (Crustacea, Portunoidea) on the subtropical coast of Brazil: a sampling outside the estuary. *Braz. J. Biol.* (2015). 75. http://dx.doi.org/10.1590/1519–6984.21513.

Anon. Crab fattening-A potential marine aquaculture industry for small scale fishermen. Feature Article 3, 2013. http://www.islf.org/index.php?option=com_content&view=article&id=64:crab-fattening-a-potential-marine-aquaculture-industry-for-small-scale-fishermen&catid=34:featuresarticles.

Anon. Hemostatic Wood Dressings https://biotextiles2013.wordpress.com/hemostatic_wound_dressings/).

Ayas, D., Özoğul, Y. The chemical composition of carapace meat of sexually mature blue crab (*Callinectes sapidus,* Rathbun, 1896) in the Mersin Bay. *Journal of Fisheries Sciences,* (2011). l, 5, 262–269.

Ayas, D., Ozogul, Y. The chemical composition of sexually mature blue swimming crab (*Portunus pelagicus,* Linnaeus, 1758) in the Mersin Bay. *Journal of Fisheries Sciences,* (2011). *5,* 308–316.

Babu, K. V. R., Chandra Mohan, S. B. K., Naik, A. T. R. Breeding biology of *Charybdis (Charybdis) feriatus* (Linnaeus) from Mangalore. *Indian J. Fish.* (2006). *53*, 181–184.

Baiduri, S. N., Akmal, S. N., Ikhwanuddin, m Mating Success of Hybrid Trials Between Two Mud Crab Species, *Scylla tranquebarica* and *Scylla olivacea*. *Journal of Fisheries and Aquatic Science*, (2014). *9*, 85–91.

Balasubramanian, C. P., Suseelan, C. Reproductive biology of the female deepwater crab *Charybdis smithii* (Brachyura: Portunidae) from the Indian seas. *Asian Fisheries Science*, (1998). *10*, 211–222.

Balasubramanian, C. P., Suseelan, C. Biochemical composition of the deep-water crab *Charybdis smithii*. *Indian J. Fish.* (2001). *48*, 333–335.

Bansal, V., Kumar, P., Sharma, S. N., Pal, O. P., Malviya, R. Applications of chitosan and chitosan derivatives in drug delivery. *Advances in Biological Research*, (2011). *5*, 28–37.

Bhatnagar, A., Sillanpää, m Applications of chitin- and chitosan-derivatives for the detoxification of water and wastewater — A short review. Advances in Colloid and Interface Science, (2009). *152*, 26–38.

Begum, M., Shah, M. M. R., Mamun, A., Alam, M. J. Comparative study of mud crab *(Scylla serrata)* fattening practices between two different systems in Bangladesh. *J. Bangladesh Agril. Univ.* (2009). *7*, 151–156.

Begum, M., Shah, M. M. R., Mamun, A., Alam, M. J. Feasibility of hi-culture of mud crab *(Scylla serrata)* and shrimp *(Penaeus monodon)*. *Bangladesh J. Fish. Res.* (2007). *11*, 189–196.

Behrman, M. The Magic Powder Made from Crab Shells That'll Keep You From Bleeding to Death. http://gizmodo.com/5806944/celox-how-crab-shells-will-stop-you-from-bleeding-to-death/

Boerner, L. K. Crab Shells Help Researchers Make New Battery Materials. http://cen.acs.org/articles/91/web/2013/06/Crab-Shells-Help-Researchers-Make.html;

Buscaino, G., Gavio, A., Galvan, D., Filiciotto, F., Maccarrone, V., de Vincenzi, G., Mazzola, S., Orensanz, J. M. Acoustic signals and behavior of *Ovalipes trimaculatus* in the context of reproduction. *Aquatic Biology* (2015). *24*, 61–73.

Cannicci, S., Dahdouh-Guebas, F., Anyona, D., Vannini, m Natural diet and feeding habits of *Thalamita crenata* (Decapoda: Portunidae). *Journal of Crustacean Biology*, (1996). *16*, 678–683.

Caulier, G., Parmentier, E., Lepoint, G., Nedervelde, F. V., Eeckhaut, I. Characterization of a population of the Harlequin crab, *Lissocarcinus orbicularis* Dana, (1852). an obligate symbiont of holothuroids, in Toliara bay (Madagascar). *Zoosymposia*, (2012). *7*, 177–183.

Chan, B. K. K. First, record of the parasitic barnacle *Sacculina scabra* Boschma, 1931 (Crustacea: Cirripedia: Rhizocephala) infecting the shallow water swimming crab *Charybdis truncata*. *The Raffles Bulletin of Zoology*, (2004). *52*, 449–453.

Chatton, E., Poisson, R. Sur l'existence dans le sang des crabs. De Péridiniens parasites: *Hematodinium perezi* n.g., n.sp. (Syndinidae). *Comptes rendus des séances de la Société de biologie et des ses filiales*, (1931). *105*, 553–557.

Cházaro-Olvera, S., Peterson, M. S. Effects of salinity on growth and molting of sympatric *Callinectes* spp. from Camaronera Lagoon, Veracruz, Mexico. Bulletin of Marine Science, (2004). *74*, 115–127.

References 377

Chisholm, J. R. S., Smith, V. J. Antibacterial activity in the haemocytes of the shore crab, *Carcinus maenas*. Journal of the Marine Biological Association of the United Kingdom, (1992). *72*, 529–542.

Christensen, S. M., Macintosh, D. J., Phuong, N. T. Pond production of the mud crabs Scylla paramamosain (Estampador) and S. olivacea (Herbst) in the Mekong Delta, Vietnam, using two different supplementary diets. *Aquaculture Research*, (2004). *35*, 1013–1024.

Chu, K. H. Morphometric analysis and reproductive biology of the crab *Charybdis affinis* (Decapoda, Brachyura, Portunidae) from the Zhujiang Estuary, China. *Crustaceana*, (1999). *72*, 648–658.

Comely, C. A., Ansell, A. D. The occurrence of black necrotic disease in crab species from the West of Scotland. *Ophelia,* (1989). *30*, 95–112.

Cores, C., Erzini, K. Aspects of the reproductive biology of Pennant's swimming crab (*Portumnus latipes*) in the south of Portugal. Resúmenes sobre el VIII Simposio MIA15, Málaga del 21 al 23 de Septiembre de 2015.

Costa, T. M., Christofoletti, R. A., Pinheiro, M. A. A. Epibionts on *Arenaeus cribrarius* (Brachyura: Portunidae) from Brazil. *Zoologia,* (2010). *27*, 387–394.

Coxworth, B. Crab shells used to produce cheaper pharmaceuticals. (2012). February 14, http://www.gizmag.com/chitin-cheaper-antiviral-drugs/21457/.

da Costa, T. V., Oshiro, L. M. Y., Flor, H. D. R. Chemical Composition of the Edible Part and the Waste of Brachyuras. *Journal of Agricultural Science and Technology,* (2012). A 2, 690–695.

Darmanto, Y. S. The effect of chitin and chitosan of crab shell on water sorption of isotherm and denaturation of myofibrils during dehydration process. *Journal of Coastal Development,* (2002). *5*, 75–83.

Dat, H.D. Description of mud crab (*Scylla* spp.) culture methods in Vietnam. In: C.P. Keenan and A. Blackshaw (eds). *Mud crab aquaculture and biology.* ACIAR Proceedings 1999 of Conference, pp. 67–71.

Davie, P. J. F., Crosnier, A. *Echinolatus* n. gen. (Crustacea, Decapoda, Portunidae) with description of two new species from the South-West Pacific. In: Richer de Forges B. and Justine J. L. (eds), *Tropical Deep-Sea Benthos.* 24. *Memoires du Museum national d'Histoire naturell,* , 193, 393–410.

Devi, K. N., Shyamasundari, K., Rao, K. H. The anatomy and histology of the midgut and hindgut of *Charybdis* (*Goniohellenus*) *truncata* (Fabricius, 1798) (Decapoda: Brachyura). *Folia Morphol (Praha)* (1989). *37*, 148–155.

Doi, W., Yokota, M., Strüssmann, C. A., Watanabe, S. Growth and reproduction of the portunid crab *Charybdis bimaculata* (Decapoda: Brachyura) in Tokyo Bay. *Journal of Crustacean Biology,* (2008). *28*, 641–651.

Eddy, F., Powell, A., Gregory, S., Nunan, L. M., Lightner, D. V., Dyson, P. J., Rowley, A. F., Shields, R. J. A novel bacterial disease of the European shore crab, *Carcinus maenas* molecular pathology and epidemiology. *Microbiology,* (2007). *153*, 2839–2849.

Ekanem, A. P., Eyo, V. O., Ekpo, I. E., Bass, B. O. Parasites of Blue Crab (*Callinectes amnicola*) in the Cross River Estuary, Nigeria. *International Journal of Fisheries and Aquatic Studies,* (2013). *1*, 18–21.

Elegbede, I. O., Fashina-Bombata, H. A. Proximate and Mineral Compositions of Common Crab Species [*Callinectes pallidus* and *Cardisoma armatum*] of Badagry Creek, Nigeria. *Poult. Fish. Wildl. Sci.* (2013). *2*, 1–5.

378 Biology and Culture of Portunid Crabs of World Seas

Elumalai, V., Selvam, D., Suresh, T. V., Pravinkumar, M., Viswanathan, C. Studies on the Prevalence of *Sacculina* spp. Infestation in *Portunus sanguinolentus* (Herbst, 1783) from Parangipettai Coastal Waters, Southeast Coast of India. *J. Biodivers endanger species,* (2012). 1:101. doi:10.4172/2332–2543.1000101

Elumalai, V., Viswanathan, C., Pravinkumar, M., Raffi, S. M. Infestation of parasitic barnacle *Sacculina* spp. in commercial marine crabs. J Parasit Dis. (2014). 38,:337–339.

Escobar-Briiones, E., Alwarez, F. Symbiosis between *Portunus spinimanus* (Decapoda, Portunidae) and *Octolasmis lowei* (Thoracica, Poecilasmatidae) from Ubatuba, Sao Paulo, Brazil. In: *Crustacean Society; Modern Approaches to the study of Crustacea,* (2002). p. 205–210.

Evgeny, R., Michel, P., Zamorov, V., Frédéric, M. The swimming crab *Charybdis smithii:* distribution, biology and trophic role in the pelagic ecosystem of the western Indian Ocean. *Marine Biology,* (2009). *156,* 1089–1107.

Fernandez-Leborans, G., Gabilondo, R. Hydrozoan and protozoan epibionts on two decapod species, *Liocarcinus depurator* (Linnaeus, 1758) and *Pilumnus hirtellus* (Linnaeus, 1761) from Scotland. Zoologischer Anzeiger – *A Journal of Comparative Zoology*, (2005). *244,* 59–72.

Fernandez-Leborans, G., Cordoba, M. J. H., del-Arco, P. G. Distribution of ciliate epibionts on the portunid crab *Liocarcinus depurator* (Decapoda: Brachyura). *Invertebrate Biology,* (1997). *116,* 171–177.

Fernandez-Leborans, G., Gabilondo, R. Invertebrate and protozoan epibionts on the velvet swimming crab *Liocarcinus puber* (Linnaeus, 1767) from Scotland. *Acta Zoologica,* (2008). *89,* 1–17.

Firat, O., Gök, G., Coğun, H. Y., Yüzereroğlu, T. A., Kargin, F. Concentrations of Cr, Cd, Cu, Zn and Fe in crab *Charybdis longicollis* and shrimp *Penaeus semisulcatus* from the Iskenderun Bay, Turkey. *Environ Monit Assess.* (2008). 147,117–23.

Flores-Albino, B., Arias, L., Gómez, J., Castillo, A., Gimeno, M.,, Shirai, K. Chitin and L(+)-lactic acid production from crab (*Callinectes bellicosus*) wastes by fermentation of *Lactobacillus* sp. B2 using sugar cane molasses as carbon source. Bioprocess Biosyst. Eng. (2012). *35,* 1193–1200.

Fortes, R. D. Preliminary Results of the Rearing of Mud Crab, *Scylla olivacea* in Brackish water Earthen Ponds. *Proceedings of International Scientific Forum. ACIAR Proceedings,* (1999). *78,* 72–75.

Fragkiadakis, G. A., Stratakis, E. K. The lectin from the crustacean *Liocarcinus depurator* recognizes O-acetylsialic acids. Comp. Biochem. *Physiol. B. Biochem. Mol. Biol.* (1997). *117,* 545–552.

Fulton, B. A., Fairchild, E. A. Nutritional Analysis of Whole Green Crab, *Carcinus maenas*, for Application as a Forage Fish Replacement in Agrifeeds. *Sustainable Agriculture Research,* (2013). *2,* 126–135.

Galil, B. S., Lutzen, J. Biological Observations on *Heterosaccus dollfusi* Boschma (Cirripedia: Rhizocephala), a Parasite of *Charybdis longicollis* Leene (Decapoda: Brachyura), a Lessepsian Migrant to the Mediterranean. *Journal of Crustacean Biology,* (1995). *15,* 659–670.

Gaudé, A. R., Anderson, J. A. Soft Shell Crab Shedding Systems. SRAC Publication No. (4306). October, 2011.

# References                                                                     379

George, C., Gopakumar, K. Biochemical Studies on Crab *Scylla Serrata*. *Fishery Technology*, (1987). *24*, 57–61.

González-Gurriarán, E., Freire, J. Sexual maturity in the velvet swimming crab *Necora puber* (Brachyura, Portunidae): morphometric and reproductive analyses. *ICES Journal of Marine Science*, (1994). *51*, 133–145.

Gordon, D. P., Wear, R. G. A new ctenostome bryozoan ectosymbiotic with terminal-molt paddle crabs (Portunidae) in New Zealand. *New Zealand Journal of Zoology*, (1999). *26*, 373–380.

Greenwood, J. G., Fielder, D. R. The zoeal stages and megalopa of *Portunus rubromarginatus* (Lanchester) (Decapoda: Portnnidae), reared in the laboratory. *J. Plankton Res.* (1979). *1*, 191–205.

Haefner, Jr. P. A., Spacher, P. J. Gill Meristics and Branchial Infestation of *Ovalipes stephensoni* (Crustacea, Brachyura) by *Synophrya hypertrophica* (Ciliata, Apostomida). *Journal of Crustacean Biology*, (1985). *5*, 273–280.

Hamasakia, K., Fukunaga, K., Kitada, S. Batch fecundity of the swimming crab *Portunus trituberculatus* (Brachyura: Portunidae). *Aquaculture*, (2006). *253*, 359–365.

Hamid, R., Masood, A., Wani, I. H., Rafiq, S. Lectins: Proteins with Diverse Applications. *Journal of Applied Pharmaceutical Science*, (2013). 3 (4 Suppl 1), S93-S103.

Hamilton, R. A. Crab shells seen as industry waste solution. *The New York Times* http://www.nytimes.com/1987/09/20/nyregion/crab-shells-seen-as-industry-waste-solution.html.

Hassan, A. B., Li, T. P., Noor, Z. Z. Coagulation and flocculation treatment of wastewater in textile industry using chitosan. *Journal of Chemical and Natural Resources Engineering*, (2011). *4*, 43–53.

Hongbin, Y., Zheng, G., Li, W., McDowell, M. T., Seh, Z., Liu, N., Lu, Z., Cui, Y. Crab Shells as Sustainable Templates from Nature for Nanostructured Battery Electrodes. Nano Lett. (2013). *13*, 3385–3390.

http://aquatrop.cirad.fr/FichiersComplementaires/Mud%20crab_Final%20report.pdf. Crab Farming Philippines (Buy/Sell).

https://en.wikipedia.org/wiki/Crab_fisheries

http://museum.wa.gov.au/explore/blogs/museummarine/creature-feature-holy-crab-crucifix-crab-charybdis-feriata-linneaus-1758.

http://nsgl.gso.uri.edu/ncu/ncug12001. Blue Crab Aquaculture in Ponds. Potentials and Pitfalls.

http://www.dfo-mpo.gc.ca/science/aah-saa/diseases-maladies/paramcb-eng.html. *Paramoeba perniciosa* (Grey Crab Disease). Fisheries and Oceans Canada

http://www.euronews.com/2014/05/12/from-crab-shells-to-raw-materials/. From crab shells to raw materials.

http://www.fao.org/fishery/culturedspecies/Scylla_serrata/en. *Scylla serrata* (Forsskål, 1755. Cultured Aquatic Species Information Programme. Fisheries and Aquaculture Department. Food and Agriculture Organization of the United Nations.

http://www.inspection.gc.ca/animals/aquatic-animals/diseases/reportable/white-spot-disease/fact-sheet/eng/1336065470439/1336068112774. White Spot Disease – Fact Sheet. Canadian food Inspection Agency.

http://www.marlin.ac.uk/species/detail/1175. Harbor crab, Liocarcinus depurator.

https://www.salcombefinest.com/news/top-5-reasons-why-eating-crab-is-good-for-you. Top 5 reasons why eating crab is good for you.

http://www.sms.si.edu/irlspec/Callinectes_ornatus.htm. Smithsonian Marine Station at Fort Pierce.

http://www.vims.edu/research/departments/eaah/programs/crustacean/research/diseases_blue_crab/index.php. Some Parasitic Diseases of Blue Crab, Virginia Institute of Marine Science.

http://www.water-technology.net/news/newscrab-shells-effective-in-purifying-contaminated-water-say-malaysian-researchers-4658258. Crab shells effective in purifying contaminated water, say Malaysian researchers

Ihwan, M. Z., Hassan, M., Ambak, M. A., Ikhwanuddin, M. Occurrence of pedunculate barnacle, *Octolasmis* spp. in blue swimming crab, *Portunus pelagicus*. (2015). https://www.researchgate.net/publication/285152882.

Ikhwanuddin, M., Khairil, I. O., Azra, M. N., Waiho, K. Biological Features of Sentinel Crab *Podophthalmus vigil* (Fabricus, 1798) in Terengganu Coastal Water, Malaysia. *Journal of Fisheries and Aquatic Science*, (2015). *10*, 501–511.

Islam, M., Shokita, S., Higa, T. 2000: Larval development of the swimming crab *Charybdis natator* (Crustacea: Brachyura: Portunidae) reared in the laboratory. *Species Diversity*, (2000). *54*, 329–349.

Islam, M. S., Kurokura, H. Gonad development and size at maturity of the male mud crab *Scylla paramamosain* (Forsskål, 1755) in a tropical mangrove swamp. *Journal of Fisheries*, (2013). *1*, 7–13.

Islam, M., Machiko, K., Shokita, S. Larval development of swimming crab, *Thalamita pelsarti* Montgomery, 1931 (Cruustacea: Brachura: Portunidae) reared in the laboratory. *Russian Journal of Marine Biology*, (2005). *31*, 78–90.

Jacoeb, A. M., Lingga, N. L. S. The Effect of Steaming on Protein and Amino Acid Charactheristic of Crab (*Portunus pelagicus*) Meat. *JPHPI* (2012). *15*, 156–162.

Janvikul, W., Uppanan, P., Thavornyutikarn, B., Krewraing, J., Prateepasen, R. In vitro comparative hemostatic studies of chitin, chitosan, and their derivatives. *Journal of Applied Polymer Science*, (2006). *102*, 445–451.

Jimmy, U. P., Arazu, V. N. The Proximate and Mineral Composition of Two Edible Crabs *Callinectes amnicola* and *Uca tangeri* (Crustecea: Decapoda) of The Cross River, Nigeria. *Pakistan Journal of Nutrition*, (2012). *11*, 78–82.

Jimoh, A. A., Ndimele, P. E., Lemomu, I. P., Shittu, U. A. The Biology of Gladiator Swim Crab (*Callinectes pallidus*) from Ojo Creek, Southwestern Nigeria. *Journal of Fisheries and Aquatic Science*, (2014). *9*, 157–169.

Jing, L; Xiao-e, C., Xu-bo, F., Hui, Y. Analysis on Flesh Rate and Muscle Nutritional Value in *Ovalipes punctatus*. http://en.cnki.com.cn/Article_en/cjfdtotal-spgy201206033.htm.

Jithendran ,K.P., Poornim, M., Balasubramanian , C.P., Kulasekarapandian ,S. Diseases of mud crabs (Scylla spp.): an overview. *Indian J. Fish.* ( 2010), 57, 55-63.

Jose, J. Seed production and farming of blue swimmer crab *Portunus pelagicus*. CMFRI Manuel Customized training Bookp. 243–248. http://eprints.cmfri.org.in/9732/1/Josileen.pdf.

Joseph, S., Ravichandran, S. Shell Diseases of Brachyuran Crabs. *Journal of Biological Sciences* (2012). *12*, 117–127.

References                                                                                              381

Kala, K. L. J., Priya, E. R., Ravichandran, S., Chandran, M. Fatty Acid Contents of Ly-ophilized and Frozen *Portunus sanguinolentus* Crabmeat. *Journal of Food Resource Science,* (2015). *4,* 43–48.

Kannathasan, A., Rajendran, K. Isolation of microbes from various tissues of marine crab Charybdis natator (herbst, 1789) (Brachyura: Portunidae). *International Journal of Recent Scientific Research,* (2010). *7,* 172–176.

Keenan, C. P., Blackshaw, A. Mud Crab Aquaculture and Biology. *Proceedings of International Scientific Forum. ACIAR Proceedings.* (1999). *78,* 216 p.

Kerch, G. The Potential of Chitosan and Its Derivatives in Prevention and Treatment of Age-Related Diseases. *Mar. Drugs,* (2015). *13,* 2158–2182.

Krishnan, T., Kannupandi, T. Morphology of the larvae and first crab of an edible estuarine crab. *Thalamita danae* Stimpson, 1858 (Portunidae) fromIindian waters. *Indian Journal of Fisheries,* (1988). *35,* 118–120.

Küçükgülmez, A., Çelik, M. Amino Acid Composition of Blue Crab (*Callinectes sapidus*) from the North Eastern Mediterranean Sea. *Journal of Applied Biological Sciences,* (2008). *2,* 39–42.

Kumari, A. S. I., Murali Shankar, A., Jaganathan, K., Soundarapandian, P. Determinations of minerals in marine crab *Charybdis lucifera* (Fabricius, 1798). *International Letters of Natural Sciences,* (2015). *45,* 1–8.

Kuris, A. M., Torchin, M. E., Lafferty, K. D. *Fecampia erythrocephala* rediscovered: prevalence and distribution of a parasitoid of the European shore crab, *Carcinus maenas. J. Mar. Biol. Ass. UK.* (2002). *82,* 955–960.

Lawal-Are, A. O. Aspects of the biology of the lagoon crab *Callinectes amnicola* (Derocheburne) in Badagry, Lagos and Lekki lagoons, Nigeria. *Proc.16th Annual Conference of the Fisheries Society of Nigeria (FISON)*2003, pp. 215–220.

Lawal-Are, A. O., Kusemiju, K. Effect of salinity on survival and growth of blue crab, *Callinectes amnicola* from Lagos Lagoon, Nigeria. *J. Environ. Biol.* (2010). *31,* 461–464.

Lawal-Are, A. O. Reproductive Biology of the Blue Crab, *Callinectes amnicola* (De Rocheburne) in the Lagos Lagoon, Nigeria. *Turkish Journal of Fisheries and Aquatic Sciences,* (2010). *10,* 1–7.

Liang, X. S., Wei, Z., Xiao Jun, Y., Ming, L. H. Analysis and comparison of nutritional quality between wild and cultured *Portunus trituberculatus. Chinese Journal of Animal Nutrition,* (2009). *21,* 695–702.

Lijauco, M. M., Prospero, O. Q., Rodriguez. E. M. Polyculture of milkfish (*Chanos chanos*) and *Scylla serrata* at two stocking densities. *SEAFDEC Aquaculture Department Quarterly Research Report,* (1980). 4,19–23.

Lira, G. M. M., Torres, E. A. F. S., Soares, R. A. M., Mendonça, S., Costa, M. F., Silva, K. W. B., Simon, S. J. G. B., Veras, K. M. A. Nutritional value of crustaceans from lagoone-estuary complex Mundaú/ Manguaba-Alagoas. *Rev. Inst. Adolfo Lutz,* (2007). *66,* 261–267.

Llewellyn, L. E., Endean, R. Toxic coral reef crabs from Australian waters. *Toxicon,* (1988). *26,* 1085–1088.

Lu, J., Gong, Y., Huang, Y., Ma, H., Zou, X., Xia, L. Nutritional Analysis and Quality Evaluation in Muscle of Crucifix Crab *Charybdis feriatus* From Three Wild Populations. *Journal of Food Research,* (2014). *3,* 27–34.

Lwin, M. M. N. Soft-shell mangrove crab farming in Southeast Asia. https://www.was.org/meetings/mobile/MG_Paper.aspx?i=34840.

Lwin, M. M. N. Polyculture of Tilapia and Seaweeds in Soft-Shell Mud Crab Ponds in Indonesia and Thailand, 2011. http://ag.arizona.edu/azaqua/ista/ISTA9/PDF's/NoeNoeMudCrabs.pdf.

Machado, G. B. O., Fabio H. C. Sanches, F, H. C., Fortuna, M. D., Tânia m Costa, T. M. Epibiosis in decapod crustaceans by stalked barnacle Octolasmis lowei (Cirripedia: Poecilasmatidae). *Zoologia* (Curitiba) (2013). *30,* 307–311.

Maheswarudu, G., Jose, J., Nair, K. R. M., Arputharaj, M. R., Ramakrishna, A., Vairamani, A., Ramamoorthy, N. Evaluation of the seed production and grow out culture of blue swimming crab *Portunus pelagicus* (Linnaeus, 1758) in India. *Indian J. Mar. Sci.* (2008). *37,* 313–321.

Mancuso, M., Zaccone, R., Carella, F., Maiolino, P., De Vico, G. First, Episode of Shell Disease Syndrome in *Carcinus aestuarii* (Crustacea: Decapoda: Portunidae) in the Volturno River. *J Aquac Res Development,* (2013). *4,* 191. doi:10.4172/2155-9546.1000191, 3p.

Mari, J., Bonami, J. W2 Virus Infection of the Crustacean *Carcinus mediterraneus*: a Reovirus Disease. *J. Gen. Virol.* (1988). *69,* 561–571.

Matozzo, V., Marin, M. G. The role of haemocytes from the crab *Carcinus aestuarii* (Crustacea, Decapoda) in immune responses: A first survey. *Marine Fish & Shellfish Immunology*, (2010). *28,* 4, 534–541.

Meena, B., Rose, S., Jayaraj, S. S., Amsath, A., Vincent, S. Microbial and haemagglutinins from the serum of estuarine Crab *Portunus sanguinolentus.* Rec. *Res. Sci. Tech.* (2011). *3,* 87–94.

Melo, G. A. S. *Manual de identificação dos Brachyura (caranguejos e siris) do litoral brasileiro. São Paulo: Plêiade.* (1996). 604 pp.

Miller, A., Inglis, G. J., Poulin, R. Comparison of the ectosymbionts and parasites of an introduced crab, *Charybdis japonica*, with sympatric and allopatric populations of a native New Zealand crab, *Ovalipes catharus* (Brachyura: Portunidae). *New Zealand Journal of Marine and Freshwater Research,* (2006). *40,* 369–378.

Morado, J. F., Dawe, E. G., Cawthorn, R. J. http://www.ices.dk/sites/pub/CM%20Doccuments/CM-2008/D/D0408.pdf. ICES CM 2008/D:04.

Moronkola, B. A., Olowu, R. A., Tovide, O. O., Ayejuyo, O. O. Determination of proximate and mineral contents of crab (*Callinectes amnicola*) living on the shore of Ojo river, Lagos, Nigeria. *Sci. Revs. Chem. Commun.* (2011). *1,* 1–6.

Morton, B. *Partnerships in the Sea: Hon Kong's Marine Symniosis*. Hong Kong University Press, Hong Kong, 1989.

Munilla, I. Henslow's swimming crab (*Polybius henslowii*) as an important food for yellow-legged gulls (*Larus cachinnans*) in NW Spain. *ICES Journal of Marine Science,* (1997). *54,* 631–634.

Muthiga, M., Mombasa, K. Edible crabs of Kenya. *Kenya Aquatica,* (1986). *3,* 61–65.

Nachimuthu, I., Kumari, S. I. A., Shankar, M. A., Jaganathan, K., Soundarapandian, P. Amino acid profile in marine crab *Charybdis lucifera* (Fabricius, 1798). *World Journal of Pharmaceutical Research,* (2015). *4,* 2345–2357.

References 383

Ng, P. K. L., Takeda, M. *Atoportunus*, a remarkable new genus of cryptic swimming crab (Crustacea; Decapoda; Brachyura: Portunidae), with descriptions of two new species from the Indo-West Pacific. *Micronesica*, (2003). 35–36, 417–430.

Norman, C. P., Reproductive Biology and evidence for hard-female mating in the brachyuran crab, *Thalassima sima* (Portunidae). *Journal of Crustacean Biology*, (1996). *16*, 656–662.

Oesterling, M. J. Soft Crabs in Closed Systems: A Virginia Success Story. http://nsgl.gso.uri.edu/vsgcp/vsgcpc00001/1996/6-shellfish_and_fish_production.pdf

Oikawa H., Fujita, T., Saito, K., Watabe, S., Satomi, M., Yano, Y. Comparison of paralytic shellfish poisoning toxin between carnivorous crabs (*Telmessus acutidens* and *Charybdis japonica*) and their prey mussel (*Mytilus galloprovincialis*) in an inshore food chain. *Toxicon*. (2004). *43*, 713–719.

Omuvwie, U., Atobatele, O. E. Growth pattern, condition factor, trace metal studies and ectoparasitic load of the blue crab, *Callinectes amnicola* from Lagos Lagoon, Badore, Ajah, Lagos, Nigeria. *Cameroon Journal of Experimental Biology*, (2013). *9*, 34–43.

Oyebisi, R., Lawal-Are, A. O., Alo, B. Comparative Study of Persistent Toxic Metal Levels in Land Crab (*Cardiosoma armatum*) and Lagoon Crab (*Callinectes amnicola*) in Lagos Lagoon. *J. Mar. Biol. Oceanogr.* (2013). *2*, 1–8.

Ö zcan, T. The swimming crab *Portunus segnis* (Forskål, 1775): host for the barnacle *Chelonibia platula* (Ranzani, 1818) from the Turkish coast. *J. Black Sea/Mediterranean Environment*, (2012). *18*, 271–278.

Patil, S. G., Patil, B. G. Chitin supplement speeds up the ethanol production in cane molasses fermentation. *Enzyme and Microbial Technology*, (1989). *11*, 38–43.

Petersen, E. H., Phuong, T. H., Dung, N. V., Giang, P. T., Dat, N. K., Vu Anh Tuan, V. A., Nghi, T. V. Bioeconomics of mud crab, *Scylla paramamosain*, culture in Vietnam. ACE Discussion Paper 2011/4. www.advancedchoiceeconomics.com.au.

Pillai, S. L., Rajapackiam, S., Sunderarajan, D. Mud crab *Scylla tranquebarica* culture in earthen pond at Tuticorin. *J. Mar. Biol. Ass. India*, (2002). *44*, 245–248.

Pinheiro, M. A. A., Fransozo, A. Reproductive Behavior of the Swimming Crab *Arenaeus cribrarius* (Lamarck, 1818) (Crustacea, Brachyura, Portunidae) in Captivity. *Bulletin of Marine Science*, (1999). *64*, 243–253.

Pinheiroa, M. A. A., Hattoria, G. Y. Growth of the speckled swimming crab, *Arenaeus cribrarius* (Lamarck, 1818) (Crustacea, Brachyura, Portunidae), in Ubatuba (SP), Brazil. *Journal of Natural History* (2006). *40*, 1331–1341.

Premarathna, A. D., Rajapakse, R. P. V. J., Pathirana, E. V. P., Karunarathna, S. C., Jayasooriya, A. P. Nutritional analysis of some selected fish and crabmeats and fatty acid analysis of oil extracted from *Portunus pelagicus*. *International Journal of Scientific & Technology Research*, (2015). *4*, 197–201.

Quinitio, E. T., Parado-Estepa, F. D., Rodriguez, E. Seed Production of Mud Crab *Scylla* spp. *Aquaculture Asia*, (2002). *7*, 29–31.

Radhakrishnan, C. K., Natarajan, R. Nutritive Value of the rab *Podophthalmus vigil* (Fabridrus). *Fish. Technol.* (1979). *16*, 37–38.

Ramesh, S., Sankar, V., Santhanam, R. *Marine Pharmaceutical Compounds*, Lambert Academic Publishing, (2012). 249 p.

Rameshkumar, G., Ravichandran, S., Chandran, K., Ajithkumar, T. T. Comparison of fatty acid profile in edible crabs Scylla serrata and *Portunus pelagicus*. *Global Journal of Environmental Research*, (2009). *3*, 42–45.

Rasheed, S., Mustaquim, J. Size at sexual maturity, breeding season and fecundity of three-spot swimming crab *Portunus sanguinolentus* (Herbst, 1783) (Decapoda, Brachyura, Portunidae) occurring in the coastal waters of Karachi, Pakistan. *Fisheries Research*, (2010). *103*, 56–62.

Raso, J. E. G., Manjon-Cabeza, E. New record of *Liocarcinus mcleayi* (Barnard, 1947), new combination (Decapoda, Brachyura, Portunidae) from South Europe. *Crustaceana*, (1996). *69*, 84–93.

Rivera, A., Santiago, K., Torres, J., Sastre, M. P., Rivera, F. F. Bacteria associated with hemolymph in the crab *Callinectes bocourti* in Puerto Rico. *Bulletin of Marine Science*, (1999). *64*, 543–548.

Rodríguez-Domínguez, G., Castillo-Vargasmachuca, S. G., Pérez-González, R., Aragón-Noriega, E. A. The size at maturity of the brown crab *Callinectes bellicosus* (Decapoda, Portunidae) in the Gulf of California. *Crustaceana*, (2012). *85*, 1513–1523.

Safaie M., Pazooki J., Kiabi B., Shokri, M. R. Reproductive biology of blue swimming crab, *Portunus segnis* (Forskal, 1775) in coastal waters of Persian Gulf and Oman Sea, Iran. *Iranian Journal of Fisheries Sciences*, (2013). *12*, 430–444.

Sankarankutty, C. A note on the abnormalities in *Thalamita integra*. Dana. http://eprints.cmfri.org.in/1608/1/Article_19.pdf.

Santhanam, R. *Nutritional Marine Life*. CRC Press (Taylor & Francis Group), (2015). 278p.

Santos, S. Symbiosis Between *Portunus spinimanus* Latreille, 1819 (Decapoda, Portunidae) and *Octolasmis lowei* (Darwin, 1852) (Thoracica, Poecilasmatidae) from Ubatuba, Sao Paulo, Brazil. Modern Approaches to the Study of Crustacea. Kluwer Acadmic / Plenum Publishers. E. Escober-Briones and F. Alvarez (Eds). pp 205–209, 2002.

Sathiadhas, J. J. R. Economics of selected coastal aquaculture. In: *Sustain Fish*. B. M. Kurup & K. Ravindran (Eds.), (2006). pp. 802–811.

Sethuramalingam, S., Radhakrishnan, C. K., Natarajan, R. Association between the bivalue, *Amussium pleuronectes* (Linne) and the crabs, *Portunus hastatoides* Fabricius and *Charybdis hoplites* Wood Mason [India]. *Indian Journal of Marine Sciences*, (1980). *91*, 68–69.

Shahidi, F., Arachchi, J. K. V., Jeon, Y. Food applications of chitin and chitosans. *Trends in Food Science & Technology*, (1999). *10*, 37–51.

Shelley, C. Capture-based aquaculture of mud crabs (*Scylla* spp.). In: A. Lovatelli and P. F. Holthus (eds). *Capture-based aquaculture. Global overview*. FAO Fisheries. Technical Paper, (2008). *508*, pp. 255–269.

Shelley, C., Lovatelli, A. *Mud crab aquaculture – A practical manual*. FAO Fisheries and Aquaculture Technical Paper. No. 567. Rome, FAO. 2011. 78 pp.

Shields, J. D. Parasites and symbionts of the crab *Portunus pelagicus* from Moreton Bay, Eastern Australia. *Journal of Crustacean Biology* (1992). 12, 94–100.

Shields, J. D., Earley, C. G. *Cancrion australiensis* new species (isopoda: Entoniscidae) found in *Thalamita sima* (Brachyura: Portunidae) from Australia. *International Journal for Parasitology*, (1993). *23*, 601–608.

Sigana, D. O. Breeding Cycle of *Thalamita crenata* (Latreille, 1829) at Gazi Creek (Maftaha Bay), Kenya. *Western Indian Ocean J. Mar. Sci.* (2002). *1*, 145–153.

References 385

Smithsonian Marine Station at Fort Pierce, http://www.sms.si.edu/irlspec/Callinectes_ornatus.htm

Sonawane, S. S., Marbate, L. K. T., Sanap, B. N., Sapkale, P. H. Embryonic development of commercially important estuarine crab, *Scylla tranquebarica* (Fabricius, 1897) along Ratnagiri coast, Maharashtra. *EM International*, (2015). *21*, 1971–1975.

Soundarapandian, P., Sivasubramanian, C., Varadharajan, D. Fattening of the Ridged Swimming Crab, *Charybdis natator* Herbst. *J. Marine Sci. Res. Dev.* (2013). 3:125. doi: 10.4172/2155–9910.1000125.

Soundarapandian, P., Varadharajan, D., Sivasubramanian, C., Irin Kumari, A. S. Amino Acid Profiles of Ridged Swimming Crab, *Charybdis natator* Herbst. *J. Aquac. Res. Development*, (2014). 5:287. doi:10.4172/2155–9546.1000287.

Soundarapandian, P., Ilavarasan, N., Varadharajan, D., Gangatharan. K. Seed Production of Commercially Important Portunid Crab, *Charybdis feriata* (Linnaeus). *J. Marine Sci. Res. Dev.* (2013). *3*, 120,6p. doi:10.4172/2155–9910.1000120

Soundarapandian, P., Varadharajan, D., Sivasubramanian, C. Mineral Composition of Edible Crab, *Charybdis natator* Herbst (Crustacea: Decapoda). *J. Bioanal. Biomed.* (2013). *5*, 99–101.

Soundarapandian, P., Varadharajan, D., Boopathi, A. Reproductive Biology of the Commercially Important Portunid Crab, *Portunus sanguinolentus* (Herbst). *J. Marine Sci. Res. Dev.* (2013). *3*, 124. doi:10.4172/2155–9910.1000

Soundarapandian, P., Roy, S., Varadharajan, D. Antioxidant Activity in Hard and Soft Shell Crabs of *Charybdis lucifera* (Fabricius, 1798). *J Aquac Res Development*, (2014). *5*, 288. doi:10.4172/2155–9546.1000288

Soundarapandian, P., Raja, D. A. Fattening of the Blue Swimming Crab *Portunus pelagicus* (Linnaeus). *Journal of Fisheries and Aquatic Science*, (2008). *3*, 97–101.

Soundarapandian, P., Varadharajan, D., Ravichandran, S. Mineral composition of edible crab *Podophthalmus vigil* Fabricius (Crustacea: Decapoda). *Arthropods*, (2014). *3*, 20–26.

Sounndarapandian, P., Dey, S. S. Proximate Composition of the Eggs of Commercially Important Crab *Portunus sanguinolentus* (Herbst). *Journal of Fisheries and Aquatic Science*, (2008). *3*, 60–65.

Sudhakar, M., Manivannan, K., Soundrapandian, P. Nutritive Value of Hard and Soft Shell Crabs of *Portunus sanguinolentus* (Herbst). *International Journal of Animal and Veterinary Advances*, (2009). *1*, 44–48.

Sudhakar, M., Raja, K., Anathan, G., Sampathkumar, P. Compositional characteristics and nutritional quality of *Podolphthalmus vigil* (Fabricius). *Asian Journal of Biological Sciences*, (2011). 4,166–174..

Suerte, N. O. Feasibility of Blue Swimming Crab *Portunus pelagicus* Linnaeus 1758 and Red Seaweed *Kappaphycus alvarezii* Doty Polyculture. in Floating Net Cages. *Mindanao Journal of Science and Technology*, (2015). *13*, 196–212.

Suganya, V., Asheeba, S. T. Isolation and characterization of astaxanthin from *Portunus sanguinolentus* (three Spotted crab), *Callinectes sapidus* (blue crab) and *Paralithodes brevipes* (Spiny king crab). *International Journal of Current Research*, (2015). *7*, 22176–22193.

Suganya, V., and Asheeba, S. T. Antioxidant and antimicrobial activity of astaxanthin isolated from three varieties of crabs. *Journal of Recent Scientific Research*, (2015). *6*, 6753–6758.

Sukumaran, K. K., Neelakantan, B. Age and growth in two marine portunid crabs, *Portunus (Portunus) sanguinolentus* (Herbst) and *Portunus (Portunus) pelagicus* (Linnaeus) along the southwest coast of India. *Indian J. Fish.* (1997). *44*, 111–131.

Sumpton, W. D., Smith, G. S., Potter, M. A. Notes on the Biology of the Portunid Crab, *Portunus sanguinolentus* (Herbst), in Subtropical Queensland Waters. *Australian Journal of Marine and Freshwater Research,* (1989). *40*, 711–717.

Takahashi, K., Kawaguchi, K. Nocturnal occurrence of the swimming crab *Ovalipes punctatus* in the swash zone of a sandy beach in northeastern Japan. *Fish. Bull.* (2001). *99*, 510–515.

Thirunavukkarasu, N., Shanmugam, A. Extraction of chitin and chitosan from mud crab *Scylla tranquebarica* (Fabricius, 1798). *International Journal on Applied Bio Engineering.* http://dx.doi.org/10.18000/ijabeg.10048.

Tom, R. A., Carroad, P. A. Effect of Reaction Conditions on Hydrolysis of Chitin by *Serratia marcescens* QMB 1466 chitinase. *Journal of Food Science*, (1981). *46*, 646–647.

Torchin, M. E., Lafferty, K. D., Kuris, A. M. Release from parasites as natural enemies: increased performance of a globally introduced marine crab. *Biological Invasions,* (2001). *3*, 333–345.

Triño, A. T., Millamena, O. M., Keenan, C. P. Monosex Culture of the Mud Crab (*Scylla serrata*) at Three Stocking Densities with Gracilaria as Crab Shelter. *Proceedings of International Scientific Forum. ACIAR Proceedings.* (1999). No. *78*, p.61–66.

Triño, A. T., Rodriguez, E. M. Mud Crab Fattening in Ponds. *Asian Fisheries Science*, (2001). *14*, 211–216.

Troell, M. Integrated marine and Brackish water aquaculture in tropical regions: research, implementation and prospects. In D. Soto (ed.). *Integrated Mariculture: A Global Review*. FAO Fisheries and Aquaculture Technical Paper, (2009). *529*, 47–131.

Venugopal, G., Razvi, S. S. H., Babu, P. P. S., Reddy, P. R., Mohan, K. M., Rao, K. P. S., Patnaik, R. R. Performance evaluation of mud crab *Scylla serrata* (Forskal, 1775) in monoculture, monosex culture and polyculture. *J. Mar. Biol. Ass. India,* (2012). *54*, 5–8.

Viswam, D. Investigation of nutritive value of crabs along Kerala coast. Final Report of Minor Research Project. p.29. 2015. http://www.sncollegecherthala.in/research%5Cdhanyaviswam.pdf.

Vivares, C. P. Parasites of decapod brachyuran crustaceans from the Gulf and Lake of Tunis. Preliminary note. *Bulletin de l'Institut National de la Sante Scientifique et Technique d'Oceanographie et de Peche de Salammbo*, (1970). *1*, 181–203.

Yadon, D. A. Cage culture and eyestalk ablation molt induction of blue crabs, (*Callinectes sapidus* Rathbun) held in Brackish water ponds. http://hdl.handle.net/1969.3/20730.

Waiho, K., Mustaqim, M., Fazhan, H., Norfaizza, W. I. W., Megat, F. H., Ikhwanuddin, m Mating behavior of the orange mud crab, *Scylla olivacea*: The effect of sex ratio and stocking density on mating success. *Aquaculture Reports*, (2015). *2*, 50–57.

Walker, G. Some Observations on the Epizoic Barnacle *Octolasmis angulata* within the Branchial Chambers of an Australian Swimming Crab. *Journal of Crustacean Biology*, (2001). *21*, 450–451.

Wang, J., Zhang, F. Y., Song, W., Fang, Y. B., Hu, J. H., Zhao, M., Jiang, K. J., Ma, L. B. Characterization of hemocyanin from the mud crab *Scylla paramamosain* and its expression analysis in different tissues, at various stages, and under *Vibrio parahaemolyticus* infection. *Genet. Mol. Res.* (2015). *14*, 16639–16651.

# References

Wang, K., Chen, F., Ma, X., Wang, S., i Chen, B., Dong, L., Shan, Z., Fan, D., Peng, H. Several newly identified immune-associated components in mud crab (*Scylla paramamosain*) and their potential anti-infection functions. http://digitalcommons.library.umaine.edu/cgi/viewcontent.cgi?article=1168&context=isfsi.

Wang, X., Chi, Z., Yue, L., g Li, J., Li, M., Wu, L. A marine killer yeast against the pathogenic yeast strain in crab (*Portunus trituberculatus*) and an optimization of the toxin production. *Microbiological Research*, (2007). *162*, 77–85.

Watanabe, T. T., Zara, F. J., Hattori, G. Y., Turra, A., Sant'anna, B. S. Biological associations of color variation in the Indo-Pacific swimming crab *Charybdis hellerii*. *An. Acad. Bras. Ciênc.* (2015). *87*, 219–232.

Wen-Jun, X., Hui, S., Han-Xiang, X. Hamish Small Preliminary study on the hematodinium infection in cultured *Portunus trituberculatus*. *Acta Hydrobiologica Sinica*, (2007). *31*, 637–642.

West, L. E. Habitat location and selection by the *sargassum* crab *Portunus sayi*: the role of sensory cues. M. S. Thesis of Florida Atlantic University Boca Raton, Florida, 2012.

Wild Fact Sheets – http://www.wildsingapore.com/wildfacts/crustacea/crab/portunidae/feriatus.htm.

Wisespongpand, P., Vareevanich, D., Khaodon, K. Nutritive value of crabs from bottom gill net as an alternative food. http://www.annualconference.ku.ac.th/cd53/04_012_P49.pdf.

Yomar-Hattori, G., Anna, B. S., Amaro-Pinheiro, M. A. Meat yield of *Callinectes bocourti* A. Milne Edwards, 1879 (Crustacea, Portunidae) in Iguape, São Paulo, Brazil. *Invest. Mar. Valparaíso*, (2006). *34*, 231–236.

Ze-Lin, W., Ngan-Kee, N., Teo, S. L. M., Parra-Velandia, F. J. Fluorescent patterns in some portunus species (Crustacea: Brachyura: Portunidae). *Contributions to Marine Science*, (2012). 135–143.

Zetlmeisl, C., Hermann, J., Petney, T., Glenner, H., Griffiths, C., Taraschewski, H. Parasites of the shore crab *Carcinus maenas* (L.): Implications for reproductive potential and invasion success. *Parasitology*, (2011). *138*, 394–401.

Zhang, J., Xia, W., Liu, P., Cheng, Q., Tahirou, T., Gu, W., Li, B. Chitosan Modification and Pharmaceutical/Biomedical Applications. *Mar. Drugs*, (2010). 8, 1962–1987.

Zmora, O., Findiesen, A., Stubblefield, J., Frenkel, V., Zohar, Y. Large-scale juvenile production of the blue crab Callinectes sapidus. *Aquaculture*, (2005). 244, 129–139.

Zohri, C. S. W. M. *Description of larval development stages of mud crab (Scylla tranquebarica) and the effectiveness of antimicrobial agents against bacterial isolates.* Masters Thesis., Universiti Malaysia Kelantan. (2011). http://umkeprints.umk.edu.my/1037/.

# INDEX

## A

Abdominal
  flab, 17, 19, 53, 73, 274, 372
  somite, 30, 132
Accessory reproductive organs, 17, 53
Accumulation, 79, 161
Acetate, 342
*Achelous*
  *depressifrons*, 30
  *spinicarpus*, 106, 354
  *spinimanus*, 107, 354
*Acineta*
  *papillifera*, 363
  *tuberosa*, 363
*Acinetides symbiotica*, 363
*Actinocyathula homari*, 363
Active
  phagocytic cells, 339
  predator, 90
Adaptation, 86, 89, 109
Adipose tissues, 333
*Aeromonas hydrophila*, 355, 356
Albinism, 353, 372
Algae, 2, 43, 84, 88, 90, 91, 213, 240, 263, 265, 277, 294, 296, 308
Allometric function, 20
Alzheimer's disease, 333, 334, 344
Ambulatory legs, 24, 29, 32, 47, 112, 115, 131, 167, 195, 202, 203, 237, 253, 259, 261
Amelioration, 333
Amino
  acid content, 324, 328
  hydroxyl functional groups, 346
Ammonia, 276, 277, 284, 358, 373
Amphipods/turbellarian, 370
Angular interspaces, 70
Anionic polysaccharides, 343
Annelids, 14, 213, 274

Antenna, 10, 21, 44, 98, 215
Antennules, 10
Antero-external angle, 25, 28, 33–35, 38, 41, 46, 60, 67, 72, 87, 257, 260, 262
Anterolateral
  borders, 25, 28, 29, 33–35, 38–40, 42, 46, 48, 51, 60, 61, 65, 67, 72, 87, 103, 114, 116, 120, 123, 127–129, 139, 150, 200, 201, 235, 236, 245, 254, 255, 257, 259–261, 267–270
  carapace spines, 68, 76
  edges, 36, 174, 189, 250
  margin, 11, 23, 24, 26, 32, 49, 57, 64, 78, 80, 81, 83, 85, 88–93, 101, 103, 142, 150, 152, 160, 163, 202, 203, 208–210, 212, 218–225, 230, 235, 237, 261, 262
  portions, 80
  spine, 41, 45, 51, 188, 202
  sulci, 81
  tooth, 51, 59, 64, 72, 140, 151, 153, 155, 170, 172, 192, 193, 197, 204–206
Antibacterial, 330, 336
Anticancer, 330, 333, 334
Anticoagulant, 6
Antidiabetic, 330
  therapeutic applications, 333
Antifungal, 330, 341
Anti-inflammatory, 330, 333
Antileukemic, 6
Antimetastatic effect, 341
Antimicrobial, 6, 71, 220, 329, 331, 334, 341, 343
  activity, 335
Antioxidant, 330, 333–335
  activity, 334
Antioxidative properties, 342

Appendages, 10, 13, 15, 16, 26, 31, 54, 78, 98, 106, 107, 109, 139, 143, 186, 229, 230, 241, 370
Apron, 10
Aquaculture, 4, 6, 7, 109, 273, 281, 300, 340
Aquatic organisms, 2, 340
*Arenaeus cribarius*, 354
Arthritis, 331, 333
Articulations, 27, 59, 64, 78, 80, 82, 86, 140, 162, 203, 258
Ascorbic acid, 344
Asexual budding, 362
*Aspergillus niger*, 146, 335
Astaxanthin pigment, 334
Asthma, 5
Atoportunus, 111, 112
Autotomy/regeneration, 22

## B

*Bacillus cereus*, 335
Bacterial
necrosis, 349, 369
shell disease, 350
Baculovirus, 368, 369
infection, 369
Balanid cirripedes, 371
*Balanus crenatus*, 363
*Barentsia matsushimana*, 353, 364
Barramundi, 69, 72, 296
Basal antennal
joint, 72, 166, 174, 175, 180, 190–192, 200, 207, 216
segment, 24, 115, 197, 253
*Bathynectes*, 3, 215–217
Bell peppers, 343
Benthic macroinvertebrates, 52
*Benthochascon*, 3, 234–236
Berried
crabs, 6, 7, 19, 276
female, 4, 52, 73, 208
Berry, 17, 19, 53, 131, 276
Big-eyed swimming crab, 39
Bioabsorbents, 345
Bioaccumulation, 79

Bioadhesion, 332
Bioadhesivity, 331, 332
Biodegradable polymer, 329, 344
Biology (portunid crabs), 10
morphology, 10
internal anatomy, 13
food/feeding, 14
behavior, 14
molting/growth, 14, 15
age/growth studies, 15
autotomy/regeneration of lost legs, 15, 16
reproduction/lifecycle, 16
sexual dimorphism/characters, 16
reproductive system, 17
mating/spawning, 17–19
fecundity, 20
lifecycle, 20–22
external anatomy, 10–13
Biomass, 2, 290, 291, 298, 300–302, 305
Biomedical applications of
chitin/chitosan, 330, 348
antioxidant/antimicrobial activities, 334–339
chitosan applications in human health, 331, 332
prevention/treatment of age-related diseases, 332–334
Biopolylmer, 347
Biosorbents, 345, 346
Bispecies culture, 306
Bitter crab disease (BCD), 357, 363, 373
Bitter crab syndrome (BCS), 357, 363
Black gill disease, 350, 373
Blastula, 77
Blue swimmer crabs, 43
Blunt-tooth swimming crab, 82
Bocourt swimming crab, 82
Body coloration, 87, 231, 233
Bone formation, 331
*Botrytis cinerea*, 335
Bottomfeeding carnivores, 43
Brachionus sp, 77
Brachyura, 90
Brachyuran true crabs, 2

Index 391

Brackish, 78, 214, 282, 293
Branchial
  chambers, 219, 356, 361, 362, 366, 371
  regions, 72, 86, 118, 123, 153, 155, 160, 194, 248
  ridge, 27
Breakage point, 15
Breast cancer, 334
Broodstock development, 52
Bryozoan, 361, 365
  growth, 365

## C

Cadmium, 142, 307, 345, 346
Calcium, 14, 43, 274, 304, 307
  salts, 14
  sources, 43
*Callinectes*
  *amnicola*, 78, 308, 355
  *bellicosus*, 81, 347
  *bocourti*, 82, 309, 355
  *danae*, 83, 309, 356
  *exasperatus*, 85
  *marginatus*, 87
  *ornatus*, 88–90, 96, 356
  *rathbunae*, 356
  *sapidus*, 82, 90, 96
  *sapidus*, 93, 96, 98, 101, 278, 290, 291, 305, 311, 334, 351, 352, 356, 367
  *similis*, 90
Camouflage, 55, 263, 266
Cancer, 5, 6, 307, 331–334, 339
*Cancrion australiensis*, 373
*Caphyra*, 3, 262–266
Caphyrinae, 3, 23
Carapace
  length, 27, 34, 37, 44, 50, 57, 78, 80, 81, 83, 88, 91, 100, 101, 106, 123, 143, 210, 216, 221, 222, 224, 227, 229, 234, 255
  width (CW), 42, 43, 53, 58, 61, 62, 70, 76, 78, 82, 106, 107, 109, 131, 157, 167, 178, 185, 203, 206, 214,

232, 233, 243, 245, 249, 251, 276, 283, 287, 291, 292, 297, 301
Carbon, 5, 6, 343, 347
Carboxymethylchitosan, 337, 342, 346
Carcininae, 3, 23
*Carcinonemertes carcinophila*, 219, 352
*Carcinonernertes mitsukurii*, 352
*Carcinus*
  *aestuarii*, 339, 357
  *maenas*, 213, 336, 352, 357–360
  *mediterraneus*, 360
Cardiac, 27, 29, 40, 42, 46, 61, 63, 64, 78, 80, 83, 84, 89, 94, 101, 103, 105, 119, 121, 123, 138, 144, 155, 158, 159, 161, 168, 182, 184, 187, 191, 192, 194, 221, 225, 227, 248, 249, 251, 261, 270
Cardiac/lateral postcardiac regions, 64
Cardio active properties, 6
Carinae, 27, 59, 126, 127, 205, 209, 216, 218, 224, 237, 238, 242, 243, 264, 265
Carpus, 34, 39, 41, 47, 59, 64, 68, 70, 76, 89, 106, 116, 118, 120–122, 125, 126, 129, 130, 136, 147, 156, 160, 161, 164, 168, 171, 173, 177, 181, 187, 194, 200, 201, 204, 210–212, 218, 221, 225, 227, 231, 236, 241, 248, 249, 255, 256, 259, 261, 268–270
Carrion feeders, 14
Cartilage, 13
Carupinae, 3, 23
Castration, 360, 371
Catechins, 344
Cationic
  nature, 331
  polymer, 332
*Catoptrus*, 3, 255, 256
Causative agents, 350–353
Cell membranes, 331
Cellulose, 332, 338, 340
Central
  confluent lobular area, 33, 114
  trapezoidal, 88, 99
Cephalothorax, 150, 151, 155, 263, 363
*Cerithidea mazatlanica*, 80

Cervical groove, 26, 116, 120, 122, 123, 129, 139, 152, 178, 183, 191, 192, 196, 201
Cestodes, 352
Characteristics of,
  caphyrinae, 262
  carcininae, 204
  carupinae, 253
  macropipinae, 235
  podophthalminae, 200
  polybiinae, 215
  portuninae, 24
  thalamitinae, 115
*Charybdis*
  *affinis*, 3, 118
  *anisodon*, 3, 120
  *annulatata*, 3
  *bimaculata*, 3, 123
  *callianassa*, 125, 361, 362
  *feriatus*, 3–5, 18, 130, 132, 314, 351
  *helleri*, 3, 136, 316, 361
  *japonica*, 3, 140, 352, 361
  *japonica*, 361
  *longicollis*, 142, 362
  *lucifera*, 3, 143, 316, 334, 351, 361
  *natator*, 145, 146, 303, 317, 361
  *smithi*, 4, 156, 319
  *truncata*, 4, 158, 362
  *vadorum*, 4, 159
  *variegate*, 4
Chela, 27, 64, 78, 83, 87–89, 92, 97, 101, 106, 153, 167, 177, 188, 211, 227–229, 232, 239, 243, 262, 270
Chelae, 16, 18, 36, 52, 78, 84, 86, 88, 89, 94, 99, 100, 152, 160, 212, 217, 230, 234, 262
Chelate
  legs, 372
  male, 52
Chelation ion exchange, 346
Chelipeds, 13, 16, 27, 41, 68, 70, 76, 80, 86, 94, 101, 105, 109, 122, 124, 142, 149, 155, 164, 175, 189, 190, 194, 195, 198, 215, 225, 229, 230, 235, 237, 247, 256, 258, 262, 263, 270, 275, 304, 355, 364, 370, 371

merus, 34, 35, 38, 41, 42, 49, 51, 52, 60, 61, 67, 118, 120, 129, 134, 152, 175, 194, 196, 199, 200, 205, 268–270
*Chelonibia patula*, 137, 354, 355, 366–368
Chemical oxygen demand (COD), 345
Chemoreception, 10, 55
Chitin hydrolysis, 340
Chitinase enzyme, 340
Chitinolytic
  activity, 357
  bacterial disease, 349, 369
Chitosan, 329–335, 337–346, 348
  coated nanoliposomes, 334
  colloidal systems, 332
  composite microgranules, 331
  glutamate, 335
  hemostatic dressings, 338
  hydrogels, 332
  lactate, 333, 335
Chromatophores, 25, 26
Chromium drug candidate, 339
Chronic inflammation, 333
Ciliates, 349, 362, 363, 370
Cinnamic acid, 344
Citric acid, 344
Clot-inducing celox, 337
Cnidarians, 2
Coagulation, 337, 344, 345
Coarsest granules, 83
Coastal
  plants, 6
  shelf waters, 22, 96, 100
  waters, 30, 83, 208, 298, 302
*Coelocarcinus*, 3, 267
*Coenophthalmus*, 3, 217
Colloidal system, 332, 345
Colon cancer cells, 334
Commensals, 14, 270, 371
Commercial
  fisheries, 1, 32, 34–36, 53, 62, 69, 75, 86, 92, 101–103, 111, 119, 121, 124, 133, 138, 141, 144, 159–161, 239
  species, 7, 62, 90, 112, 232
  tablet dosage form, 331

Index 393

Common name, 120, 126, 152, 156, 227, 256, 271
*Conchacineta constricta*, 363
Conchoderma virgatum, 55
Consumable species of crabs, 2
    caneridae, 2
    gecarcinidae, 2
    geryonidae, 2
    lithodidae, 2
    macidae, 2
    ocypodidae, 2
    portunidae, 2
    potamidae, 2
    xanthidae, 2
Conspicuous spine, 29, 59
Continental shelf areas, 57
Convex dorsal surface, 85
Copepoda, 353
Copious intake, 15
Copper, 307, 308, 333, 345, 346
Copulation, 16, 18, 19, 69, 76, 77, 110, 131, 240
Coral atoll, 2
Cortez swimming crab, 81
*Corynophrya anisostyla*, 363
Cosmetics, 330
Coxa, 26, 230
Coxal spines, 21
Crab
    exploitation practices, 4
    immune response, 71
    larvae, 22, 100
    meat, 1, 4, 5, 307, 308
    megalopa, 21
    shell wastes, 337, 340, 344–348
    stage, 44
    zoea larvae, 21
Cradle-carrying position, 18
Crannies, 14
Crocodiles, 69, 72
Crustaceans, 2, 14, 43, 52, 55, 57, 72, 80, 90, 91, 96, 100, 131, 142, 157, 171, 204, 208, 219, 221, 228, 238, 240, 292, 351, 353, 356, 357, 363, 370
Cucumbers, 343
Culture ponds, 6, 281, 369

Cypris larva, 362
Cytomegalo virus promoter sequence, 332

## D

Dactylus, 27, 41, 45, 47, 59, 67, 106, 111, 163, 167, 182, 195, 202, 208, 210, 215, 216, 218, 219, 222, 224, 234, 258, 267
*Dasyatis brevicaudata*, 45
Dead-tooth-swimming crab, 82
Decapoda, 2, 81, 86, 336
Deep-sea blue swimming crab, 87
Deformities, 353, 372
Degree of acetylation, 347
Delicate swimming crab, 25
Demersal, 36, 78, 80, 101
Dental medicine, 331
Dentiform angles, 39
Denuded specimens, 33, 67
Depletion of oxygen, 342
Depolymerized chitosan oligomers, 332
Deproteinization, 347
Destabilization, 345
Detritivorous, 14
D-glucosaminic acid, 338, 339
Diabetes hepatitis, 331
Diabetes mellitus, 333
Diabetic rats, 333
Diagnostic character, 12
Digestive
    enzymes, 13, 94
    system, 10, 13, 94
Dinoflagellates, 349, 357, 363, 371
Direct economic value, 6
Distal
    spine, 25, 27, 33, 34, 38–41, 46, 49, 64, 67, 87, 104, 114, 163, 205
    spiniform lobule, 28
Distinct spines, 70
*Dolfusiella martini*, 352
Dorsodistal margin, 65
Dorsoventrally flattened, 29, 36, 44, 46, 231
Drug-chitosan mixture, 331

## E

Ecdysis, 14
Echinoderms, 90, 238
*Echinolatus*, 3, 210–212, 230
Ecofriendly/biodegradable nature, 342
Economic heavy metal remover, 346
Ectoparasitic barnacles, 355
Ectosymbiont, 361, 365
Effective immune methods, 70
Egg-bearing females, 20
Ejaculatory ducts, 17, 53
Electroplating operations, 344
Elucidation, 6
Embryology, 20, 100
Embryonic development, 17, 77, 110, 131, 369
Endangered species, 83, 197
Endopodite, 17, 19, 23, 97, 98
Endopods, 44
*Ephelota gemmipara*, 363
Epibionts, 354, 356, 364
Epibranchial, 40, 42, 48, 51, 61, 72, 78, 82–84, 104, 123, 127, 130, 141, 156, 166, 171, 178, 182, 183, 191, 192, 260, 270, 361
    line, 78, 116, 120, 122, 129, 201
    region, 26, 233, 234
    teeth, 29
Epistome, 39
Epizoite, 361
Epizootics, 357
Erosion/melanization, 369
*Escherichia coli*, 146, 335
Esophagus, 10
Estuaries, 22, 29, 42, 68, 80, 83–85, 87, 88, 93, 95, 96, 99, 100, 141, 206, 239
Estuarine-marine ecosystem, 2
*Euphylax*, 3, 202, 203
Euryhaline, 79, 95, 215
Exoskeleton, 11, 14, 18, 19, 69, 189, 329, 357, 361, 362, 369
External morphology, 17, 53
Extracellular chitinase production, 340
Extruded eggs, 19
Extrusion, 19

Eye placode, 77

## F

Fatty acid content, 328
*Fecampia erythrocephala*, 359
Fecundity, 20, 43, 58, 61, 77, 79, 107, 108, 124, 131, 146, 157, 206, 208, 219, 238, 240
Female
    crab, 13, 17, 18, 52, 58, 73, 119, 189, 273, 276, 283, 284, 295, 300, 301, 362, 366, 371
    reproductive system, 17, 53
Fertilization, 6, 16
    rates, 73
Filamentous bacteria, 350, 369
First
    anterolateral tooth, 64
    *zoea*, 44, 97
Fisheries, 1, 3, 4, 6, 45, 58, 79, 81, 82, 85, 88, 98, 146, 172, 215, 218, 240, 271
Flatface swimming crab, 30
Flesh-colored blotches, 54
Floating sargassum, 54
Flocculation, 344, 345
Fluorescing patterns, 12
Food
    applications, 348
    feeding, 22, 43, 52, 55, 57, 60, 69, 72, 76, 80, 84, 86, 90–92, 94, 100, 102, 107, 109, 131, 157, 171, 219, 289, 291, 300
    ingredient, 330
    pharmaceutical industries, 334
    reserves, 13
Foraminifera, 90
Foraminiferous species, 2
Freshwater, 78, 100, 298, 300
Fringing hairs, 28, 33, 41, 67
Frontal
    lobe spines, 68, 70, 76
    postfrontal patches, 59
Fungal
    development, 343

Index 395

diseases, 370
growth, 335, 342
*Fusarium* sp., 351

### G

Gastric
  mill ossicles, 14
  regions, 40, 72
Gastroenteritis, 356
Gastropods, 72, 80, 214
Gastrula, 77
Gene delivery, 332
Genus triticella, 365
Gills, 13, 372
Gladiator swimming crab, 34, 86
Glucose, 307, 333, 357, 358
Glutaraldehyde, 343
Glutathione dependent antioxidant
  system, 333
Gonad
  development, 57, 70
  mature, 58
  maturing, 58
Gonochoric, 81, 86, 241
Gonophores, 18, 19
Gonopod, 16, 18, 24, 83, 115, 124, 205
Gonopores, 80, 81, 83
Goose barnacles, 55
Gram-positive/gram-negative bacteria,
  335
Granular
  eminences, 42, 61
  ridges, 37, 40, 47, 59, 126, 150,
    153–155, 162
Granulate, 78, 80, 83, 84, 88, 92, 94,
  99, 101, 129, 192, 200, 221, 225, 259,
  261, 268, 361
Granulated ridges, 32, 64, 89, 101, 103
Granulation, 28, 67, 80, 89, 99, 136,
  153, 242, 245
Granulose, 11, 23
Green crab, 81, 213, 214, 357, 363
Grow-out culture, 289, 372
Growth rate, 15, 43, 52, 79, 95, 109,
  119, 228, 284, 288, 289, 301, 340

Gumbo crab, 81

### H

Habitat, 55, 68, 198, 210, 271, 286
Haemagglutinin (HA), 53
Haemocyanin, 357
Haemocyte infiltration, 357
Haemocytes, 336, 339, 357
Haemolymph, 334, 358, 368, 370
*Haliphthoros* sp., 351
Hard shell blue crab, 81
  crab farming, 306
Hatchery conditions, 77, 370
Hatching, 17, 61, 62, 73, 77, 131, 146,
  157, 186, 277, 278
Hazardous-waste disposal problem, 344
Healing, 16, 331
*Hematodinium* sp, 365, 371
Hemichannel-associated transmembrane
  protein, 71
Hemocyanin, 71
*Hemophilus paracuniculus*, 355
Hemoproteins, 342
Hemorrhage, 338
Hemostasis effects, 337
Hemostatic wound dressing, 337
Hepatic region, 58
Hepatopancreas, 13, 94, 119, 159, 189,
  350, 358, 361, 368–370
  cells, 358
  epithelium, 369
*Heterosaccus*
  *dollfusi*, 362
  *lunatus*, 361
High-density lipoprotein (HDL), 308,
  340
Hirsute, 47, 59, 112, 115, 140, 162, 241,
  243, 245
Histopathological alteration, 349, 372
Histozoic microsporidian, 371
Human
  astrocytoma cells, 334
  health applications, 348
Hybridization, 77
*Hydroides norvegica*, 353, 364, 365

Hydroids, 55
Hydrophilic drugs, 331
Hypercholesterolemia, 333
Hypersaline, 84, 100
Hypo-branchial chamber, 366
Hypocholesterolemic
  effect, 339
  properties, 330
Hyposaline, 84, 86, 89, 95

# I

Icosahedral virus, 369
Immune associated components, 70, 71
Immuno-enhancing effect, 341
Inconspicuous teeth, 39
Industrial
  agricultural processes, 345
  applications, 329, 348
Inflammation, 5, 333
Inner
  spine, 57, 64, 142, 156
  supraorbital lobe, 37, 47, 121, 148, 182, 240
Intermolecular electrostatic repulsion, 343
Intermolt period, 356
Internal carapace width (ICW), 70
Intertidal zone, 57, 83, 165, 171, 173, 185, 194, 214, 225, 239, 244
Intestine, 13
Intramolecular hydrogen bonds, 343
Intromittent copulatory organ, 16
Invertebrates, 84, 86, 92, 100, 109, 157
Iridescent patches, 31
Iron, 142, 307, 333, 339, 342
Ischium, 26, 41, 44, 51, 111, 163
Isopoda, 353
Isoprenaline-induced oxidative stress, 333

# J

Juvenile crab, 73, 291, 352
Juveniles, 12, 15, 42, 43, 51, 52, 68, 69, 89, 94, 99, 100, 142, 157, 171, 181, 228, 275, 278, 279, 291, 354, 372

# L

Lachrymal flow, 332
Lactate, 335, 342
*Lagenidium* sp., 351
Lagoon, 2, 82, 89, 298, 302
Lanceolate, 78, 83, 92, 94, 182, 208, 210–212, 215, 218, 222, 224, 231, 234, 237, 253, 262
Lancer swimming crab, 36
Larval
  abundance, 22, 96
  development, 44, 72, 73, 77, 110, 131, 146, 186, 207
  rearing, 277, 306
  release, 22, 96, 100
  stages, 44, 45, 131
Late maturing, 17, 53, 157
Lateral
  borders convex, 47
  epibranchial spine, 39
  frontal teeth, 32, 153
  spines, 21, 31, 44, 54, 84, 86, 98, 106, 107, 109, 110, 111, 132, 146
Lecanicephalid cestode, 352
Lepas anatifera, 55
Lessepsian migrant crab, 362
*Libystes*, 259, 260
Life cycle, 22, 157, 354
*Liocarcinus*
  *depurator*, 219, 220, 352, 357, 362–364
  *puber*, 349, 351–353, 363, 364
*Lissocarcinus*, 267, 268, 270
Lithium-ion batteries, 348
Liver gluconeogenesis, 333
L-malic acid, 344
Longevity, 43
Longitudinal granulated ridges, 64
*Loxothylacus*
  *carinatus*, 365
  *nierstrasz*, 362
  *texanus*, 356
Luciferase reporter gene, 332
Luminescent
  bacterial disease, 370

Index 397

vibriosis, 350
*Lupocycloporus aburatsubo*, 67

# M

Macropipinae, 3, 23
Male
  abdominal segments, 24
  first pleopod, 65, 166, 169, 174, 176, 180, 184, 190–192, 200, 201
  reproductive system, 13, 17, 52
Mandible, 44, 97, 98
Mangroves, 42, 68–70, 72, 83–85, 88, 286, 288, 298
Manus, 41, 64, 105, 136, 141, 156, 158, 160, 161, 164, 182
Marbled swimming crab, 87
Marine
  benthos, 2
  breeding crabs, 70
  ecosystems, 1, 5
  food web, 5
  habitats, 1
*Maritrema subdolum*, 357–359
Mating, 17, 18, 52, 56, 69, 72, 76, 86, 96, 240
  activity, 76, 77
  behavior, 18, 52, 56, 69, 86
Mature female abdomen, 80, 81, 83, 94, 101
Maxillipeds, 23, 25, 27, 28, 33–35, 38, 41, 44, 46, 49, 60, 67, 87, 97, 98, 106, 114, 141, 163, 207, 211, 245, 253, 257, 260
Median, 32, 34, 37, 46, 59, 67, 87, 103, 105, 116, 120, 121, 123, 125, 127, 134, 136, 138, 141, 144, 145, 147– 149, 158, 159, 161, 162, 166, 171, 173, 174, 178, 181, 182, 185, 187, 188, 196, 197, 222
  frontal lobe, 63, 193
  teeth, 67, 122, 139, 152, 243
Megalopa, 21, 44, 45, 51, 73, 75, 77, 96, 98, 110, 111, 131, 146, 186, 207, 215, 240, 274, 276, 277, 279
  metamorphoses, 44

  stage metamorphoses, 44
  stages, 51
Megalopae, 51, 92, 95, 274
Megalopal stage, 20, 22, 96, 100, 146, 207
Merus, 39, 49, 50, 59, 64, 65, 112, 116, 120, 124, 129, 140, 141, 147, 154, 167, 173, 181, 186, 188, 202, 204, 206, 211, 212, 236, 245, 249, 254, 255, 258, 260, 265, 266
Mesenteric lymphadenitis, 356
Mesobranchial, 29, 40, 42, 61, 78, 80, 83, 84, 89, 94, 101, 103, 105, 119, 121, 125, 138, 144, 148, 158, 159, 163, 168, 175, 178, 182, 184, 187, 191, 192, 197, 242
Mesogastric, 29, 32, 42, 46, 48, 51, 61, 63, 64, 78, 94, 101, 103, 105, 116, 120, 122, 123, 129, 130, 135, 141, 159, 178, 182, 196, 201, 231, 249
Metacercariae, 361
Metagastric, 40, 42, 48, 51, 61, 64, 86, 88, 89, 99, 102–105, 130, 141, 159, 182
  adults, 84, 89
    ridges, 42, 48, 51, 61, 102, 104, 141
Metal
  anions, 347
  cations, 347
  plating industry, 344
Metamorphosis, 22, 96, 100, 157
Metazoan parasitic infections, 352
Metazoans, 349, 370
*Metschnikowia bicuspidata*, 368
Microparticulate drug-carrier, 332
*Microphallus claviformis*, 357, 359
Milky
  appearance, 358
  disease, 350, 358, 365, 368, 371, 373
Mineral content, 328
Mobility, 2, 370
Modern industrialized fishing fleets, 1
Molariform denticles, 41
Molluscs, 2, 14, 43, 57, 69, 72, 76, 80, 84, 86, 90, 91, 94, 102, 109, 142, 213,

214, 219, 221, 228, 239, 244, 285, 294, 296
Molting, 10, 14, 15, 22, 43, 45, 69, 79, 92, 95, 96, 100, 131, 213, 238, 240, 283, 288, 292, 295, 303–305, 360, 369, 371, 372
Monosex culture, 280, 301, 306
Monospecies culture, 283, 306
Morphological changes, 77
Mortality, 70, 279, 294, 359, 368–371
Mucin, 332
Mucoadhesive recipient, 331
Mud crab reovirus (MCRV), 369
Multidentate, 11, 23
Muscle
    food products, 342
    necrosis virus, 369

# N

*N*(2-hydroxy)propyl-3-trimethylammo-nium chitosan chloride, 337
*N, O*carboxymethylchitosan (NOCC), 337, 342
N-acetylneuraminic acid (NANA), 336
N-carboxymethylchitosan, 342, 346
Necrosis, 368, 373
Neonatal streptozotocininduced diabetic mice, 333
Neurodegenerative diseases, 333, 334
Nitrogen, 5, 284
Noninfested individuals, 356
Nonovigerous females, 354
Nonrenewable biological substances, 347
Nutraceuticals, 330, 333
Nutritional
    health value, 7
    qualities, 2
    values, 308, 328

# O

Obscure maroon, 86
Obsolescent granules, 83
O-carboxymethylchitin films, 343
*Octolasmis*

*angulata*, 361, 362, 366
*lowei*, 354, 356, 366
*tridens*, 366, 367
*warwickii*, 366, 367
Ointments, 332
Omega-3 fatty acids, 4, 308
Omnivorous, 5, 25, 52, 131, 208, 304
Oocyte, 61, 62
Opalescent hemolymph, 368
Opercular plates, 371
Ophthalmic
    chitosan gels, 332
    drug delivery, 332
Opportunistic, 14, 43, 91, 100, 239, 370–372
    omnivores, 14
    pathogens, 356
Oral
    administration, 334
    bioavailability, 332
    drug delivery, 331
    mucous wounds, 331
Ornate blue crab, 89
Osteo
    conductive, 331
    properties, 331
Outer
    orbital angle, 39, 152
    ventrodistal spine, 64
*Ovalipes catharus*, 239, 352, 361
Ovigerous
    crabs, 58
    female, 19, 57, 58, 66, 84, 106, 146, 164, 192, 194, 219, 220, 276, 278, 354
Oviposition, 62
*Ovlaipes catharus*, 353
Oxalic acid, 344
Oxidative
    enzyme activities, 339
    stability, 342
    stress, 333

# P

Paddle-like dactyl, 21

# Index 399

Pairs of ovaries/oviducts, 17, 53
*Paramoeba perniciosa*, 357
Parasites, 14, 94, 353, 355, 362,
  365–368, 370, 371, 373
  diseases, 370
  trematodes, 361
Parasitization, 367
Partially N-acetylated chitosan (PNAC),
  337
*Pasteurella multocida*, 355, 356
Pathogen infection, 70
Pathological changes, 357
Peak spawning seasons, 7
Pectinase treatment, 343
Penultimate
  segment, 34, 46, 87, 118, 119, 121,
  125, 126, 130, 135, 137, 138, 141,
  144, 147, 149, 160, 161, 164, 167,
  168, 171, 173, 175–177, 179, 181,
  186–188, 194–197, 230
  somite, 48
Peptides, 71, 331
Pereiopods, 41, 44, 45, 57, 97, 98, 106,
  134, 174, 190, 208–210, 216, 218,
  221, 222, 225, 227, 229, 231, 234,
  259, 261, 262, 271, 363
Periodontal surgery, 331
Permeability-enhancing properties, 332
Permeation enhancer, 331
Pharmaceutical
  applications, 329
  tablets, 331
Phenoloxidase, 336
Phosphorus, 5, 307
Phylum arthropoda, 14
Phytotoxicity, 335
Picolinate chromium complex, 339
Pigmentation, 73, 369
Pincers, 13, 14, 34, 133, 203, 246, 283,
  301, 305
Piniform, 88
Pink crab disease (PCD), 357, 363
Planktonic larvae, 17, 215
Plant matter, 80, 214
Plasma recalcification time (PRT), 337
Platyhelminthes, 352

Pleopod buds, 44, 98, 132
Pleopods, 13, 16, 17, 19, 53, 73, 85, 110,
  214, 251–253, 305, 362, 371
Plume-like filaments, 13
Plumose seta, 29, 44, 46, 47, 98, 132
Pneumonia, 356
Podophthalminae, 3, 23
*Podophthalmus*, 3, 204–206, 319, 351
  *vigil*, 4
Polybiinae, 3, 23
Polychaetes, 2, 90, 107, 208, 219, 238,
  240, 294, 370
Polychlorinated biphenyls (PCB), 346
Polyculture, 280, 293–297, 306
Polyelectrolytes, 343
Polygonal patterning, 68, 70, 76
Polymeric excipients, 332
Polyphenolic compounds, 344
*Polypocephalus moretonensis*, 352
*Pomatoceros triqueter*, 353, 364
Portunid crabs, 2–5, 6, 9, 11, 13, 14, 16,
  17, 20, 21, 124, 273, 293, 334, 348,
  349, 353
*Portunion maenadis*, 353, 357
*Portunus*
  *argentatus*, 24
  *binoculus*, 26, 365
  *gibbesii*, 31
  *gladiator*, 4, 34, 86
  *gracilimanus*, 4, 32
  *granulatus*, 33
  *haani*, 4, 34, 86, 321
  *hastatoides*, 4, 35, 365
  *hastatus*, 36, 365
  *macrophthalmus*, 39
  *nipponensis*, 40, 257
  *pelagicus*, 2, 4, 5, 12, 14, 15, 17, 19,
  20, 42, 44, 45, 276, 289, 302, 321,
  351, 352, 366
  *pseudohastatoides*, 46
  *pseudotenuipes*, 47
  *pubescens*, 48
  *pulchricristatus*, 49
  *rubromarginatus*, 50
  *sanguinolentus*, 4, 53, 334

*sanguinolentus*, 2, 6, 12, 15, 20, 51, 273, 323, 334, 351, 367
*segnis*, 57, 367, 368
*tenuipes*, 48, 60
*trituberculatus*, 2, 4, 61, 273, 326, 368
*tuberculosus*, 62
*ventralis*, 63
*vossi*, 64
*xantusii*, 66
*yoronensis*, 66
Postcardiac regions, 29
Postcopulation, 69
Post-copulatory guarding, 19, 69, 77, 131
Posterior/posterolateral surfaces, 40
Postero-distal border, 28, 38, 42, 49, 52, 60, 61
Posterolateral
    junction, 25, 28, 32–35, 38, 40–42, 46, 49, 51, 52, 60, 61, 67, 87, 114, 116, 120, 123, 129, 155, 162, 201
    margin, 11, 23, 27, 41, 110, 111, 139, 150, 152, 156, 160, 198, 212, 214, 233
    spine, 51
Postlarvae, 72, 296
Postlateral junction, 59, 126
Pre-copulatory
    guarding, 76, 77, 131
    position, 69
Predation, 45, 94
Predators, 5, 14, 45, 55, 69, 72, 96, 158, 239, 287
Premolt
    guarding, 18
    stage, 18, 305
Proanthocyanidins, 344
*Profilicollis botulus*, 352, 357, 359
Pro-inflammatory cytokine, 334
Prophenoloxidase activating system, 336
Propodi, 24, 88, 115, 231, 270
Propodus, 59, 88, 94, 101, 119, 121, 122, 125, 126, 134, 137, 138, 141, 144, 145, 148, 149, 158, 160, 161, 163, 164, 167, 168, 171, 173, 175,

177, 179, 181, 182, 187, 188, 194, 196, 197, 202, 205, 224, 234, 236, 248
Proteins, 307, 308, 329, 339, 345
*Proteus vulgaris*, 335
Protistans, 349
Protobranchial, 29
Protogastric, 64, 103, 105, 116, 120, 123, 129, 139, 201, 231, 249
Proximal
    somites, 37
    tooth, 88
*Pseudomonas*
    *cepacia*, 355, 356
    *mallei*, 355, 356
    *putrefa sciens*, 355
    *putrefasciens*, 356
*Psychrobacter immobilis*, 336
Pubescence, 24, 28, 33, 34, 42, 46, 48, 64, 67, 87, 114, 156, 188
Purplish-vinaceous, 80
Pyrrolidine carboxylate salts, 342

## Q

Quadridentate, 67
Quadrilobate, 28, 191, 210, 228, 229

## R

Radionuclides, 347
Rat myocardium, 333
Recalcification, 43
Red
    blue crab, 82
    legged swimming crab, 26, 29
    sternum syndrome, 350, 373
Reflex action, 15
Regeneration, 6, 16, 331
Reovirus, 360, 369
Reproduction, 22, 43, 72, 84–86, 90, 95, 96, 100, 109, 124, 131, 146, 215, 234, 367
Reproductive
    biology, 52
    capacity, 20
    females, 85
Resources management, 4

Index 401

Respiration, 13, 94, 109, 343, 371
Rhizocephalan cirripedes, 361, 365
Rhomboid, 36
*Rhzopus stolonifer*, 335
Rostral spine, 44
Rostrum, 21, 44, 110, 111
Rotifer, 77
Rudimentary buds, 44
Rugose swimming crab, 85
Rust spot shell disease, 372, 373

## S

*Saccharomyces*
  *cerevisiae*, 341
  *uvarum*, 341
*Sacculina*
  *carcini*, 221, 360
  *carinata*, 365
  *granifera*, 366
  *scabra*, 362
Sacculinid parasites, 354
Salinity range, 79, 298, 300
Sand
  mud substrates, 80
  cleansers, 14
Saprophagous species, 90
*Sargassum* spp, 55
Scavengers, 43
Sclerotinia rot (carrot), 335
*Scylla*
  *olivacea*, 4, 68, 76, 285, 292, 293,
    326, 372
  *paramamosain*, 4, 69, 70, 290, 292,
    293
  *serrata*, 2, 4, 5, 19, 71, 76, 273, 284,
    285, 294, 295, 297, 300, 301, 345,
    349–351, 369, 372
  *tranquebarica*, 75–7, 330, 372
Sea spiders, 55
Seafood industries, 348
Seagrass, 14, 42, 56, 89, 290
Seaweed, 14, 55, 266, 284, 297
*Second zoea*, 44, 97
Secondary sexual characteristics, 366
Seed production/larval rearing,

callinectes sapidus, 278, 279
portunid crabs, 273, 306
portunus pelagicus, 276–278
scylla serrata, 273–276
Segmentation, 21, 254
Selenium, 307, 333
Seminal receptacles, 17, 19, 53
Semipelagic, 2
Septicemia, 356
*Serratia marcescens*, 340
Serum
  cholesterol, 340
  triacylglycerol, 333
Setae, 19, 21, 29, 44, 45, 47, 65, 73, 74,
  97, 98, 109–111, 131, 132, 146, 198,
  211, 213, 232, 256
Setation, 21, 65, 97, 132
Severance, 15
Sexual
  characters, 16
  dimorphism, 16, 124, 271
  maturity, 52, 53, 58, 70, 73, 109, 124,
    137, 206, 228
Sharp-tooth swim crab, 87
Shell diseases, 349, 373
  bacterial diseases, 349, 350
  fungal diseases, 355
  non-infectious/other diseases,
    353–373
  parasitic infections, 351, 352
  viral diseases, 350,351
*Shigella flexneri*, 355, 356
Short spine, 32, 163
Shrimps, 14, 78, 157, 215, 238, 274,
  277, 280, 281, 294–296
  culture, 4, 283, 369
Sinuous, 16, 65, 99, 141, 159, 162, 174,
  193, 194, 212, 250, 261, 269
Sixth thoracic segment, 17, 19
Slow-moving
  invertebrates, 43
  prey, 14
Soft
  shell crab farming, 306
  shelled crab, 335
Somites, 30, 37, 44, 132, 234

Spawning, 17, 19, 58, 73, 90, 96, 100, 110, 124, 131, 146, 208, 240, 245, 274
Spear-shaped swimming crab, 35
Species
description, 271
diversity, 2
identification, 12
profile, 271
Spermatheca, 17, 19, 53
Spermatophores, 19
Sphistin, 71
SpHyastatin, 71
Spiniform
elevations, 40
granules, 33, 232
Spinules, 35, 42, 49, 52, 61, 65, 66, 86, 116, 118–122, 124–127, 129, 130, 137, 144, 145, 147–149, 153, 154, 160–162, 171, 173, 179, 180, 182, 184, 187, 190, 192, 194, 197, 201, 202, 205, 232, 248, 251
Squamiform
arrangement, 59
markings, 25, 34, 46, 60, 87, 114, 120, 125, 138, 139, 155, 158, 161, 178, 184
*Staphylococcus aureus*, 146, 335
Starch, 332
Stationary bottom-dwelling animals, 69, 72
Sternal
cornuae, 51
spines, 21
Sternites, 17, 19, 83, 85, 88, 92, 156
Sternum, 80, 127, 135, 190, 230, 256
Stocking density, 69, 274, 277, 279, 286, 287, 289, 290, 292, 295, 297, 305
Stomach, 13, 157, 171, 208
Straight margins, 70
Strawberries, 335, 343
Stump tooth swimming crab, 82
Subdistal spine, 67, 103–105
Subhexagonal, 24, 115, 207, 210–212, 230, 253
Submedian red spots, 27
Sub-regional elevations, 39

Substantial commercial fishery, 58
Subterminal spine, 24, 59, 97, 115
Superoxide anion production, 339
Supraorbital
lobe, 37, 118, 122, 126
margin, 39
Sustainable fishery, 7
Swimmerets, 13, 16, 371
Swimming
crabs, 2, 4, 5, 14, 23, 171, 297
legs, 13, 30, 73, 80, 103–105, 304
pereiopods, 57

## T

*Tagelus affinis*, 80
Tartaric acid, 344
Telson, 21, 29, 44, 47, 78, 80, 81, 83, 84, 88, 89, 92, 94, 97–99, 101, 127, 131, 132, 146, 157, 163, 199, 210–212, 231, 250, 258, 262
Testes, 13
Tetraoxypyrimidine, 339
Tetraphyllidean cestode, 352, 366
*Thalamita*
*coeruleipes*, 168, 361, 373
*cooperi*, 169, 373
*Thalassia testudinum*, 55
The United States Food and Drug Administration (USFDA), 339
Thiolated Chitosan Nanoparticles, 334
*Third zoea*, 44, 97
*Thlamita sima*, 373
Thoracic
appendages, 44, 98, 366
sternites IV and V, 80, 81, 84, 89
Tomentose, 28, 29, 46, 134
Tomentum, 37, 116, 120, 123, 139, 229, 230
Toxic metals, 79
Toxicity, 353
Transmucosal drug carriers, 332
Transpiration loss, 343
Transverse gastric ridges, 27
Triacylglycerol, 340
*Triticella capsularis*, 353, 361, 365

Index
403

Truncate, 28, 121, 123, 126, 128, 130, 139, 145, 155, 156, 173, 176, 197, 201
*Trygonorhina fasciata guanerius*, 45
Tuberculate, 40, 81, 237, 247, 248, 258
Tunicates, 55
Turbidity, 289, 344, 345, 373
    reduction, 345
Turtles, 55, 69, 72

## U

United States Environmental Protection Agency (USEPA), 346
Uropod, 21

## V

Value-added products, 329
Vas differentia, 17, 53
Vascular plants, 2
Ventral posterior end, 44, 132
*Vibrio*
    *campbellii*, 370
    *fischeri*, 370
    *fluvialis*, 355, 370
    *harveyi*, 350, 370
    *nereis*, 370
    *vulnificus*, 370
Virosomes, 368
Visual organs, 10
Vitamin C deficiency, 373

## W

Walking legs, 13, 150, 213, 229, 258, 269
Warmed-over favor (WOF), 342
Warrior swimming crab, 81
Waste
    management, 340
    water treatment, 344
White spot disease, 351, 373
White spot syndrome virus (WSSV), 351, 368, 369, 370

## X

*Xaiva*, 3, 209
Xanthium gum, 332

## Y

*Yersinia pseudotuberculosis*, 355
Y-shaped carina, 59

## Z

Zinc, 142, 307, 333
Zoeae, 17, 19, 61, 62, 77, 132, 207, 274, 276, 277
Zoeal stages, 20, 22, 44, 73, 96, 98, 100, 110, 131, 146, 186, 277
*Zoothamnium* sp., 363
Zygomycetes, 335